Fluorescent Nanodiamonds

Fluorescent Nanodiamonds

Huan-Cheng Chang
Institute of Atomic and Molecular Sciences, Academia Sinica
Taiwan, Republic of China

Wesley Wei-Wen Hsiao
FND Biotech, Inc., Taiwan, Republic of China

Meng-Chih Su
Department of Chemistry, Sonoma State University, CA, USA

Registered Offices
John Wiley & Sons, Inc., 111 River Street, Hoboken, NJ 07030, USA
John Wiley & Sons Ltd, The Atrium, Southern Gate, Chichester, West Sussex, PO19 8SQ, UK

Editorial Office
9600 Garsington Road, Oxford, OX4 2DQ, UK

For details of our global editorial offices, customer services, and more information about Wiley products visit us at www.wiley.com.

Wiley also publishes its books in a variety of electronic formats and by print-on-demand. Some content that appears in standard print versions of this book may not be available in other formats.

Library of Congress Cataloging-in-Publication Data

Names: Chang, Huan-Cheng, author. | Hsiao, Wesley Wei-Wen, author. |
    Su, Meng-Chih, author.
Title: Fluorescent nanodiamonds / by Huan-Cheng Chang (Institute of Atomic and Molecular Sciences, Academia
    Sinica, Taiwan, R.O.C.), Wesley Wei-Wen Hsiao (Institute of Atomic and Molecular Sciences, Academia Sinica,
    Taiwan, R.O.C.; FND Biotech, Inc., Taiwan, R.O.C.), Meng-Chih Su (Department of Chemistry, Sonoma State
    University, CA, U.S.).
Description: Hoboken, NJ : Wiley, 2019. | Includes bibliographical references and index. |
Identifiers: LCCN 2018021226 (print) | LCCN 2018022679 (ebook) | ISBN 9781119477105 (Adobe PDF) |
    ISBN 9781119477044 (ePub) | ISBN 9781119477082 (hardcover)
Subjects: LCSH: Nanodiamonds. | Nanostructured materials. | Diamonds. | Nanomedicine. | Imaging systems.
Classification: LCC TP873.5.D5 (ebook) | LCC TP873.5.D5 C53 2018 (print) | DDC 549/.27–dc23
LC record available at https://lccn.loc.gov/2018021226

Cover Design: Wiley
Cover Image: Permission from Sotoma, S., Epperla, C.P., and Chang, H.-C. (2018) Diamond nanothermometry.
    ChemNanoMat 4 (1), 15–27. © (2018) John Wiley & Sons Ltd.

Set in 10/12pt Warnock by SPi Global, Pondicherry, India
Printed in Singapore by C.O.S. Printers Pte Ltd

10 9 8 7 6 5 4 3 2 1

To our wives and children
Karen H.-J. Chen *and* **Alan Y.-S. Chang**
Pei-Fen Lu *and* **Yu-Ching Hsiao**
Sukha C. Su
for their firm belief and tireless support
and to
Academia Sinica
on the 90th anniversary

Contents

Preface

Fluorescent nanodiamond (FND) is a carbon-based nanomaterial with exceptional physicochemical properties and extraordinary functional capacities. Invented in 2005, FND evolved from the studies of nanometer-sized diamond particles and color centers in bulk diamonds. The invention was followed by an ever-growing list of applications in diverse areas of science and engineering, which has led to fast increasing interests among researchers worldwide. While all is going well and fine, it brings us to the question:

Why this book now?

There are two versions of answers to this question: a comprehensive answer obviously requires reading the entire book and the brief one (perhaps, also closer to our initial thoughts) is shared here.

As illustrated multiple times in the book, the FND's superior biocompatibility coupled with its long emission lifetimes makes it a unique drug carrier with an added benefit that its movement in living organisms can be tracked in three dimensions and of real time. Using FND as a platform for drug delivery falls in the hardcore of the much-anticipated development in *nanomedicine*, which actively looks for new solution in target-selective and site-specific therapeutic treatments at the molecular level. This is the medicine of the twenty-first century and, when fully developed, FND will have a great role to play in the field. The foreseeable impact on human physiology and health is so profound that it may change the way the therapeutic treatment is delivered for future generations.

It has become a trend that new science grows from a sector in the overlapping areas of knowledge subdomains, complimentary and/or collectively benefited and strengthened by all participating disciplines. This is certainly the case for the FND studies, originally rooted in basic sciences such as physics, chemistry, and materials science; it also covers (to a lesser extent) bioengineering, electrical engineering, and optoelectronics. Accordingly, FND can be adapted in just about any areas in science and applied to a wide range of research and industry. With a further and in-depth understanding of FND, it is expected that the use of this nanomaterial will inspire more new thinking in physics, chemistry, and biology as well as innovative designs for more potential applications.

So far, to the best of our knowledge, there is no single book that has provided a coherent presentation of all the subject materials necessary for a consistent understanding of FND. This book takes a systematic approach, beginning with the basic principles, to cover the status of current FND research and future outlook. Also included in the book is a collection of application examples in some specific areas, intended to inspire our readers to devise new applications for their own needs. We find creative use of FND every day. As we all know, creativity grows out of imagination; and when it comes to imagination, sky is the limit.

Like any other cutting-edge sciences, FND is still relatively young in the scientific community. Much more can be achieved if there are more people to know about it (and, hence, be benefited by it). Throughout the years, we have enjoyed working with FND in a variety of aspects, been rewarded with many exciting discoveries, and learned a great deal of what FND is capable of doing for us humans and for our environment as well. It has been (and still is) such a fascinating adventure for us. With passion, we put together in this book the materials that we have learned from working with FND, which, we hope, will bring the same (or, more) excitement to our readers, ultimately, inviting more contributions to the growth of this field. Who knows what wonders FND may bring to us tomorrow?

This book intends to be a comprehensive treatment of the subjects with FND up to date (2018), with only a brief review of nanodiamonds in Chapter 2. Further discussions on nanodiamonds can be found elsewhere. Categorically, the content materials are grouped into two parts, namely, *basics* (Part I, Chapters 1–5) and *specific topics* (Part II, Chapters 6–14). A list of the titles of all 14 chapters in the book is given here as a quick view of the content in the book:

- Chapter 1: Introduction to Nanotechnology
- Chapter 2: Nanodiamonds
- Chapter 3: Color Centers in Diamond
- Chapter 4: Surface Chemistry of Nanodiamonds
- Chapter 5: Biocompatibility of Nanodiamonds
- Chapter 6: Producing Fluorescent Nanodiamonds
- Chapter 7: Single Particle Detection and Tracking
- Chapter 8: Cell Labeling and Fluorescence Imaging
- Chapter 9: Cell Tracking and Deep Tissue Imaging
- Chapter 10: Nanoscopic Imaging
- Chapter 11: Nanoscale Quantum Sensing
- Chapter 12: Hybrid Fluorescent Nanodiamonds
- Chapter 13: Nanodiamond-Enabled Medicine
- Chapter 14: Diamonds in the Sky

These topic-oriented chapters, not necessarily independent, explore multiple aspects of FND, comprising structural configurations, fundamental optical and magnetic properties, various fabrication techniques, and specific applications involving the state-of-the-art instrumentation, with ample illustrations, case studies, practical examples, and historical perspectives. Putting it all together in a book provides our readers a full landscape of what is out there in this field today and plenty of opportunities for future growth.

It is our goal that the book will serve a broad audience, including beginners such as upper division (junior or senior level) undergraduate students majoring in science and engineering, and graduate students of related majors. This book can be adopted as a textbook or a reference for a one-semester-long special topic course in biology, chemistry, physics, materials science, and other engineering areas. The exhausted list of citations at the end of each chapter should provide our readers sufficient resources either to patch up their background or for their further study of some specific subjects. The book's *bottom-up* approach, emphasizing the understanding of the basics first, is equally suited for experienced researchers or industry professionals who have needs to work with nanomaterials or related biotechnology. While they may find the book helpful in refreshing and enhancing the necessary backgrounds, we expect these readers will be particularly interested in the current progress on cell tracking, nanoscale sensing, drug delivery, and/or any of the contemporary techniques still under development. The book supplies plenty of materials extracted from the primary literature concerning the specific areas of interest, which we believe will adequately prepare our readers before they launch into their own research.

Therefore, a recommended use of the book is to begin with Part I for a fundamental understanding of the working principles, followed by a thorough reading throughout the chapters in Part II, including the subjects of cell tracking, nanoscale imaging, and quantum sensing, which are arguably the most important use of FND at the moment. Experienced researchers in the nanocarbon areas may wish to skip the chapters in Part I and choose only topics from Part II that seem to appeal to them as they extend their interests further into new directions of the FND development. Of course, there are several other ways to use the book, all tailored to fit the individual's special needs, and our readers will be the ones to make that ultimate judgment, deciding for themselves.

Finally, we fully understand that a book like this represents mainly a working progress, while much advancement is being made each and every day. By no means should we claim that the book has covered *all* areas of the field in FND. In fact, we are confident that the book probably has missed to include some exciting or even important work that may be of particular interest to some of our readers. However, we made a strategic decision in an early stage of the writing to include materials that we are familiar with and those that can clearly be presented in a cohesive and coherent context. We believe a well-thought-out presentation will be the best way to help our readers learn, hopefully, more effectively. If you, the reader, find it otherwise, please send us your comments and advices how we may improve to achieve this goal. We will appreciate your input.

Welcome to the world of FND and enjoy your reading of the book.

Taipei
California
August 2018

Huan-Cheng Chang
Wesley Wei-Wen Hsiao
Meng-Chih Su

Acknowledgements

The authors gratefully acknowledge their colleagues and collaborators including Profs. Ming-Shien Chang, Wen Chang, Jui-I Chao, Chia-Chun Chen, Kuei-Hsien Chen, Yit-Tsong Chen, Bing-Ming Cheng, Chia-Liang Cheng, Bon-chu Chung, Jim-Min Fang, Wunshain Fann, Chau-Chung Han, Cheng-Hsiang Hsiao, Chia-Lung Hsieh, Patrick C. H. Hsieh, Hsao-Hsun Hsu, Jui-Hung Hsu, Jyh-Chiang Jiang, Hsien-Ming Lee, Sheng-Chung Lee, Te-Chang Lee, Yuan-Tseh Lee, Chung-Leung Li, Tsung-Lin Li, Tsong-Shin Lim, Cheng-Huang Lin, Chih-Kai Lin, Chun-Hung Lin, Jiing-Chyuan Lin, Sheng Hsien Lin, Thai-Yen Ling, Tzu-Ming Liu, Chung-Yuan Mou, Ker-Jar Song, Kwok Sun, Juen-Kai Wang, Chih-Che Wu, Yi-Chun Wu, Alice L. Yu, John Yu, Tsyr-Yan Yu, and Yueh-Chung Yu in Taiwan for their kind support and assistance.

We also thank all group members in the Biophysical Chemistry Laboratory at the Institute of Atomic and Molecular Sciences, Academia Sinica, for their important contributions to this work over the past 24 years. They include Be-Ming Chang, Che-Wei Chang, Cheng-Chun Chang, Yu-Tang Chang, Yuan-Chang Chang, Chao-Sheng Chen, Kowa Chen, Oliver Y. Chen, Yen-Wei Chen, Yi-Ying Chen, Chi-An Cheng, Chandra Prakash Epperla, Orestis Faklaris, Chia-Yi Fang, Feng-Jen Hsieh, Chih-Wei Huang, L.-C. Lora Huang, Yuen Yung Hui, Ming-Wei Kang, Xianglei Kong, Shan-Jen Kuo, Yung Kuo, Chien-Hsun Lee, Hsiao-Wen Lee, Hung-Cheng Li, Yingqi Li, Chung-Lun Lin, Hsin-Hung Lin, Ko-Wei Lin, Yu-Chun Lu, Yi-Wen Mau, Nitin Mohan, T.-Thanh-Bao Nguyen, Lei Pan, Dinh Minh Pham, Shu-Yao Sheu, Hualin Shu, Shingo Sotoma, Long-Jyun Su, Pei-Chang Tsai, Yan-Kai Tzeng, V. Vaijayanthimala, Tse-Luen Wee, Shih-Hua Yeh, Shu-Jung Yu, and Bailin Zhang.

Across the Pacific Ocean, we thank Dr. Ian Jones for his enthusiastic support. Our appreciation also goes to the students at Sonoma State University who have helped to improve the manuscript, including Shaanti Trikha, Claire Sylva, Megumi Hallberg, and Amir Arshi.

Part I

Basics

1

Introduction to Nanotechnology

In his 2017 budget request to the US Congress, President Obama included $1.4 billion for the *National Nanotechnology Initiative* (NNI), an overarching program coordinating some 20 federal agencies and departments in all activities of nanotechnology R&D, policies, and regulation [1]. For the same time period, President Obama's budget for the entire *National Science Foundation* (NSF) was $7.964 billion, a 6.7% increase from the previous year's, which would cover the vast majority of research programs in all areas of science in the United States, including a $400 million contribution to the NNI's budget [2], a share of about 18% of the NSF's budget. All combined, the United States alone has invested nearly $24 billion since the inception of NNI in 2001 with its annual budget tripled during the same period of time. Today, similar national nanotechnology programs exist in more than 60 countries worldwide. At the other end, the global market for nanotechnology products was reported to be about $26 billion in 2014 alone, an impressive growth from the previous year's $23 billion, and was predicted to reach $62 billion by 2019 based on a compound annual growth rate of 19.8% [3]. With such a vast investment worldwide and a double-digit annual growth rate in its global gross revenue, nanotechnology has been coined as the technology of the twenty-first century. Yet, the century is still young and so is the nanotechnology industry.

So, how was it all started? What have happened to nanotechnology throughout the years? And, where are we now? We will survey the now and then in the development of nanotechnology in this chapter, with specific interest in the fields of nanocarbons, paving the way to further discussions of nanodiamonds and fluorescent nanodiamonds in the following chapters.

1.1 Nanotechnology: From Large to Small

1.1.1 Feynman: Plenty of Room at the Bottom

Nanotechnology, a term first adopted by Norio Taniguchi of Tokyo University of Science in 1974 to describe the thin film deposition in his semiconductor processing work [4], is defined by NNI as "science, engineering, and technology conducted at the nanoscale, which is about 1 to 100 nanometers." Building from the small size, nanotechnology has developed across a wide open landscape in science and engineering, including physics, chemistry, biology, materials science, and many other hybrid new fields. Despite all its

Fluorescent Nanodiamonds, First Edition. Huan-Cheng Chang, Wesley Wei-Wen Hsiao and Meng-Chih Su.
© 2019 John Wiley & Sons Ltd. Published 2019 by John Wiley & Sons Ltd.

blooming development, the original ideas of nanoscience and nanotechnology can be traced back to Richard Feynman's speech at Caltech in 1959 on his vision of miniaturization in materials [5]. After half a century now, the tiny world that Feynman so vividly painted in his speech is becoming ever more realistic than the world we presently live in with plenty more development still to come.

Feynman gave his first views on nanotechnology over a talk entitled "There's Plenty of Room at the Bottom" on 29 December 1959 at the annual meeting of the American Physical Society. In this talk, as anyone would have done in promoting a new product, Feynman tried to address the fundamental questions in this new field like: Why making things small? Can it be done? If so, how? What good will come out of things being small? Where and how to use it, including models, examples, challenges, etc.? Feynman first used an illustration of writing an entire 24 densely packed volumes of the *Encyclopedia Britannica* on a head of a pin to describe his ideas of being small. Making any material in such a small size was neither perceivable nor imaginable at that time. If we were present at Feynman's talk, in retrospect, we would have witnessed that there were less than 2000 computers in the United States and each IBM computer could take up more than a floor's space and, with accessories and service facilities, could easily occupy an entire building. We would also see the first ever moon landing by a no-man spacecraft, Luna 2, and the invention of the first laser by Theodore Maiman and the first appearance of COBOL, a high-level business-oriented computer language dominating throughout the rest of the twentieth century and into the beginning of the new millennium. As for our financial investment, Hewlett-Packard stock was only just accepted by New York Stock Exchange for trading and the first integrated circuit chip was sold for a $120 market price. Looking back, Feynman's idea of "manipulating and controlling things on a small scale" was both revolutionary and largely fictional under such environment. No doubt, there was a lot of convincing to do in this speech.

But, first, why? Why make things small? How small is small?

Feynman answered the questions right from the beginning of his talk in his own words: it is precisely because "enormous amounts of information can be carried in an exceedingly small space." For example, if we were to use some 100 atoms (e.g. a $5 \times 5 \times 5$ cube of metal atoms) for one "bit" of information, we could store the information of all 24 million volumes of books that the world had accumulated up to 1959 in a metal cube of 1/200 of an inch wide, a tiny dust barely visible by human eyes. In addition to storing massive information in a small space, we want nanomaterials to work for us. Feynman used cellular biosystems as a model for all man-made nanomaterials to follow. These nanomaterials, for example, should be small in size, active, mobile, and each one has its own specific functions. Moreover, they must be able to manipulate themselves so as to transmit information from one onto another, again, like that occurring in biosystems, not simply to store information. Imagine the possibilities, if we would just stop for a moment and look around, what it could do for us once we learn how to manufacture nanomaterials: from personal health care to agriculture, construction, communication, and explorations in new frontiers both on our earth and in outer space.

There will be plenty of good things coming from making things on such small scale. The most significant improvement is, perhaps, the speed. Using computer-aided facial recognition as an example, Feynman quickly pointed out that with all necessary materials and powers available at his present time, a computer of the size of Pentagon would have been needed to complete the task and it would have taken an impatiently long time to process the information. Furthermore, "because of its large size, there is finite time required to get the information from one place to another." After all, the information transfer cannot go faster than the speed of light. Even half a century ago, Feynman was correct to predict that "when our computers get faster and faster and more and more elaborate, we will have to make them smaller and smaller." Interestingly, his vision of making smaller computers has been realized in part by today's state-of-the-art lithography technologies, including electron beam lithography and extreme ultraviolet lithography, to increase the integration density of silicon chips [6].

Taking a daring step forward, Feynman considered bolts and nuts for making small machines and, just for the fun of it, "nanocars." He looked at the precision in machining, stress and expansion, strength of materials, magnetic force fields, lubrication, heat dissipation, etc., and concluded that with a new design of engine and redesign of the electrical parts, it is possible to make cars on a small scale. He even proposed that one of the ways to use these nanomachines was to perform medical surgery inside our own body. It was a wild idea as he admitted, but little did he know that similar concepts are also being implemented 50 years later today to develop nanocarriers for drug delivery.

As with any initiatives, Feynman acknowledged challenges and obstacles existing along the way. The first and foremost one was the electron microscope, which was capable of resolving only 10 Å, the very best at the time. In order to carry out the work described in his speech, Feynman would need electron microscopes improved by at least 100 times in resolving power. Consequently, he made a public plea in this speech both to his audience in the meeting and the physics community at large for redesigning new high-power electron microscopes. It was not until 20 years later when Gerd Binnig and Heinrich Rohrer invented the *scanning tunneling microscope* (STM) in 1982 [7] and the *atomic force microscope* (AFM) in 1986 [8] before Feynman's death in 1988. The invention of both STM and AFM, together with the design of the first electron microscope by Ernst Ruska, was recognized by the Nobel Prize in Physics 1986 [9]. These three types of microscopes have become standard instruments today in the studies of nanoscience and nanotechnology.

A wonderful inspiration as Feynman's speech was, it did not cause an immediate rush in the science community to make nanoscale materials, mainly because of the lack of appropriate tools (and, perhaps equally deficient, the knowledge of making) at the time. Nevertheless, the significance of Feynman's vision precipitated gradually throughout time and today anyone who studies the history of nanoscience and nanotechnology will regard this speech a milestone marking the birth of a new field. The field, which affects every aspect of human life from electronics, energy, medicine, and cosmetics, to food and agriculture, virtually writes a new page in human civilization. As our historian friends would be happy to remind us, history and civilization are frequently made by circumstantial occasions and incidents, which might be as simple as a speech. It may even not necessarily be a large audience or the speaker may not be aware at the time of the prevailing outcome that would follow in the long years. Mark Antony's "Friends, Romans, countrymen, lend me your ears" in replying to Marcus Brutus's assassination of Julius Caesar, Abraham Lincoln's "The Gettysburg Address," and Martin Luther King's "I have a dream" are few examples of well-known speeches that had profound impacts on the course of human history. Now, we add Feynman's speech to this list.

1.1.2 Nanotechnology Today

Not until mid-1980s, when STM and AFM became available and C_{60} was discovered [10], did the field of nanoscience and nanotechnology finally receive a long-awaited boost and set off to a healthy growth by way of developing new nanomaterials as well as new applications. A nice historical timeline is kept current by NNI on its official website [11]. In 2001, recommended by members of his science advisory committee, the US President Bill Clinton inaugurated the NNI involving 20 federal departments, independent agencies, and commissions working together to meet the mission aiming at the creation of "a future in which the ability to understand and control matter at the nanoscale leads to a revolution in technology and industry that benefits society" [12]. NNI is managed by the *National Science and Technology Council* (NSTC), a cabinet-level council reporting directly to the US President. The components of federal agencies and departments in NNI include NSF, NIH, NASA, DOE, DOD, EPA, FDA, etc. Therefore, it is a concerted effort across all areas of government institutions to develop a new field that arguably is becoming the signature technology of the twenty-first century.

In 2008, after two years' comprehensive work, NNI published a report entitled "Strategy for Nanotechnology-Related Environmental, Health and Safety Research," which is the first official EH&S document dealing with nanomaterial management and safety policies in the United States [13]. The document covers five primary categories: (i) instrumentation, metrology, and analytical methods, (ii) nanomaterials and human health, (iii) nanomaterials and the environment, (iv) human and environmental exposure assessment, and (v) risk management methods, which have become the operational standards for nanotechnology industry and research. Five years later, in 2013, the project began its second run of strategic planning and EH&S review and revision, which is still in working progress at the present time. The most exciting current event, perhaps, is the "Nanotechnology-Inspired Grand Challenge," launched on 20 October 2015, calling for the scientific community to collaborate in creating "a new type of computer that can proactively interpret and learn from data, solve unfamiliar problems using what it has learned, and operate with the energy efficiency of the human brain" [14]. That is, a humanized computer or an artificial intelligence machine. According to the NNI reports, this *Grand Challenge* has generated broad interest within the nanotechnology community, including federal and state agencies, and private sectors.

In other parts of the world, on 12 May 2004, the European Commission adopted the communication "Towards a European Strategy for Nanotechnology" [15], which brought the discussion on nanoscience and nanotechnology to an institutional level as an integrated and responsible strategy planning. In the following year, the European Commission adopted the action plan "Nanosciences and Nanotechnologies: An action plan for Europe 2005–2009" [16], which defined a series of articulated and interconnected actions for immediate implementation based on the strategy approved in the previous year. At the same time, the Britain's Royal Society and the Royal Academy of Engineering published "Nanoscience and Nanotechnologies: Opportunities and Uncertainties" [17], cautioning the risks in the potential health, environmental, social, ethical, and regulatory issues with this new field. The concerns with the general public's environmental, health, and safety raised in this document may have triggered NNI's response with a two-year study and its final 2008 EH&S document, which eventually set the global policies on nanomaterial-operating protocols and related safety regulations.

1.1.3 The Bottom-Up Approach

Our discussion so far seems to have focused narrowly on the size miniaturization and, indeed, there are virtues of the size reduction in improving the performance of devices. Personal electronics *3C* (computer, communication, and consumer electronics), for example, are getting smaller and smaller each day now and yet they also deliver faster and better quality results ever than before. So, is it that nanotechnology is all about things being small? Though many people may have emphasized over the "nano" part of the technology, this field is far more than the size. It is with the concept of building functional materials and devices from the bottom up. Manufacturing things from the atomic and molecular levels, so-called *nano-manufacturing*, requires carefully designed scale-up of nanomaterials, devices, and systems. The quality of materials can be greatly enhanced through the processes of nano-manufacturing, leading to stronger, light-weighted, durable, and flexible products that, in most cases, also possess unique features to meet their specific needs.

How to manufacture materials from the bottom up? An extreme approach is to design, synthesize, and assemble functional molecules through organic chemistry. A fascinating example in this regard is the synthesis of molecular machines made of parts that are mobile relative to each other. The machines that have been demonstrated to work at the molecular level include a mechanically interlocked architecture consisting of two inter-locked macrocycles (known as the *catenane*), a dumbbell-shaped molecule threaded through a macrocycle (known as the *rotaxane*), and a molecular motor consisting of bis-helicene connected by an alkene double bond [18]. The first nanocar ever built, as envisioned by Feynman in 1959, was made up from oligo(phenylene ethynylene) with alkynyl axles and four spherical C_{60} wheels as in a conventional car [19]. In recognition of their achievements in molecular nanotechnology, the Nobel Prize in Chemistry 2016 was awarded jointly to Jean-Pierre Sauvage, Fraser Stoddart, and Bernard Feringa "for the design and synthesis of molecular machines" [18].

Another effective approach is molecular self-assembly, by which large molecules automatically arrange themselves in an orderly three-dimensional structural configuration with the secondary interactions (hydrogen and polar bonding) and tertiary interactions (van der Waals forces) in various parts of the molecules. Self-assembled nanomaterials, which are typically grown from the surface of a nicely prepared substrate, normally show well-defined and regular patterns in its three-dimensional conformational structure and therefore greatly enhance the material strength against any external stress or shear forces. Some other common practices in nanomanufacturing include *chemical vapor deposition* to create high-performance thin films, *dip pen lithography* using AFM to write nanoprints, and *roll-to-roll processing* to produce high-volume rolls of nanowires, etc. [11]. The number of new manufacturing processes is increasing as more nanomaterials are discovered.

A new aspect in manufacturing nanomaterials is the entering of quantum effects. When the dimension of a substance is reduced to 100 nm or less, in the range of the so-called *quantum realm*, there will be physical properties that behave at a different level than the classic Newtonian physics due to the quantum effects, as predicted in

Feynman's speech. Scientists are finding more and more nanomaterials displaying a set of unique optical, magnetic, mechanical and/or electrical properties that are different than those of the same materials in the bulk. Some of these physical features have added benefits in establishing new ways of conducting research or even potentially creating a new technology all by itself. Using these unique properties at the nanoscale, new frontiers have been opened in science which would not have been possible otherwise in the past and new research fields have been launched that may lead to an entirely new landscape of science and technology in the next generations. Fluorescent nanodiamond presented in this book is an example of such nanomaterials.

A final note, before we leave this discussion of nanotechnology, is that some in the general public may have been misled to believe that nanotechnology is a niche field. It is not. Nanotechnology is truly an interdisciplinary and multipurpose field, as its impact will reach virtually all industries and research areas, which will be demonstrated in ample occasions later in the book.

1.2   Nanocarbons: Now and Then

The second half of this chapter will focus on nanomaterials made up by only carbon atoms, i.e. the nanocarbon family. It seems that we have already known the chemistry of carbon quite well, mostly in the form of graphite, through its long association with humans in history. Carbon is a simple element, predominantly consisting of the same number (6) of protons and neutrons, and therefore C-12 (98.9% natural abundance) is radioactively stable. The other two naturally occurring isotopes are C-13 (1.1%) and C-14 (~1 ppt), of which C-14 is radioactive with a half-lifetime of 5730 years and has been routinely used to date archeological findings [20]. With its relatively low mass, carbon is in the 6th place of all the 100 plus elements known today with an atomic number of $Z = 6$. In the universe, carbon ranks the fourth element of cosmic abundance, after hydrogen, helium, and oxygen, and counts for about 20% by weight of all life stocks and humans on earth, next only to oxygen (60%) [21]. Grouped together with silicon, germanium, tin, and lead (the 4A group) on the Periodic Table of Elements, the elemental carbon may be considered chemically inert compared with the neighboring nitrogen and oxygen groups because of its thermodynamic stability. Also because of this stability, graphite solid is often used as a reference state for just about all thermodynamic measurements and data. Of course, such inertness is changed completely when carbon is combined with hydrogen, oxygen, and/or nitrogen atoms, and a whole world of interesting reactions will occur, as we have learned some in our college organic chemistry courses that strong carbon chains are the backbones of all organic molecules.

So, why carbon now?

Carbon is arguably the most common element that supports and interacts with a vast variety of materials since the dawn of human civilization, much like a dear old friend. What is it so special about carbon among all elements on the planet Earth that scientists and researchers are now calling it the "carbon age" [22, 23]? Specifically, why are the carbon nanomaterials? Perhaps, there is something that we never knew about this old friend and it deserves a close look particularly in the nanometer regime. Here, we trace the development of carbon nanomaterials to explore their molecular structures and intrinsic properties, which have shown great potentials to many innovative practices and applications both in research laboratories and as commercial products.

1.2.1 Classification

Carbon nanomaterials have four major allotropes: *fullerene, carbon nanotube* (CNT), *graphene*, and *nanodiamond* (ND), all consisting of only carbon atoms and each has its own characteristic nanostructure. These four forms of carbon are often referred to as zero-dimensional (fullerene), one-dimensional (CNT), two-dimensional (graphene), and three-dimensional (ND) nanomaterials, mainly based on their molecular structures and electronic properties. Graphite is normally viewed as graphene layers stacking up in three dimensions and hence not an allotrope of carbon. Compared with the other three allotropes, ND is unique in that its constituent carbon atoms are all present in sp^3 electronic configuration, except those on the surface. The multiple forms of carbon present a rich class of solid-state materials that are both environmentally friendly and sustainable among other distinct features of their own.

In an attempt to maintain a focused theme, we will discuss fullerenes, CNTs, and graphenes here and leave NDs to the next chapter.

1.2.2 Fullerenes

A brief look into the history and development of carbon nanomaterials here seems useful to reacquaint ourselves with some milestones in the development of new forms of carbon, with an emphasis on their unique structures and general features. A good place to start is at the time when carbon nanoparticles first entered the center stage of science as the team of Harold Kroto, Richard Smalley, and Robert Curl boldly introduced to the world their soccer ball structure for the C_{60} molecule, now recognized as *buckminsterfullerene* (or *fullerene* in short), in 1985 [10].

The discovery of C_{60} stemmed from the research of Kroto, who investigated the origins of long linear carbon chain molecules in the interstellar medium [24]. These unusual, long, and flexible molecules in skies, originally hypothesized to be created in the atmospheres of carbon-rich red giant stars, were suspected to link with a long-standing puzzle in astronomy – the carriers of the mysterious diffuse interstellar bands (see more details in Chapter 14). Through an introduction by Curl, Kroto applied the cluster beam apparatus developed by Smalley to perform experiments in searching for new and stable carbon molecules and/or clusters that may exist in the outer space. Their experimental results indicated the presence of an odd molecule containing 60 carbon atoms with a relatively strong signal in the time-of-flight mass spectra. The molecule, C_{60}, formed readily after annealing in the gas phase and exhibited extraordinary stability. Its molecular structure, however, took the scientists some extra effort to finally settle on an unprecedented model of all 60 carbons enclosed in totally symmetric spheroidal geometry (Figure 1.1) [24]. An immense (and somewhat unexpected) passion caught up almost immediately in the scientific community to search for members of the fullerene family, to learn their chemistry as well as the applications of these new buckyballs [25]. Indeed, C_{70}, C_{76}, C_{84}, and many other fullerene siblings were found in no time, all sharing a set of unique features as outlined here.

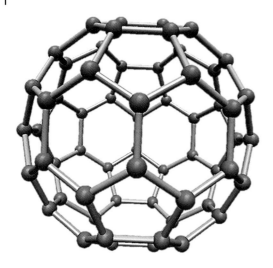

Figure 1.1 A molecular model of C_{60}.

Fullerene is a carbon molecule comprising a hollow cage with each carbon atom connected to its three immediate neighbors in a ring configuration. There are two types of rings fused together to form the fullerene cage: five (pentagonal) and six (hexagonal) members. As D'Arcy Thompson had pointed out at the turn of the 1900s that no system of hexagons alone can enclose a space [26], it would need exactly 12 pentagonal rings in combining with a varying number of hexagonal rings to piece together a cage. This is known as the *Euler's theorem* in geometry and topology, which states that the number of vertexes (V), edges (E), and faces (F) of a *simple* polyhedron are related by the formula:

$$V - E + F = 2. \tag{1.1}$$

The word "simple" here refers to "one piece without holes."

For C_{60}, as shown in Figure 1.1, the 60 carbon atoms give $V = 60$. There are three bonds per atom and therefore a total of $3 \times 60 = 180$ bonds, yielding 90 edges (each edge shared by two atoms). Following the Euler's formula, $60 - 90 + F = 2$, C_{60} has 32 faces (32 rings), i.e. 20 hexagons and 12 pentagons. Therefore, fullerenes are defined as "polyhedral closed cages made up entirely of n three-coordinate carbon atoms and having 12 pentagonal and $(n/2 - 10)$ hexagonal faces, where $n \geq 20$." [27]. So, C_{70} has 25 hexagons, C_{84} has 32, and so forth. Other polyhedral and closed cages that are made up entirely of n three-coordinated carbon atoms but may contain more than just five- or six-membered rings are known as quasi-fullerenes. No two pentagons can share the same edge and, consequently, every carbon atom must belong simultaneously to one pentagonal and two hexagonal rings. Therefore, while C_{60} shapes as a true spheroid, other fullerenes such as C_{70} appear as an oblong figure resembling more like a rugby ball, rather than a soccer ball.

Fullerenes are stable molecules. C_{60}, for example, weighs only 720 amu (or 720 Da) in a tiny size of about 1 nm in diameter. In an early study of the chemistry of fullerenes, both C_{60} and C_{70} were discovered to become superconducting when doped with alkali metals and showed a record-high temperature by the time: 18 K [28]. On the other hand, when placed under favorable conditions (mostly through photocatalyzed reactions), fullerenes may become reactive with incoming nucleophiles, converting carbon's sp^2-hybridization to sp^3 through nucleophilic additions [29]. During the reactions, the hexagonal benzene rings remain largely intact as the nucleophiles mainly interact with the sp^2-electrons on the pentagonal rings. The changes in hybridization, accompanied by a conformational change from the curved planar to a three-dimensional structure, are believed to relax the overall surface strain and therefore provide necessary stability for the final products.

In 1996, the Nobel Prize in chemistry was awarded to Robert Curl, Harold Kroto, and Richard Smalley "for their discovery of fullerenes" [30]. It is clear now as we look back to the history that the impact of C_{60} is not merely the discovery of a new form of carbon in material; rather, it has opened our eyes to see the infinitive possibilities of new structures that can be constructed by familiar elements, carbon and beyond, and the applications never imaginable before are now possible to humans. After all, we may not really know carbon, an old friend of ours, as well as we have thought.

1.2.3 Carbon Nanotubes

Around 1990, while exploring new forms of fullerenes, Smalley suggested the possible existence of the tubular fullerene: a straight segment of a carbon tube capped, perhaps, by two hemispheres of C_{60} on both ends of the tube [31]. Smalley's idea may or may not directly link with the grand entry of CNTs into the scientific world, but for sure when Sumio Iijima reported the first observation of *multiwalled carbon nanotubes* (MWCNTs) [32] in the following year (1991), the world met yet another wave of awes and wonders of a new form of carbon.

Producing MWCNTs by arc-discharge evaporation of graphite and examining the products with electron microscopy, Iijima was able to show clearly the presence of concentric nanotubes comprised of 2–50 layers of carbon sheets. At the center, the smallest hollow tube has a diameter of 2.2 nm with additional layers separated from each other by 0.34 nm, matching exactly the distance between two adjacent layers of bulk graphite. Two years later (1993), *single-walled carbon nanotubes* (SWCNTs) were made and identified for the first time in laboratories by Iijima and Ichihashi [33] and, independently, by Bethune et al. [34]. SWCNT is now recognized as rolling a one-atom-thick, sp^2-hybridized carbon sheet into a cylindrical configuration with diameters ranging from a few Angstroms to a few nanometers, all depending on the particular methods and conditions of synthesis. The length typically varies from hundreds of nanometers to several micrometers.

Almost all of CNT's intrinsic properties are derived from its unique structure, which at the first glance looks like "chicken wire," all composed of six-membered rings. We use SWCNT here as an example to describe parameters necessary for defining the nanotube structure. As depicted in Figure 1.2a [35], a SWCNT can be pictured as a section cut off from a graphene sheet, represented by dashed lines on the figure and rolled over to seal on both edges seamlessly so that any point along one edge will match an equivalent point on the other edge perfectly, thereby forming a cylindrical tube. Depending on how the section is cut, the relative orientation of the cuts with respect to the sheet will determine the structure of CNT as well as many of the nanotube's physical properties. By convention, the orientation of these cuts and the size of the tube (in diameter) are all defined by a pair of numbers, called *chiral indices (n,m)* [36], as described below.

(a)

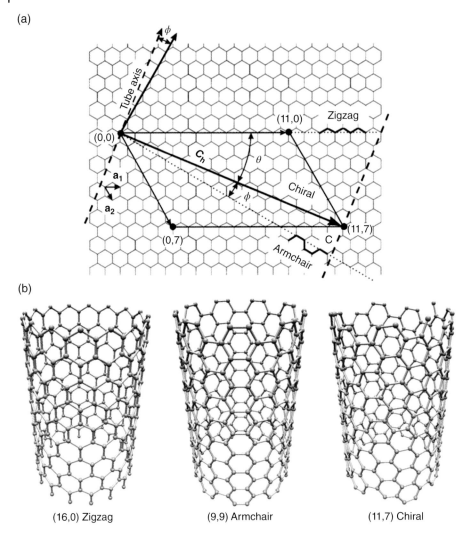

(b)

(16,0) Zigzag   (9,9) Armchair   (11,7) Chiral

Figure 1.2 (a) Formation of SWCNT from a carbon sheet showing the unit vectors (a_1 and a_2), the chiral vector (C_h), the chiral indices (n,m), and the chiral angle (θ). The wrapping angle is $\phi = 30° - \theta$. In this particular example, by rolling up the sheet along C_h such that the origin (0,0) coincides with point C, a nanotube indicated by indices (11,7) is formed. (b) Three representative SWCNT structures in similar diameter: zigzag, armchair, and chiral. *Source:* Adapted with permission from Ref. [35]. Reproduced with permission of Nature Publishing Group.

The first and foremost structural parameter to describe the nanotube is the *chiral vector*, C_h, which follows along the circumference of the tube and is perpendicular to the wall or the tube axis (Figure 1.2a). The chiral vector connects two equivalent sites on the graphene sheet and is defined as $C_h = na_1 + ma_2$, where a_1 and a_2 are unit vectors of the hexagonal honeycomb lattice and n and m are integers. The vector also defines a *chiral angle*, θ, which is the angle between C_h and the zigzag direction of the sheet. In a sense, the zigzag vector originates at point (0,0) and intersects only joining points of the rings, defining the orientation of the entire carbon sheet, whereas the chiral vector results from the CNT alone. The angle θ is a function of the chiral indices as

$$\theta = \tan^{-1}\left[\frac{\sqrt{3}\,m}{2n+m}\right] \tag{1.2}$$

and the diameter (d) is also related to the chiral indices by

$$d = \frac{\sqrt{3(n^2 + nm + m^2)}d_{cc}}{\pi}, \tag{1.3}$$

where d_{cc} is the C—C bond length (0.145 nm). For the example given in Figure 1.2a, the chiral indices are (11,7), which makes it a chiral SWCNT, with a chiral angle of $\theta = 23°$ and a diameter of $d = 1.26$ nm. For SWCNTs with a zigzag (n,0) or armchair (n,n) structure, the corresponding chiral angles are $\theta = 0°$ and 30°, respectively.

Evidently, the topology of a nanotube can be well characterized by this pair of numbers, (n,m), which in turn defines the unique symmetry of CNT as either chiral or achiral (i.e. armchair and zigzag in Figure 1.2b). As it turns out, the chirality is closely tied to the electronic properties of the nanotubes. Derived from their electronic energy gaps, CNTs can be metallic or semiconducting, depending on their chirality and diameters [36–38]. In addition, the huge aspect ratio (tube length/diameter) of CNT makes it an ideal one-dimensional system, in which the charge carrier scattering is drastically reduced. It has been found that the one-dimensional carbon crystalline lattice can conduct electricity at room temperature with virtually no resistance, a phenomenon known as ballistic transport, where electrons can move freely throughout the structure without any scattering from atoms. Furthermore, the lack of interface states as those existing at the silicon/silicon dioxide interface provides a greater flexibility to the fabrication process. It is one of the most promising nanomaterials to realize molecular electronics [39].

The strong carbon–carbon bonds and their resultant network make CNTs 20 times stronger (tensile strength) than steel but only 1/16 as dense [40]. As a result, CNTs have found their commercial use in lightweight applications, such as bicycle frames and golf clubs. With an enormous specific surface area, greater than 1600 m^2 g^{-1}, SWCNTs show an extraordinary surface adsorption capability comparable or, in some cases, better than activated carbon as an adsorbent [41]. The discovery of MWCNTs and SWCNTs together greatly boosted the intensity of the CNT research, leading to an exponential growth of scientific publications each year. In the following 10 years since 1991, published CNT papers reached a total number of nearly 19 000 [41], setting a phenomenal record for a single material in public's interests.

1.2.4 Graphenes

As all we know, graphenes may have been around us for ages. Nevertheless, it was only in 2004 that Konstantin Novoselov and Andre Geim elegantly isolated single- and few-layer of "suspended" graphenes from highly oriented pyrolytic graphite using a Scotch tape [42]. Graphene is characterized as individual or few-layer stacked sheets of

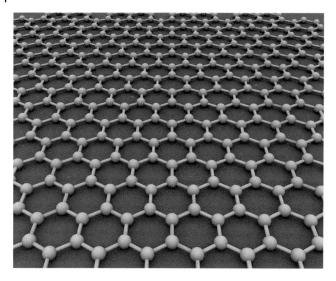

Figure 1.3 A molecular model of graphene. The almost perfect web is only one atom thick. *Source:* Reprinted with permission from Ref. [50].

sp^2-hybridized carbon, where the number of sheets does not exceed 10. The structure of graphene can be referred to as infinite polycyclic aromatic hydrocarbons containing an infinite number of benzene rings fused together (Figure 1.3). Graphene is the first two-dimensional nanomaterial known to exist in suspended form, defying previous conventional knowledge that this two-dimensional material would have been too thermodynamically unstable to exist [43]. In fact, with an intrinsic strength of 130 GPa, the Young's modulus per layer of 350 N m^{-1}, and a breaking strength of 42 N m^{-1}, graphene is one of the world's strongest materials ever discovered and warrants a title as *super carbon* [44].

The honeycomb structure of graphene is formed by an infinite number of benzene rings where each carbon atom uses three of its four valence electrons to form sp^2-hybridized covalent bonds with three coplanar neighboring carbons. The fourth valence electron occupies the carbon's p_z orbital that forms sp^2-bonds shared equally in three directions, leading to a bond order of 1 and 1/3. Delocalized sp^2-electrons now spread over on a continuous layer of honeycomb constructed by short and rigid covalent bonds, and together they provide graphene an extraordinary stability to withstand a great deal of mechanical strain and stress, as indicated by its unusually high Young's modulus value [44, 45]. On a smooth flat surface like this, any injected charge carriers can run freely at an incredibly high speed, while at the same time the graphene sheet experiences rapid lattice vibrations (i.e. phonons). Both effects contribute to the excellent electrical and thermal conductivity of this nanomaterial. Furthermore, because all carbon atoms are identical and symmetrically arranged on the honeycomb plane, graphene is nonpolar and hydrophobic except the edges. Therefore, the nanomaterial has a very poor solubility in water or even regular polar solvents used in the laboratories, which actually imposes challenges in processing graphene functionalization [46].

As a two-dimensional crystalline membrane, graphene possesses a set of unique physical properties, collectively, surpassing any other materials known today. These properties include an exceedingly large specific surface area ($2630 \, \mathrm{m^2 \, g^{-1}}$), low density (<$1 \, \mathrm{g \, cm^{-3}}$), ultrahigh charge mobility (>$2 \times 10^5 \, \mathrm{cm^2 \, V^{-1} \, s^{-1}}$), excellent electrical conductivity ($106 \, \mathrm{S \, cm^{-1}}$) and thermal conductivity (>$5000 \, \mathrm{W \, m^{-1} \, K^{-1}}$), an uniform broadband optical absorption (ultraviolet to far infrared), superb mechanical strength, and unusual flexibility [47]. So far, researchers have steered graphenes into an array of areas including some of the most popular subjects such as energy conversion and storage in fuel cells (e.g. supercapacitors), ultrafast field-effect transistors, transparent conductors, carbocatalysis (i.e. green chemistry), chemical sensors, polymer composites, … and the list goes on. It is expected that with more research done, we will get to learn more about the full potential of graphenes and what the nanomaterials can do for the human society.

Graphenes are in the early stages of macroscale manufacturing for commercial use. They can be applied as conductive reinforcement coatings on Kevlar fibers and in the fabrication of large-area transparent electrodes and low-loss high-permittivity composites, only to name a few [48]. Used in energy storage devices, graphenes support energy capacity that is cost-effective far beyond other materials (in terms of energy per weight and cost). With all these remarkable electronic properties, graphenes are the top candidate to replace silicon-based electronics, which is approaching its own material limit. When realized, the carbon-based electronics promises to perform at a lightning speed with superb capacity and, now more pressingly than ever, truly green, both socially and economically.

Finally, as an endnote, Andre Geim and Konstantin Novoselov received the Nobel Prize in Physics 2010 "for groundbreaking experiments regarding the two-dimensional material graphene" [49].

References

1 United States National Nanotechnology Initiative (2016). NNI supplement to the President's 2017 Budget. https://www.nano.gov/node/1573 (accessed 16 April 2018).
2 National Science Foundation (2016). FY 2017 Budget Request. https://www.nsf.gov/about/budget/fy2017 (accessed 16 April 2018).
3 McWilliams, A. (2014). Nanotechnology: a realistic market assessment. https://www.bccresearch.com/market-research/nanotechnology/nanotechnology-market-assessment-report-nan031f.html (accessed 16 April 2018).
4 Taniguchi, N. (1974). On the basic concept of nanotechnology. *Proceedings of the International Conference on Production Engineering*, Tokyo, Part II (Japan Society of Precision Engineering).
5 Feynman, R.P. (1959). There's plenty of room at the bottom. http://calteches.library.caltech.edu/1976/1/1960Bottom.pdf. (accessed 16 April 2018).
6 Ito, T. and Okazaki, S. (2000). Pushing the limits of lithography. *Nature* 406: 1027–1031.
7 Binnig, G., Rohrer, H., Gerber, C., and Weibel, E. (1982). Surface studies by scanning tunneling microscopy. *Phys Rev Lett* 49: 57–61.
8 Binnig, G. and Quate, C.F. (1986). Atomic force microscope. *Phys Rev Lett* 56: 930–933.
9 The Nobel Prize in Physics 1986. *Nobelprize.org*. Nobel Media AB 2014. https://www.nobelprize.org/nobel_prizes/physics/laureates/1986 (accessed 16 April 2018).

10 Kroto, H.W., Heath, J.R., O'Brien, S.C. et al. (1985). C_{60}: Buckminsterfullerene. *Nature* 318: 162–163.

11 United States National Nanotechnology Initiative (2018). Nanotechnology timeline. http://www.nano.gov/timeline (acessed 16 April 2018).

12 National Science and Technology Council Committee on Technology (2014). National Nanotechnology Initiative Strategic Plan. https://www.nano.gov/sites/default/files/pub_resource/2014_nni_strategic_plan.pdf (acessed 16 April 2018).

13 The National Science and Technology Council (2008). Strategy for Nanotechnology-Related Environmental, Health and Safety Research. https://www.nano.gov/node/254 (acessed 16 April 2018).

14 United States National Nanotechnology Initiative (2015). Nanotechnology-inspired grand challenges. http://www.nano.gov/grandchallenges (accessed 16 April 2018).

15 European Commission (2004). Towards a European strategy for nanotechnology. https://ec.europa.eu/research/industrial_technologies/pdf/policy/nano_com_en_new.pdf (accessed 16 April 2018).

16 European Commission (2005). Nanosciences and nanotechnologies: an action plan for Europe 2005–2009. https://ec.europa.eu/research/industrial_technologies/pdf/policy/action_plan_brochure_en.pdf (accessed 16 April 2018).

17 The Royal Academy of Engineering (2003). Nanoscience and nanotechnologies: opportunities and uncertainties. http://www.nanotec.org.uk/finalReport.htm (acessed 16 April 2018).

18 The Nobel Prize in Chemistry 2016. *Nobelprize.org.* Nobel Media AB 2014. https://www.nobelprize.org/nobel_prizes/chemistry/laureates/2016 (accessed 16 April 2018).

19 Shirai, Y., Osgood, A.J., Zhao, Y. et al. (2005). Directional control in thermally driven single-molecule nanocars. *Nano Lett* 5: 2330–2334.

20 Greenwood, N.N. and Earnshaw, A. (1997). *Chemistry of the Elements.* 2e. Butterworth-Heinemann.

21 Croswell, K. (1995). *Alchemy of the Heavens.* Anchor Book.

22 Roston, E. (2009). *The Carbon Age.* Walker & Co.

23 Feaver, A. (2010). The carbon age: dark element, brighter future. https://www.cnet.com/news/the-carbon-age-dark-element-brighter-future (acessed 16 April 2018).

24 American Chemical Society (2010). Discovery of fullerenes. National historic chemical landmark. https://www.acs.org/content/acs/en/education/whatischemistry/landmarks/fullerenes.html (acessed 16 April 2018).

25 Kroto, H.W. and Walton, D.R.M. (eds.) (1993). *The Fullerenes: New Horizons for the Chemistry, Physics and Astrophysics of Carbon.* Cambridge University Press.

26 Thompson, D.W. (1969). *On Growth and Form.* Cambridge University Press.

27 Godly, E.W. and Taylor, R. (1997). Nomenclature and terminology of fullerenes: A preliminary study. *Pure Appl Chem* 69: 1411–1434.

28 Hebard, A.F., Rosseinsky, M.J., Haddon, R.C. et al. (1991). Superconductivity at 18 K in potassium-doped C_{60}. *Nature* 350: 600–601.

29 Haddon, R.C. (1993). Chemistry of the fullerenes: the manifestation of strain in a class of continuous aromatic molecules. *Science* 261: 1545–1550.

30 The Nobel Prize in Chemistry 1996. *Nobelprize.org.* Nobel Media AB 2014. http://www.nobelprize.org/nobel_prizes/chemistry/laureates/1996 (accessed 16 April 2018).

31 AZoNano (2004). Carbon nanotubes – history and development of carbon nanotubes (Buckytubes). http://www.azonano.com/article.aspx?ArticleID=982 (accessed 16 April 2018).

32 Iijima, S. (1991). Helical microtubes of graphitic carbon. *Nature* 354: 56–58.

33 Iijima, S. and Ichihashi, T. (1993). Single shell nanotubes of 1-nm diameter. *Nature* 363: 603–605.

34 Bethune, D.S., Kiang, C.H., Devries, M.S. et al. (1993). Cobalt-catalyzed growth of carbon nanotubes with single-atomic-layer walls. *Nature* 363: 605–607.

35 Wilder, J.W.G., Venema, L.C., Rinzler, A.G. et al. (1998). Electronic structure of atomically resolved carbon nanotubes. *Nature* 391: 59–62.

36 Saito, R., Fujita, M., Dresselhaus, G., and Dresselhaus, M.S. (1992). Electronic structure of chiral graphene tubules. *Appl Phys Lett* 60: 2204–2206.

37 Hamada, N., Sawada, S.I., and Oshiyama, A. (1992). New one-dimensional conductors: graphite microtubules. *Phys Rev Lett* 68: 1579–1581.

38 Carlson, L.J. and Krauss, T.D. (2008). Photophysics of individual single-walled carbon nanotubes. *Acc Chem Res* 41: 235–243.

39 Avouris, P. (2002). Molecular electronics with carbon nanotubes. *Acc Chem Res* 35: 1026–1034.

40 Collins, P.G. and Avouris, P. (2000). Nanotubes for electronics. *Sci Am* 283: 62–69.

41 Kondratyuk, P. and Yates, J.T. Jr. (2007). Molecular views of physical adsorption inside and outside of single-wall carbon nanotubes. *Acc Chem Res* 40: 995–1004.

42 Novoselov, K.S., Geim, A.K., Morozov, S.V. et al. (2004). Electric field effect in atomically thin carbon films. *Science* 306: 666–669.

43 Novoselov, K.S., Jiang, D., Schedin, F. et al. (2005). Two-dimensional atomic crystals. *Proc Natl Acad Sci USA* 102: 10451–10453.

44 Savage, N. (2012). Materials science: super carbon. *Nature* 483: S30–S31.

45 Lee, C., Wei, X.W., Kysar, J.W., and Hone, J. (2008). Measurement of the elastic properties and intrinsic strength of monolayer graphene. *Science* 321: 385–388.

46 Quintana, M., Vazquez, E., and Prato, M. (2013). Organic functionalization of graphene in dispersions. *Acc Chem Res* 46: 138–148.

47 Allen, M.J., Tung, V.C., and Kaner, R.B. (2012). Honeycomb carbon: a review of graphene. *Chem Rev* 110: 132–145.

48 James, D.K. and Tour, J.M. (2013). Graphene: powder, flakes, ribbons, and sheets. *Acc Chem Res* 46: 2307–2318.

49 The Nobel Prize in Physics 2010. *Nobelprize.org*. Nobel Media AB 2014. https://www.nobelprize.org/nobel_prizes/physics/laureates/2010 (accessed 16 April 2018).

50 AlexanderAlUS (2010). Own Work, CC BY-SA 3.0. https://commons.wikimedia.org/w/index.php?curid=11294534 (accessed 16 April 2018).

2

Nanodiamonds

Now the final allotrope of nanocarbon–*nanodiamonds* (NDs), and before that, there are diamonds. Famous or not, every diamond comes with a story, some publicly known and others remain in private talks. This is the story of the French *Regent diamond*.

2.1 Ah, Diamonds, Eternal Beautiful

Many famous diamonds today, including the French Regent diamond, originally came from India, each shadowed by a long trail of history [1]. Behind the flawless beauty of diamonds lies the dark side of human nature, reflecting personal selfishness and greediness, social and political power struggles, theft, robbery, murders, and even wars. The Regent diamond has it all. The original gem was said to have been discovered in 1701 by a slave working in a diamond mine at Kistna of Southern India. It presented as the opportunity of lifetime for the slave to gain his freedom. He hid the 410 carats stone, weighing 82 g, in a deep wound cut in his leg before fleeing to the south coast in search of a quick getaway to a new land and new life altogether. Shrouded by the agony of being caught and the risks of exposing his valuable treasure, he soon hooked up with an English skipper, entrusting the seaman with his valuable stone in exchange for a stealthy sail immediately. No one ever knew the fate of the slave, but the seaman came back alone with the stone in his pocket, which he sold quickly to a diamond merchant, Jamchund, for about £1000. Like a drunken sailor, he squandered the money as fast as it had come in and was last seen "hanging himself in remorse."

Jamchund, well known for his dealings in the eastern region of India, and now holding the stone, proceeded to negotiate with the few potential buyers available to him, including the English governor of Fort St. George, Thomas Pitt, who eventually bought it at a price of £20 400. After Mr. Pitt's purchase of the stone, suspicions lingered so strongly as to how the transaction actually took place, that upon returning to England in 1710 he felt obliged to send a public letter to the chief editor of the *European Magazine* clarifying the early history of the diamond and his negotiations with Jamchund. The story as told was never clear whether it was the fame of the *Pitt diamond* (as it was called at the time) or Mr. Pitt's own reputation as a bold, temperamental, and sometimes lawless businessman that has fostered such sustained scandalous suspicions. Whatever it may have been, the criticism haunted Mr. Pitt for the remainder of his life.

Fluorescent Nanodiamonds, First Edition. Huan-Cheng Chang, Wesley Wei-Wen Hsiao and Meng-Chih Su.
© 2019 John Wiley & Sons Ltd. Published 2019 by John Wiley & Sons Ltd.

By this time the diamond had been cut by a skillful jeweler in London named Harris, in a period over two years time (from 1704 to 1706), rendering a 136.8 carat diamond. It was a *brilliant* cut, which was a particular form of cut maximizing the reflecting facets in a cone shape, thereby channeling maximum light through the top of the diamond. In a broadly square shape, 1.2 inch by 1 inch, the diamond was cushioned in a 0.75-inch depth and appeared to be perfectly white with a faintly pale blue "of the first water." Its flawless cut presented brilliant speckles in a breathtaking beauty that no other diamond could match and, to this day, is still regarded as the finest diamond in the world.

In 1717, assisted by the powerful Scotch financier John Law, Mr. Pitt sold the diamond for a price of £135 000 to Philippe II, the Duke of Orleans "Regent of France," ending Mr. Pitt's enduring ordeal of seeking a Royal buyer. France, at that time under Louis XV, purchased the diamond upon the suggestion of Mr. Law that "France ought to possess a gem the finest ever seen in Europe." From then on the diamond assumed its new title, the "Regent," the name carried till this day.

The Regent made its public debut in 1721, when Louis XV wore it to attend a party at the Turkish embassy, where it immediately became the center of attention among the European royals and nobles. The next year at his coronation (25 October 1722), Louis XV had the Regent temporarily mounted on his new crown as the lead gem at the front center. Later, the King began to wear the Regent on his hat, a habit he kept for the rest of his life at the height of French art and culture in Europe. However, under the glamorous surface was the hardship wrought upon France by years of social unjustness and wars, in particular the *Seven Years War*, as so vividly characterized by historian Arthur Tilley [2]:

> At this time the fable of the four cats became current: the thin cat was the people, the fat cat the financiers, the one-eyed cat the ministry, and the blind cat the King who saw nothing and refused to see anything.

The King himself lived a good life and died in 1774, preceded by the death of his own son nine years earlier. The throne was therefore passed on to one of the grandsons, Louis XVI, who also inherited the Regent and used it again as the lead gem at his coronation, on 11 June 1775. Interestingly, Louis XVI also carried on his grandfather's habit of wearing the gem on his hat. Was it a sign of glorification for his Grandpa or a final homage to the glory of the monarchy that was fading fast? History could never tell. Later, Queen Marie Antoinette was also seen to wear the Regent occasionally on her black velvet hat. In his reign, Louis XVI attempted some unsuccessful reforms bringing France into further financial hardship especially for those in the working class, fueling more social unrest. The dissatisfaction and anger of peasants finally erupted in the 1789 French Revolution. Louis XVI and Queen were both captured and later guillotined on 10 January 1792, which brought an end to the uninterrupted French monarchy of more than 1000 years.

One year earlier in 1791, a commission was convened by the National Assembly of the *French First Republic* including the most experienced jewelers in Paris at the time to inventory the French royal crown jewels, with the Regent leading a list of regalia including 9547 diamonds. The collection of the regalia was put on display for the first time to the general public at Garde-Meuble (the *Public Treasury*). Early in September of 1792, as a precaution due to the growing violence in nearby Paris, the display was closed and all treasures safely stored away in the halls of Garde-Meuble. On the morning of

17 September 1792, security guards discovered that the Regent and 11 cabinets full of the regalia had been stolen overnight – the robbery of the century had just been committed, which quickly sent shock waves throughout France and the rest of the world. As expected, an enormous effort was instituted to hunt for the thieves and every rock was turned in the entire country to look for the lost jewelry. In the end, the robbery was never solved.

One year later, in the midst of chaos with the Revolution, the Regent and some other regalia items were miraculously recovered from the attic of a house in the Allee des Veuves, Champs-Elysees, at a tip from an anonymous letter addressed to the *Commune*. Rumors circulated wildly about this remarkable and highly unexpected reappearance of the royal jewelry, yet to this date, history could never reveal the true identity of the mysterious author(s) of the letter. The general consensus held by historians seemed to cite the great difficulties of reselling a diamond of such high profile as the main reason for the return of the Regent, or, is that because the fate of the Regent was already inter-twined with the French politics? After all, this was not the only time the Regent was returned.

It was the Prussians who made the next return of the Regent after Napoleon's defeat in the *Battle of Waterloo*, that lead to his exile to the island of Elba in 1814. Before that, following years of terrors and political power struggles of the *France First Republic*, Napoleon staged the *Coup of 18 Brumaire* in 1799 and seized the power from reign of the *Directory*, declaring essentially a dictatorship over France. Victories of his military campaigns in Europe (the *Napoleonic wars*) and the profits thereby brought to France had empowered Napoleon to become *Emperor of the French* by 1804. Napoleon appeared to have a special preference for the Regent among all other royal gems, and had placed the Regent in the pommel of his sword as depicted in the portraits by Jacques-Louis David, First Painter to the Emperor. Later, it was removed to decorate his two-edged sword: A wish of glorifying the Emperor's undefeated might or a hope for a long-lasting prosperous France? Unfortunately, neither came true. Shortly after losing the battle at Waterloo, Napoleon signed abdication of the throne and was exiled to Elba Island. His son and second wife, Marie Louis, went back to her home country, Austria, taking the Regent with her. Marie's father, the *Habsburg Emperor* Francis II of Austria, who apparently was well acquainted with the value of the Regent and its history within the French royalty, returned the gem to France to avoid any unnecessary misunder-standing or, even worse, some regrettable disputes that could arise because of the gem. Obviously, the *Habsburg Emperor* valued peace for his own people, no matter how brief it might be at that time, above the possession of a gem.

Once back in the collection of French crown jewels, the Regent was mounted again on the crowns of Louis XVIII at his coronation (1814), as well as Charles X (1825) and Napoleon III (1852), during which time the people in France endured years of repaying the war debts caused by the Napoleonic military campaigns. Power struggles in the parliament and royal politics often delayed or derailed the much-needed economic reforms that were necessary for reconstructing the nation. The parliament was filled with elites and wealthy nobles, suspicious of corruption and the abuse of power, which had widened the gaps between the rich and poor sending labor-class people further into poverty and suffering. People's resentment and frustrations continued to mount until finally they erupted into the *Second Revolution* in 1848. As the result, Charles X abdicated from the throne and the *French Second Republic* was established with Napoleon III, a nephew of Napoleon, the former *Emperor of French*, as the President who subsequently became the *Emperor of France* himself four years later.

Under Napoleon III's government, France finally began to rebuild and grow in prosperity. The long-awaited overhaul of the French banking systems and economic reforms significantly improved the country's competiveness in export and trade. Industrialization greatly improved agricultural efficiency, resulting in surplus that finally eliminated the lasting cyclical famines. Urban infrastructure in Paris and other major cities in France were upgraded or reconstructed, attracting increasing number of foreign visitors from all over Europe. Napoleon III was skillful in international relations: Forging alliances with Britain, curbing Russia's aggressiveness, signing favorable trading treaties, and expanding overseas business and colonies. France was emerging as a major player again in the Europe's political landscape. Ironically, while the first Napoleon had left France enormous amounts of national debts after his short-lived military glory, it was his nephew who actually revived France's fortune.

In 1870, Napoleon III lost the Franco-German War and was captured by Prussians in the *Battle of Sedan*. The French parliament immediately proclaimed its *Third Republic* forcing Napoleon III and his wife, Empress Eugenie, into exile in England. The former Emperor died three years later. On official records, the last royalty to wear the Regent was Empress Eugenie, who was reported to have the gem on her Grecian diadem probably before her exile. Since 1887, the Regent has been in the collection of the *French Royal Treasury*, still on display at the Louvre in Paris today [3].

The Regent has witnessed many changes in the world since its birth, but there are few things that never changed and one of them is that the Regent is still the world's finest diamond and of eternal beauty.

2.2 Diamonds: From Structure to Classification

Traditions surrounding diamonds across so many cultures have imbued them with mystical aura. Their ability to endure down through the ages sets them apart from most other substances. But beneath all that history of glamour and status, diamonds are a natural material with a set of unique physical properties, capable of equally critical engineering and industrial utility, and now are making their contributions at a microscopic level to the wellbeing of humanity. We start this section with a discussion on the structure of diamond crystals from a scientific perspective.

2.2.1 Structure

The crystal structure of diamond is symmetric in all three dimensions, known as the diamond cubic lattice. It is the same structure shared by silicon, geranium, and some other alloys, in addition to carbon. Figure 2.1 shows a unit cell of the diamond cubic *Bravais lattice*, in which every carbon atom is represented by a sphere in a three-dimensional matrix that can be repeated infinitely to fill the entire space [4]. A way to look at this crystal structure is by dividing the unit cell into eight smaller cubes, which can be further separated into two groups. The cube of the first group consists of five carbon atoms in a sp^3 tetrahedral configuration with four carbons occupying four corners and the last one at the center of the cube. These carbon-filled cubes are connected by two

Figure 2.1 Unit cell of the diamond cubic crystal structure with a lattice constant of 3.57 Å. Carbon atoms are presented in gray spheres.

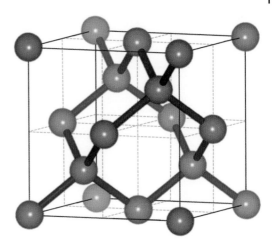

pairs of carbon atoms: One pair located in the lower half of the unit cell aligned diagonally, while the other pair in the upper half also aligned diagonally. These two pairs are staggered in perpendicular from each other. The cube of the other group, however, contains no carbon atoms at the center and, as a result, there are a total of eight carbon atoms $[1 \times 4+(\frac{1}{8}) \times 4 \times 8]$ in the unit cell.

Another way to view the diamond crystal structure is that it is composed of two interpenetrating face-centered cubic lattices aligned in the direction of the body diagonal and displaced by ¼ length of the diagonal. All carbon atoms in the lattice are bonded tetrahedrally with a bond angle of 109.5° and a uniform carbon–carbon bond length of $d_{cc} = 0.154$ nm throughout. Since the bond length corresponds to a translation between the two basis points, (0,0,0) and (1,1,1), it can be related to the width of the unit cell (called *lattice constant, a*) as $d_{cc} = a[(\frac{1}{4})^2 + (\frac{1}{4})^2 + (\frac{1}{4})^2]^{1/2}$, which leads to $a = 3.567$ Å for diamond, a value confirmed by X-ray diffraction measurement. The total number of carbon atoms in such a unit cell is also 8, calculated as $1 \times 4+(\frac{1}{2}) \times 6+(\frac{1}{8}) \times 8$. William Henry Bragg and William Lawrence Bragg were the first to determine the crystal structure of diamond in 1913, and were awarded the Nobel Prize in Physics 1915 "for their services in the analysis of crystal structure by means of X-rays" [5].

With eight carbon atoms squeezed in a 3.567-Å cube, diamond has a mass density of 3.515 g cm^{-3} at room temperature and a number density of 1.76×10^{23} atoms cm^{-3}, which is the highest of any material. Given this exceptionally compact structure, diamond shows several superlative physical properties not found in other carbon materials (Table 2.1) [6]. First, diamond is the hardest known material to date, about 40 times harder than stainless steel. Second, diamond has the highest thermal conductivity of any bulk material, 2000 W m^{-1} K^{-1} at 300 K, about 4 times more than copper. Third, diamond possesses the largest refractive index ($n = 2.41$ in the visible) of all dielectric materials, about 1.6 times as large as that of quartz. Because of these unique properties, together with their remarkable chemical inertness, diamond grits and powders have found a wide range of industrial applications over decades [7]. In 2015, the global rough diamond production from the world's 54 largest diamond mines combined was estimated to be over 100 million carats, or 20 million grams [8], a high demand that comes at a steep price of harsh environmental hazards associated with the diamond mining [9].

Table 2.1 Typical physical properties of diamond.

Property	Value	Unit
Hardness	10 000	$kg\,mm^{-2}$
Strength, tensile	>1.2	GPa
Strength, compressive	>110	GPa
Density	3.52	$g\,cm^{-3}$
Young's modulus	1220	GPa
Thermal expansion coefficient	0.0000011	K^{-1}
Thermal conductivity	20.0	$W\,cm^{-1}\,K^{-1}$
Debye temperature	2200	K
Optical index of refraction (at 591 nm)	2.41	Dimensionless
Bandgap	5.45	eV

Source: From Ref. [6].

Table 2.2 Classification of diamonds.

Type	Feature	Color	Note
Ia	Nitrogen atom concentration up to 0.3%, aggregated nitrogen	Colorless	Most natural diamonds
Ib	Nitrogen atom concentration up to 500 ppm, singly substitutional nitrogen	Yellow, orange	Rare in nature, almost all HPHT diamonds
IIa	No or few nitrogen atoms	Colorless	Rare in nature, the "purest" diamonds
IIb	No or little nitrogen, boron-doped	Blue, gray	Extremely rare in nature, p-type semiconductor

Source: From Ref. [13]. Reproduced with permission of Elsevier.

2.2.2 Classification

A chemically pure and structurally perfect diamond is transparent and colorless, like a drop of pure water. This is because diamond is a wide bandgap semiconductor with an energy gap of $E_g = 5.45$ eV (Table 2.1) between valence and conduction bands (or 5.47 eV in [10]). Colors may arise from the presence of chemical impurities and/or structural defects in the diamond matrix. In gem and jewelry industry, *color, clarity, cut,* and *carat weight* (known as *4Cs*) are four well-accepted standards in describing the gem-quality of diamond [11]. Scientists have been interested in the origin of diamond color not only to gain a better control of the color for market values, but also to learn the factors responsible for these interesting physical properties suitable for applications in electronics, optics, and even medicine.

Nitrogen is the most common impurity present in the crystal lattice of natural diamond [12]. In fact, the content of nitrogen both in quantity and structural form has been used to classify diamond materials (Table 2.2) [13]. Type I diamond contains up to 0.3% (3000 ppm) nitrogen and type II contains none or little nitrogen. Type I is further divided

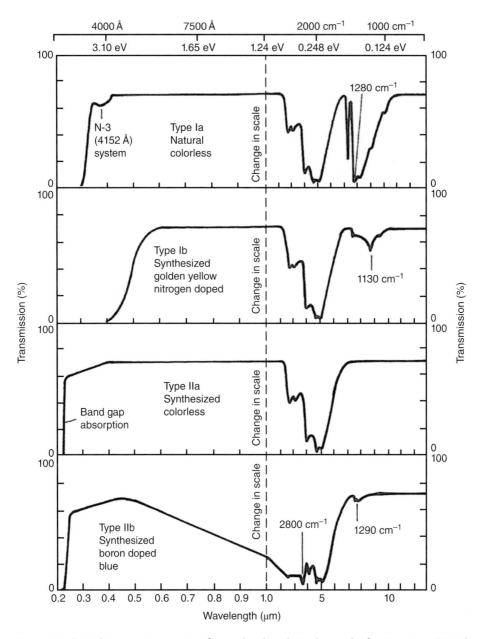

Figure 2.2 Optical transmission spectra of natural and synthetic diamonds of various types. Note the change in scale at 1 μm. *Source:* Adapted with permission from Ref. [14].

into two groups, Ia and Ib, depending on the nitrogen concentration and how these atoms are incorporated into the crystal lattice. The nitrogen atoms in type Ia diamond exist as aggregates, while in type Ib as atomically dispersed nitrogen entities. About 98% of natural diamonds belong to type Ia, containing nitrogen aggregates, with a strong optical absorption below 300 nm and thus colorless (Figure 2.2) [14], whereas the nitrogen atoms

Figure 2.3 Diamond crystallites synthesized by the HPHT method. The particles, diameter approximately 400 μm, contain typically 100 ppm nitrogen.

in type Ib diamond absorb strongly at a longer wavelength (i.e. 400 nm), imparting yellow color to diamond (Figure 2.3). There are also two groups of diamond in class II: Type IIa for pure diamond and type IIb for boron-doped diamond. Type IIa diamond makes up 1–2% of all natural diamonds and is optically transparent with its transmittance spanning from 225 to 2000 nm (Figure 2.2).

To analyze the diamond composition in laboratories, the commonly used analytical instruments include thermal gravimetric analysis, mass spectrometry, infrared spectroscopy, Raman spectroscopy, ultraviolet-visible spectroscopy, X-ray diffraction, optical microscopy, electron microscopy, etc. Mid-infrared spectroscopy (4000–400 cm^{-1} or 2.5–25 μm) is arguably the most useful tool to characterize the nitrogen concentration without causing damages to the crystals (cf., Section 3.1). As we will see throughout the later chapters, there are many examples and problems that cannot be solved without these instruments.

2.3 Diamond Synthesis

Being an allotrope of carbon in the sp^3 configuration, diamond in fact is a metastable phase at normal temperature and pressure (Figure 2.4) [15]. It is less thermodynamically stable than graphite by 2.90 kJ mol^{-1} at 0 °C and 1 atm (i.e. the standard conditions for temperature and pressure). As a result, natural diamonds are typically formed in the lithospheric mantle at a depth of more than 140 km on Earth [16], discovered only after being forced to the earth surface via volcanic activities. At room temperature, diamond is converted to graphite at a very slow rate and notable structural changes only occur at temperatures above 1700 °C in a high vacuum [17]. For diamonds without defects inside the crystal lattice, graphitization starts at surface and proceeds gradually toward the core [18, 19].

Attempts to make diamonds artificially have a long history back to 1797, the year that diamond was first discovered to comprise only carbon atoms [20]. A famous example of the early attempts is the synthesis of diamonds by heating charcoal with iron inside a

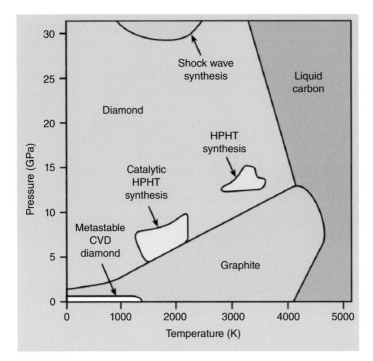

Figure 2.4 Phase and reaction diagram of carbon. The regions where the three synthesis methods, HPHT, CVD, and detonation (or shock wave), discussed in the text are indicated in the figure. *Source:* Adapted with permission from Ref. [15]. Reproduced with permission of John Wiley & Sons.

carbon crucible at temperatures above 3000 °C, followed by rapid cooling of the molten iron in water [21]. The reproducibility of this method, however, was poor. It was not until the 1950s that scientists were able to produce diamonds reliably in the laboratory, recreating the physical construction of what was once thought a singularity of nature. The first approach in this synthesis was to mimic nature by applying a high pressure (either static or transient) to graphite under a high temperature condition. Low-pressure and low-temperature methods were later developed to produce diamonds that is physically and chemically identical to those found in nature. However, the synthesis under such conditions often yielded a mixture of diamond, graphite, and amorphous carbon. Luckily, methods have been developed to overcome this problem by controlling the crystal formation kinetics such that the growth rate of diamond is higher than those of graphite and amorphous carbon [22]. Currently, the three commonly used methods for industrial production of synthetic diamonds are: *High-pressure high-temperature* (HPHT), *chemical vapor deposition* (CVD), and detonation.

2.3.1 HPHT

The first reproducible synthesis of "man-made diamonds" with the HPHT method was demonstrated by a research group at General Electric in 1955 [23]. The method used a hydraulic press to generate a pressure in the range of 5–11 GPa while maintaining the temperature in the range of 1200–2200 °C. A special design of this ultra-high pressure belt

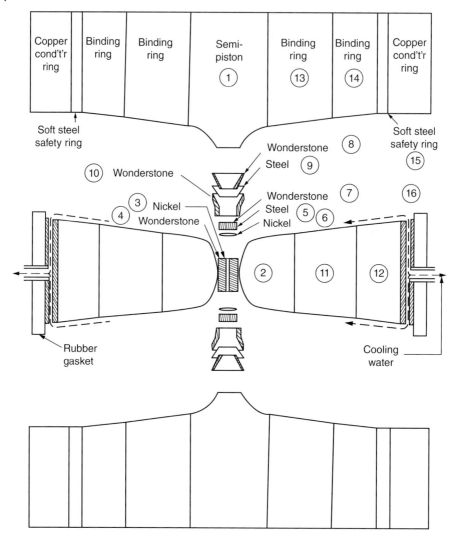

Figure 2.5 Ultrahigh pressure belt apparatus used by General Electric for the HPHT diamond synthesis. *Source:* Reprinted with permission from Ref. [24]. Reproduced with permission of AIP Publishing LLC.

apparatus (Figure 2.5) was that both the chamber and pistons received lateral support from stressed binding rings, enabling the extreme conditions to be maintained for hours [24], which was a critical step in the synthesis. The addition of "solvent-catalyst" consisting of nickel, cobalt, and iron also helped dissolve carbon, allowing the graphite-to-diamond conversion to proceed at lower temperatures and pressures as indicated in Figure 2.4.

The diamond crystals grown by this sophisticate HPHT technique are of high quality and do not contain non-diamond carbon structures (e.g. diamond-like carbon, graphite, and amorphous carbon). Their sizes range from tens of nanometers to about 1 cm. A shortcoming of this method is that the diamonds grown are of type Ib with typical 100–200 ppm of single substitutional nitrogen atoms incorporated into the crystal lattice and thus appear yellowish (Figure 2.3). These man-made diamonds can be easily distinguished from natural diamonds by measuring the concentrations of nitrogen contaminants with infrared spectroscopy (Table 2.2). Although the HPHT diamonds have not played a significant role in the mainstream diamond trade, they are monocrystalline and useful for a variety of industrial processes, including cutting and machining of mechanical components as well as polishing and grinding of optics.

2.3.2 CVD

Diamond growth through CVD was first reported in the 1960s [21]. The method grew diamonds by applying low-pressure carbon-containing gases on a chosen substrate. However, the rate of the growth in early experiments proved to be slow. This was mainly caused by the deposition of graphite, leading to the formation of mixed sp^3/sp^2 phases [25]. The breakthrough in the CVD diamond synthesis came from the discovery that the presence of a large amount of hydrogen atoms in the gas mixtures helped etch away sp and sp^2 carbon atoms, enhancing the diamond growth [26]. The CVD-based diamond growth has become an active and extensive area of research since the 1980s and stimulated industrial manufacture and use of diamond materials in many applications.

A typical source of carbon for CVD diamond growth is methane; however, other carbon-containing gases also work. The growth requires a means of activating precursor molecules diluted in hydrogen, which is commonly performed through thermal methods (e.g. a hot filament) or in microwave plasma (Figure 2.6) [27]. The typical concentration of methane in the gas mixture is 0.5–2%. The growth pretreatments such as mechanical abrading, ultrasonic seeding, and ion bombardment are often used to prepare nucleation centers on the surface of the substrate, which is usually made of copper or silicon [25]. The CVD technology allows for the deposition of thin polycrystalline diamond films on areas up to 100 cm in diameter. Compared with HPHT, the method is less cost-effective but is more versatile in producing diamond films of various kinds. For example, the addition of gases such as trimethyl boron to the gas mixtures facilitates the incorporation of boron defects into the diamond (i.e. doping of diamond), which in turn allows precise tuning of the physical properties (e.g. bandgap and electrical conductivity) of such diamond materials. The CVD films can later be milled to gain particles of various sizes and structures. Furthermore, diamond films with varying isotopic compositions of ^{12}C and ^{13}C are also synthesizable with this method [28].

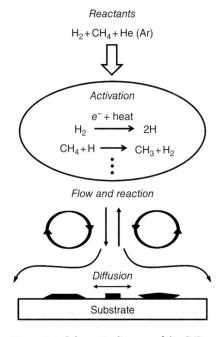

Figure 2.6 Schematic diagram of the CVD diamond growth mechanism. Methane is used as the carbon source in this example. *Source:* Adapted with permission from Ref. [27].

(a)　　　　　　　　　　　(b)

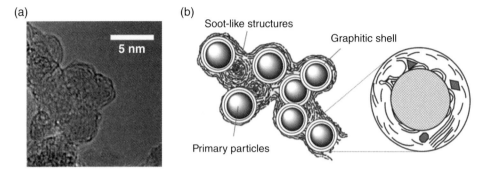

Figure 2.7　(a) High-resolution TEM image of DNDs. The particles are surrounded by graphitic and soot-like materials. (b) Structure model of the DND agglomerates. *Source:* Reprinted with permission from Ref. [31]. Reproduced with permission of John Wiley & Sons.

2.3.3　Detonation

In 1963, a group of scientists in the Soviet Union discovered tiny crystallites of diamond particles in soot produced by detonating an oxygen-deficient TNT/hexogen composition in an inert media [29]. The average size of the primary particles of these so-called "*detonation nanodiamonds* (DNDs)" was only about 5 nm as a result that the shock waves, which compressed carbon into diamonds, generated by the explosion lasted only a fraction of a microsecond (Figure 2.4). For several reasons, including security measures in the USSR and a lack of interest in nanotechnology at that time, the application of this type of diamonds remained under-exploited until recently [30].

DNDs have the smallest particle sizes among all synthetic diamonds. However, the transient explosion conditions in the reaction chamber inevitably cause severe agglomeration of the material, leading to an effective size of approximately 200 nm. *Transmission electron microscopy* (TEM) has revealed that the particles are linked by covalent bonds among surface functional groups and also by the soot structures surrounding every primary particle (Figure 2.7a). Therefore, thorough purification of the material is required in order to obtain graphite-free NDs. Figure 2.7b shows a model explaining the unusually tight aggregation [31]. To isolate the primary particles, elaborate post-processing deagglomeration and purification are often undertaken to break the covalent linkage between the disordered (sp and sp^2) carbon atoms on the surfaces. A proven method is that of Osawa [32] using wet ball milling with ZrO_2 microbeads. The resulting DND particles have a fairly uniform size distribution with a mean diameter of 4–5 nm. They are available in kilogram quantities in the form of colloidal sols, gels, and assemblies, suitable for various types of applications.

2.4　Nanodiamonds: A Scientist's Best Friend

While humans have valued the beauty of diamonds throughout history, as the story of the Regent vividly described earlier in this chapter, there are other virtues that diamonds have long been recognized equally valuable as a resourceful material. Rare in quantity, diamonds are the hardest material ever known to exist on the planet. What most people may not know, or has not fully recognized or utilized, is the other unique properties that make diamond distinctive from other materials. They include: (i) diamond is chemically inert, nontoxic, and biocompatible; (ii) the surface of diamond is readily derivatizable with a variety of functional groups; (iii) the structural defects in diamond are highly fluorescent and photostable. Leveraging these properties has opened doors to new developments and applications of the materials, particularly nanoscale diamonds, in biology and medicine [33].

As the diamond size decreases, the immediate question is: Do NDs share the same physical properties of diamonds given the same molecular structure in both? If not, to what extent ND begins to deviate its behaviors from diamonds? It is likely, as we have experienced with other nanomaterials, additional properties may occur and we will need to know the causes and how to control them effectively. We will discuss the size of NDs here and other size effects in the next chapters.

NDs belong to a broad family of nanocarbon including fullerenes, nanotubes, and graphenes, as discussed in Section 1.2. NDs can be loosely classified by size: Nanocrystalline (10–100 nm) or ultrananocrystalline (<10 nm) [34]. DNDs belong to the latter; they are polycrystalline and composed of a sp^3-hydridized diamond core coated with a sp^2-hybridized graphite shell or amorphous carbon (Figure 2.7a). The typical carbon content in these particles is in the range of 90%, in addition to 2% nitrogen atoms as impurities [35]. As a result, DNDs are not so suitable for optical applications but are more useful as drug-delivery devices because of their small size (Chapter 13). HPHT-NDs, in contrast, are monocrystalline with high optical transparency. These nanomaterials are produced by crushing and/or ball milling of micrometer- or millimeter-sized diamond crystallites synthesized by HPHT methods and then separated by centrifugal force. Although their size is larger (typically 10–100 nm) and their size distribution is considerably broader than DNDs, they contain less sp^2 carbon on surface and thus can be purified more easily in acids. Moreover, the HPHT-ND particles can host a high-density ensemble (>10 ppm) of fluorescent centers for applications as bioimaging agents (Chapter 6).

The smallest possible diamonds are the *diamondoids*, which are naturally present in crude oil [36]. With a formula of $C_{10}H_{16}$, adamantane is the simplest diamondoid molecule. It is a cycloalkane with a structural configuration similar to a segment of the diamond lattice (Figure 2.8). The molecule has a diameter of only approximately 0.5 nm, given a C—C bond length of 0.154 nm and a C—H bond length of 0.109 nm. A wide range of diamondoids consisting of 1–10 adamantane cages have been discovered and structurally identified in the laboratories with various spectroscopic techniques [37]. Larger

Figure 2.8 Structures of adamantane (red), diamantane (green), and triamantane (blue) as segments of the diamond lattice.

Figure 2.9 Diamond nanoparticles (white powders in vial) extracted from a piece of the carbonaceous Murchison meteorite. *Source:* Reprinted with permission from Ref. [53].

diamondoids, such as cyclohexamantane ($C_{26}H_{30}$) with a size of more than 1 nm [38], are particularly interesting because they bridge the gap between hydrocarbon molecules and hydrogenated diamond nanocrystals. Compared to NDs with nonuniform size and irregular shape, these cage compounds are advantageous in having well-defined structures and have attracted considerable attention of organic chemists to synthesize novel diamondoid-based chemicals and materials. Recent developments of the techniques to derivatize diamondoids with various functional groups have expanded the research on these unique structures and the applications of these cage compounds in pharmaceutics, molecular electronics, and many other areas [39].

NDs that fill the size gap between hydrocarbon molecules and hydrogenated diamond nanocrystals, surprisingly, came from the outer space. In 1987, scientists discovered a new type of natural diamonds called *meteoritic nanodiamonds*, isolated from acid dissolution residues of primitive carbonaceous meteorites (Figure 2.9) [40]. The median diameter of these NDs was 2.6 nm, about 10–1000 times smaller than dust grains in interstellar space [41]. The concentration of the ND grains found in the Orgueil meteorite, a meteorite discovered in southwestern France in 1864, was up to 1400 ppm [42]. Further detailed analysis of the microstructures of the meteoritic NDs with high-resolution TEM suggested that the predominant mechanism for the presolar diamond formation may be a vapor deposition process, rather than high-pressure compression by shock waves [43]. Another notable finding of these studies is that ND is highest in abundance among all stardust particles (including SiC, graphite, and oxides) in meteorites [44]. This has led to a suggestion that NDs as small as 3 nm are thermodynamically more stable than graphite of similar size [45, 46]. It has also triggered an active search for interstellar NDs by astronomical observations in conjunction with laboratory studies. We will continue the discussion of this interesting topic in Chapter 14.

The recent emergence of *fluorescent nanodiamonds* (FNDs) has sparked a new era of technological innovation with NDs for cell labeling, imaging, and tracking applications [47–50]. This technological advancement is made possible by high-yield fabrication of these fluorescent carbon nanoparticles with excellent biocompatibility and unique optical properties (Chapter 6). A special attribute of FNDs is their bright and stable fluorescence originating from the nitrogen-vacancy color centers (Chapter 3). When exposed to green-yellow light, the FNDs emit non-photobleaching tissue-penetrating red photons, making them well suited for bioimaging applications. Contemporary research studies in the field have been focusing on applying the surface-functionalized FNDs for nanoscale imaging and quantum sensing, as to be elaborated in Chapters 10 and 11, respectively. Moreover, the conjugation of FNDs with other nanoparticles to form hybrids has opened up exciting new horizons for this novel nanomaterial (Chapter 12).

Today, we are witnessing a rapid increase of interest in the use of ND particles in a broad range of science and technology areas. These include tribology, quantum information, catalysis, nanoscale sensing, biomedicine, and many more [51, 52]. We will provide comprehensive discussions of FNDs and their biomedical applications in the following chapters.

References

1 Streeter, E. (1882). *The Great Diamonds of the World*. G. Bell & Sons.
2 Tilley, A. (1922). *Modern France. A Companion to French Studies*. Cambridge University Press.
3 Mabille, G. (2001). Diamond, known as the "Regent". http://www.louvre.fr/en/oeuvre-notices/diamond-known-regent (accessed 16 April 2018).
4 Kittel, C. (2005). *Introduction to Solid State Physics*, 8e. Wiley.
5 The Nobel Prize in Physics 1915. *Nobelprize.org*. Nobel Media AB 2014. https://www.nobelprize.org/nobel_prizes/physics/laureates/1915 (accessed 16 April 2018).
6 Yoder, M.N. (1994). The vision of diamond as an engineered material. In: *Synthetic Diamond: Emerging CVD Science and Technology* (ed. K.E. Spear and J.P. Dismukes), 3–17. Wiley.
7 Tomlinson, P.N. (1992). Applications of diamond grits and composites. In: *The Properties of Natural and Synthetic Diamond* (ed. J.E. Field), 637–666. Academic Press.
8 Zimnisky, P. (2015). Global rough diamond production estimated to hit over 135M Carats in 2015. Kitco Commentary, Kitco.
9 The Environmental Literacy Council (2015). Diamond mining. https://enviroliteracy.org/land-use/mineral-resources/diamond-mining (accessed 16 April 2018).
10 Clark, C.D., Dean, P.J., and Harris, P.V. (1964). Intrinsic edge absorption in diamond. *Proc R Soc A* 277: 312–329.
11 Dundek, M. (2009). *Diamonds*, 3e. Nobel Gems Publications.
12 Kaiser, W. and Bond, W.L. (1959). Nitrogen, a major impurity in common type I diamond. *Phys Rev* 115: 857–863.
13 Field, J.E. (1992). Applications of diamond grits and composites. In: *The Properties of Natural and Synthetic Diamond* (ed. J.E. Field), 669. Academic Press.
14 Pankove, J.I. and Qui, C.H. (1994). Optical properties and optoelectronic applications of diamond. In: *Synthetic Diamond: Emerging CVD Science and Technology* (ed. K.E. Spear and J.P. Dismukes), 401–418. Wiley.
15 Bundy, F.P. (1980). The P,T phase and reaction diagram for elemental carbon, 1979. *J Geophys Res* 85: 6930–6936.
16 Tappert, R. and Tappert, M.C. (2011). *Diamonds in Nature: A Guide to Rough Diamonds*. Springer.

17 Howes, V.R. (1962). The graphitization of diamond. *Proc Phys Soc* 80: 648–662.

18 De Vita, A., Galli, G., Canning, A., and Car, R. (1996). A microscopic model for surface-induced diamond-to-graphite transitions. *Nature* 379: 523–526.

19 Adiga, S.P., Curtiss, L.A., and Gruen, D.M. (2010). Molecular dynamics simulations of nanodiamond graphitization. In: *Nanodiamonds: Applications in Biology and Nanoscale Medicine* (ed. D. Ho), 35–54. Springer.

20 Tennant, S. (1797). On the nature of the diamond. *Philos Trans R Soc Lond* 87: 123–127.

21 Hazen, R.M. (1999). *The Diamond Makers*. Cambridge University Press.

22 Spear, K.E. and Dismukes, J.P. (ed.) (1994). *Synthetic Diamond: Emerging CVD Science and Technology*. Wiley.

23 Bundy, F.P., Hall, H.T., Strong, H.M., and Wentorf, R.H. (1955). Man-made diamonds. *Nature* 176: 51–55.

24 Hall, H.T. (1960). Ultra-high-pressure, high-temperature apparatus: the "Belt". *Rev Sci Instrum* 31: 125–131.

25 May, P.W. (2000). Diamond thin films: a 21st-century material. *Phil Trans R Soc A* 358: 473–495.

26 Angus, J.C. and Hayman, C.C. (1991). Low-pressure, metastable growth of diamond and 'diamondlike' phases. *Annu Rev Mater Sci* 241: 913–921.

27 Butler, J.E. and Woodin, R.L. (1993). Thin film diamond growth mechanisms. *Phil Trans R Soc Lond A* 342: 209–224.

28 Anthony, T.R. and Banholzer, W.F. (1992). Properties of diamond with varying isotopic composition. *Diam Relat Mater* 1: 717–726.

29 Danilenko, V.V. (2004). On the history of the discovery of nanodiamond synthesis. *Phys Solid State* 46: 595–599.

30 Vul', A. and Shenderova, O. (2014). *Detonation Nanodiamonds: Science and Applications*. Pan Stanford.

31 Krueger, A., Ozawa, M., Jarre, G. et al. (2007). Deagglomeration and functionalisation of detonation diamond. *Phys Stat Sol A* 204: 2881–2887.

32 Osawa, E. (2008). Monodisperse single nanodiamond particles. *Pure Appl Chem* 80: 1365–1379.

33 Mochalin, V.N., Shenderova, O., Ho, D., and Gogotsi, Y. (2012). The properties and applications of nanodiamonds. *Nat Nanotechnol* 7: 11–23.

34 Williams, O.A. (2011). Nanocrystalline diamond. *Diam Relat Mater* 20: 621–640.

35 Shenderova, O.A., Vlasov, I.I., Turner, S. et al. (2011). Nitrogen control in nanodiamond produced by detonation shock-wave-assisted synthesis. *J Phys Chem C* 115: 14014–14024.

36 Marchand, A.P. (2003). Diamondoid hydrocarbons – delving into nature's bounty. *Science* 299: 52–53.

37 Dahl, J.E., Liu, S.G., and Carlson, R.M. (2003). Isolation and structure of higher diamondoids, nanometersized diamond molecules. *Science* 299: 96–99.

38 Dahl, J.E.P., Moldowan, J.M., Peakman, T.M., and Carlson, R.M. (2003). Isolation and structural proof of the large diamond molecule, cyclohexamantane ($C_{26}H_{30}$). *Angew Chem Int Ed* 42: 2040–2044.

39 Schwertfeger, H., Fokin, A.A., and Schreiner, P.R. (2008). Diamonds are a chemist's best friend: diamondoid chemistry beyond adamantine. *Angew Chem Int Ed* 47: 1022–1036.

40 Lewis, R.S., Ming, Y., Wacker, J.F. et al. (1987). Interstellar diamonds in meteorites. *Nature* 326: 160–162.

41 Lewis, R.S., Anders, E., and Draine, B.T. (1989). Properties, detectability and origin of interstellar diamonds in meteorites. *Nature* 339: 117–121.

42 Huss, G.R. and Lewis, R.S. (1995). Presolar diamond, SiC, and graphite in primitive chondrites: abundances as a function of meteorite class and petrologic type. *Geochim Cosmochim Acta* 59: 115–160.

43 Daulton, T.L., Eisenhour, D.D., Bernatowicz, T.J. et al. (1996). Genesis of presolar diamonds: comparative high-resolution transmission electron microscopy study of meteoritic and terrestrial nano-diamonds. *Geochim Cosmochim Acta* 60: 4853–4872.

44 Davis, A.M. (2011). Stardust in meteorites. *Proc Natl Acad Sci U S A* 108: 19142–19146.

45 Nuth, J.A. (1987). Small-particle physics and interstellar diamonds. *Nature* 329: 589.

46 Badziag, P., Verwoerd, W.S., Ellis, W.P., and Greiner, N.R. (1990). Nanometre-sized diamonds are more stable than graphite. *Nature* 343: 244–245.

47 Yu, S.J., Kang, M.W., Chang, H.C. et al. (2005). Bright fluorescent nanodiamonds: no photobleaching and low cytotoxicity. *J Am Chem Soc* 127: 17604–17605.

48 Chang, Y.R., Lee, H.Y., Chen, K. et al. (2008). Mass production and dynamic imaging of fluorescent nanodiamonds. *Nat Nanotechnol* 3: 284–288.

49 Wu, T.J., Tzeng, Y.K., Chang, W.W. et al. (2013). Tracking the engraftment and regenerative capabilities of transplanted lung stem cells using fluorescent nanodiamonds. *Nat Nanotechnol* 8: 682–689.

50 Hsiao, W.W.W., Hui, Y.Y., Tsai, P.C., and Chang, H.C. (2016). Fluorescent nanodiamond: a versatile tool for long-term cell tracking, super-resolution imaging, and nanoscale temperature sensing. *Acc Chem Res* 49: 400–407.

51 Greentree, A.D., Aharonovich, I., Castelletto, S. et al. (2010). Twenty-first century applications of nanodiamonds. *Opt Photonics News* 21: 20–25.

52 Arnault, J.-C. (ed.) (2017). *Nanodiamonds: Advanced Material Analysis, Properties and Applications*. Elsevier.

53 Beatty, K. (2003). A solar source for diamond dust? *Sky & Telescope* (23 July 2003). http://www.skyandtelescope.com/astronomy-news/a-solar-source-for-diamond-dust (accessed 16 April 2018).

3

Color Centers in Diamond

In this chapter, we will discuss the key features of nanodiamonds that are relevant to the development of *fluorescent nanodiamonds* (FNDs) and their many uses in bio-technology and beyond. Up till now, we have surveyed nanotechnology in previous chapters starting with its historical development and focusing, in particular, on the nanocarbon materials. With a general introduction to nanodiamonds (Chapter 2), we have learned some methods commonly used to prepare man-made diamonds and the fundamental properties shared by all diamonds and nanodiamonds alike. In this and the following two chapters, we are about to find out that nanodiamonds can glow (Chapter 3), support (Chapter 4), and is safe to use (Chapter 5). These unique features have made nanodiamonds a superior material for a fast-growing list of applications at present time and in the coming future.

First comes an optical property that puts a touch of color on diamond. After all, who can turn away from the speckles of a diamond with an irresistible color destined to capture our hearts? It is only fitting to start with color centers in diamond.

Color centers are crystal defects that absorb light in a spectral region where the crystal itself has no absorption [1]. The term "color center" is derived from the German word *Farbenzentren*, which was first discovered in alkali halide crystals by Pohl and cowork-ers in the 1930s [2]. It has later been confirmed both experimentally and theoretically that the Farbenzentren (or F-center) is a crystalline vacancy with captured unpaired electron(s) that absorbs light in the visible region, thus giving various colors to the crys-tals. Similar to the F-centers in alkali halides, color centers can be found in diamond as well [3], although their formation occurs under far more extreme conditions due to the chemical inertness and mechanical rigidity of the material (Chapter 2). This chapter begins with an introduction to the optical properties of diamond, followed by a brief review of the history of diamond coloration by radiation damage and a comprehensive discussion of some important spectral characteristics of vacancy-related color centers.

3.1 Nitrogen Impurities

As discussed in the previous chapter, a pure diamond is optically transparent with its transmittance extending from 225 to 2000 nm (Figure 2.2). The diamond crystal shows only infrared (IR) absorption bands at 2–7 μm resulting from two- and three-phonon excitation processes [4]. The absence of one-phonon absorption is due to the high

Fluorescent Nanodiamonds, First Edition. Huan-Cheng Chang, Wesley Wei-Wen Hsiao and Meng-Chih Su.
© 2019 John Wiley & Sons Ltd. Published 2019 by John Wiley & Sons Ltd.

symmetry of diamond's crystal structure, of which C–C vibrations are IR-inactive. Such vibrational modes, alternatively, appear as a sharp Raman scattering peak at $1332\,\mathrm{cm}^{-1}$ (or $7.508\,\mu\mathrm{m}$), serving as a fingerprint for easy identification of diamond from other carbon-based materials [5]. Therefore, for any colored diamond, its optical absorption and emission must be contributed by impurities and structural defects in the crystal matrix.

Nitrogen is the major impurity in diamond [6]. The nitrogen concentration of diamond is typically measured by inert gas fusion analysis [7], which involves fusing the sample material in a graphite crucible at high temperatures and then determining the amount of the nitrogen gas released by a mass spectrometer. Quantitative as is, the method is destructive. IR absorption spectroscopy, on the other hand, is nondestructive and it has been applied as a tool to characterize the nitrogen concentrations [8]. Figure 3.1a presents a typical IR spectrum of type Ib bulk diamond, obtained by direct absorption methods [9]. Two prominent features are observed at 1130 and $1344\,\mathrm{cm}^{-1}$ and they are attributed to the localized vibrational modes of C–N bonds [10, 11]. Studies have established the correlations between the absorption coefficients (μ in units of cm^{-1}) of these two bands and the concentration ($[N^0]$ in units of ppm) of neutral nitrogen atoms dispersed in diamond, giving [10–12]

$$\left[N^0\right] = 37.5\mu_{1344} = 25\mu_{1130}. \tag{3.1}$$

Similarly, for type Ia bulk diamond, a prominent feature is found at $1282\,\mathrm{cm}^{-1}$ (Figure 3.1b) [13], which is related to the localized vibrational modes of the –N–C–N– moieties in the N aggregates, N_A, whose concentrations can be estimated as [14]

$$\left[N_A\right] = 16.5\mu_{1282}. \tag{3.2}$$

In both cases, the absorption coefficient is defined as

$$\mu = \frac{2.303A}{L}, \tag{3.3}$$

where A is the measured absorbance at the specified wavenumber and L is the sample thickness (in cm).

Though useful in quantifying $[N^0]$ and $[N_A]$ in bulk diamonds, Eqs. (3.1) and (3.2) are not suited for diamond powders because the sample thickness L (and, thus, absorption coefficient μ) is ill-defined. A way to circumvent this problem is to use the two-photon absorption band of diamond at approximately $5\,\mu\mathrm{m}$ as the internal standard (Figure 3.1) [15]. The absorption band shows a characteristic dip at $2120\,\mathrm{cm}^{-1}$ and the depth of this dip is directly proportional to the effective sample thickness. Su et al. [16] have reported that the $[N^0]$ in type Ib diamond powders can be estimated by

$$\left[N^0\right] \approx 203\left(\frac{\Delta\mu_{1130}}{\Delta\mu_{2120}}\right) = 203\left(\frac{\Delta A_{1130}}{\Delta A_{2120}}\right), \tag{3.4}$$

where $\Delta\mu_{1130}$ is the absorption coefficient difference corresponding to the height of the hump at $1130\,\mathrm{cm}^{-1}$ (the "mountain") and $\Delta\mu_{2120}$ is the absorption coefficient difference between the linear line drawn across the two peaks at 2033 and $2160\,\mathrm{cm}^{-1}$ and the dip at $2120\,\mathrm{cm}^{-1}$ (the "valley") (Figure 3.1). Hence, $\Delta\mu_{1130}/\Delta\mu_{2120}$ is the mountain/valley ratio. Similarly, for type Ia diamond, one has

(a)

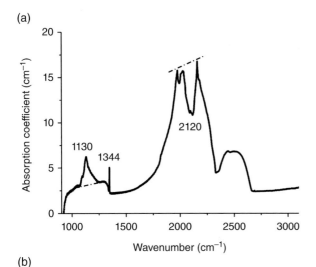

(b)

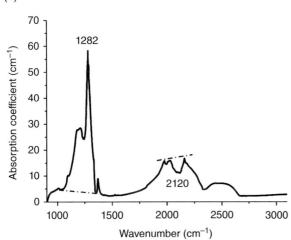

Figure 3.1 (a) IR absorption spectrum of a synthetic diamond crystal with $[N^0] = 109$ ppm. The ratio of the height of the hump at 1130 cm^{-1} versus the depth of the dip at 2120 cm^{-1} was measured to determine the nitrogen concentration. (b) IR absorption spectrum of a natural diamond crystal with a nitrogen concentration of $[N_A] = 900$ ppm. The ratio of the height of the peak at 1282 cm^{-1} versus the depth of the dip at 2120 cm^{-1} was measured to determine the nitrogen concentration. *Source:* Adapted with permission from Refs. [9, 13]. (a) Reproduced with permission of American Chemical Society and (b) Reproduced with permission of Elsevier.

$$[N_A] \approx 91\left(\frac{\Delta\mu_{1282}}{\Delta\mu_{2120}}\right) = 91\left(\frac{\Delta A_{1282}}{\Delta A_{2120}}\right), \qquad (3.5)$$

where $\Delta\mu_{1282}$ is the absorption coefficient difference corresponding to the height of the peak at 1282 cm^{-1}.

The major advantage of using these modified equations is that the measured $[N^0]$ or $[N_A]$ is independent of the sample thickness since it involves only relative absorbance in the calculations. Additionally, as a ratio, the result is virtually free of any baseline shift typical of every spectrum obtained experimentally. The method is readily applicable for diamond crystallites measured by *diffuse reflectance infrared Fourier transform spectroscopy* (DRIFTS) [17], a technique widely used to determine IR photons scattered from fine particles [18]. For instance, with DRIFTS, the nitrogen content of submillimeter diamond crystallites prepared under high-pressure and high-temperature conditions was successfully measured to be $[N^0] = 100-300$ ppm [16], classified as type Ib diamonds. The method also confirmed that increasing the nitrogen content to more than 600 ppm could be achieved by doping the solvent catalyst with inorganic nitrogen-containing substances in high concentrations [19].

3.2 Crystal Defects

Crystal defects are the second source responsible for diamond color. The discovery of diamond coloration through radiation damage can be traced back to the turn of the twentieth century. In 1896, French scientist Henri Becquerel accidentally discovered radioactivity while performing phosphorescence experiments with uranium salts. Inspired by this finding as well as the discovery of X-ray by Wilhelm Röntgen in 1895, Marie and Pierre Curie continued the studies for different uranium minerals and eventually discovered two new elements, *polonium* (Po, $Z = 84$) and *radium* (Ra, $Z = 88$) in 1898. Becquerel and the Curies shared the Nobel Prize in 1903 for their work in discovering spontaneous radioactivity and the new radiation phenomena [20]. A year later, in a paper presented to the *Royal Society of London*, the English scientist William Crookes reported that the exposure of diamonds to radiation generated by radium bromide for several months could give the gem a bluish green to green color. The discovery, also by serendipity, had caused considerable interest and excitement in the gemological community because the color alternations were permanent, unlike painting or coating on the gems in the early days for added market value [21].

So, what has caused diamond to change color?

Radium bromide used by Crookes contains radium, which is a highly radioactive element. It is now known that all isotopes of radium are radioactive and the most stable isotope is Ra-226 with a half-lifetime of 1600 years [22]. The isotope slowly decays to radon, which is a stable inert gas, emitting alpha particles along with the process. Research studies by physicists during the time period between 1910 and 1940 had concluded that the color change of diamond was mainly due to the alpha radiation from radium. The color appeared to be confined to a shallow layer on the diamond's surface. More studies of the effect found that the green or blue-green color could be changed further to various shades of color ranging from yellow to brown by thermal annealing. It was this permanent color enhancement through irradiation combined with high-pressure high-temperature annealing that opened the modern era of diamond treatment. An excellent review on the historical accounts of diamond treatments has been given by Overton and Shigley [21].

Alpha particles are not the only type of radiation useful for treating diamonds. Exposure to high-energy sources such as electrons, neutrons, protons, helium ions, or gamma rays can also change the diamond color in a similar fashion [23]. It is well understood now that the color change is due to structural damages that occur when the radiation passes through the diamond crystalline. Specifically, the penetrating particles collide with the diamond lattice, knocking off the carbon atoms from their originally occupied positions to create vacant sites in the lattice matrix. Since the vacancies can

Figure 3.2 An irradiation-treated green diamond. *Source:* Reprinted with permission from Ref. [21]. Reproduced with permission of Springer.

exist in both neutral and negatively charged states, they give rise to a broad range of absorption in the visible and near-infrared regions of the spectrum, yielding a blue-to-green color (Figure 3.2). Subsequent heating of the irradiated diamonds above 500 °C in an inert atmosphere changes the blue green color to yellow, brownish, or other hues [21]. More details are discussed later in this chapter.

3.3 Vacancy-Related Color Centers

3.3.1 GR1 and ND1

Systematic investigations of radiation damage in natural and synthetic diamonds were first carried out in the 1950s [24, 25]. Since then, a wealth of information has been collected through optical spectroscopy and electron paramagnetic resonance studies for the color centers [3, 26]. Nearly all radiation-damaged diamonds show an absorption band associated with the neutral vacancy defect (cf., Figure 3.3a for the molecular structure [27]), designated as V^0, in their ultraviolet-visible (UV-Vis) spectra [28]. The neutral vacancy, known as the GR1 center where GR stands for "general radiation" (Table 3.1), is one of the best characterized optical defects in diamond [8]. It has a T_d point symmetry and exhibits a narrow *zero-phonon line* (ZPL) at 741 nm (or 1.673 eV) accompanied by a broad phonon sideband of a maximum around 620 nm in the UV-Vis absorption spectrum (Figure 3.4a and b). When exposed to orange-red light, the center emits near-infrared photoluminescence at approximately 800 nm (Figure 3.5) [29] with a quantum yield of only 1.4% at low temperature (Table 3.2) [30]. Although GR1 is not a good chromophore for fluorescence imaging because of such a low quantum yield, it can be produced with a high number density (>100 ppm) by extensive irradiation without the need of thermal annealing. Davies [8] has provided a calibration constant that correlates the concentration ($[V^0]$ in cm^{-3}) of the neutral vacancy and the integrated absorption coefficient (A_{GR1} in cm^{-1}) of its ZPL at 80 K to be (Table 3.3):

$$A_{GR1} = 1.2 \times 10^{-16} \left[V^0 \right]. \tag{3.6}$$

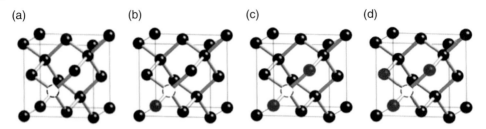

Figure 3.3 Structures of vacancy-related color centers in diamond: (a) V⁰, (b) NV, (c) H3, and (d) N3. The carbon atoms, nitrogen atoms, and vacancies are denoted by black spheres, dark red spheres, and blue dashed circles, respectively. *Source:* Reprinted with permission from Ref. [27]. Reproduced with permission of John Wiley & Sons.

Table 3.1 Nomenclature for zero-phonon lines of color centers in diamond.

Letter	Significance	Range
GR	General radiation (induced in all diamonds by irradiation)	GR1 (1.67 eV) to GR8 (3.00 eV)
H	Heat treatment (preceded by irradiation)	H1 (0.18 eV) to H18 (3.56 eV)
N	Natural diamond	N1 (1.50 eV) to N9 (5.26 eV)

Source: From Ref. [26].

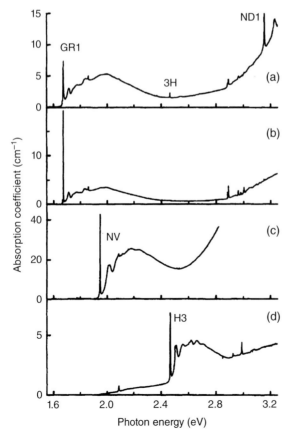

Figure 3.4 UV-Vis absorption spectra of natural and synthetic diamonds after irradiation at room temperature with 2-MeV electrons: (a) type Ia, (b) type IIa, (c) type Ib with additional annealing, and (d) type Ia with additional annealing. All the spectra were recorded at liquid-nitrogen temperature. *Source:* Reprinted with permission from Ref. [28]. Reproduced with permission of American Physical Society.

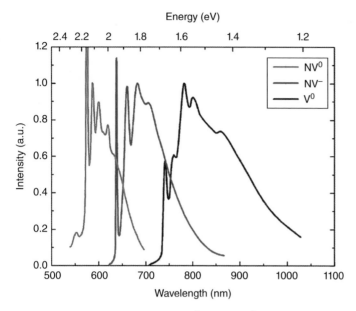

Figure 3.5 Luminescence spectra of NV⁰, NV⁻, and V⁰ in diamond at 77 K. The characteristic ZPLs are located at 575, 637, and 741 nm, respectively. *Source:* Reprinted with permission from Ref. [73]. Reproduced with permission of John Wiley & Sons.

Table 3.2 Spectroscopic properties of vacancy-related color centers in bulk diamonds.[a]

Center	Point group	ZPL (nm)	τ (ns)	QE (%)	References
V⁰ (GR1)	T_d	741	2.55	1.4	[29, 30]
V⁻ (ND1)	T_d	394	—	—	[24]
NV⁻	C_{3v}	637	11.6	99	[33, 34, 57]
NV⁰	C_{3v}	575	19	—	[32, 35]
N–V–N (H3)	C_{2v}	503	16	95	[43, 44]
N₃+V (N3)	C_{3v}	415	41	29	[44, 46]
SiV⁻	D_{3d}	738	1.28	—	[48, 49]

[a] Listed are zero-phonon line (ZPL), emission lifetime (τ), and quantum efficiency (QE) at low temperatures.

Table 3.3 Calibrations of the optical absorption in zero-phonon lines of some vacancy-related color centers in bulk diamonds.

Line	Calibration
1	$A_{GR1} = (1.2 \pm 0.3) \times 10^{-16}$ [V⁰]
2	$A_{ND1} = (4.8 \pm 0.2) \times 10^{-16}$ [V⁻]
3	$A_{NV^-} = (1.4 \pm 0.35) \times 10^{-16}$ [NV⁻]
4	$A_{H3} = (1.0 \pm 0.35) \times 10^{-16}$ [H3]
5	$A_{N3} = (8.6 \pm 2) \times 10^{-17}$ [N3]

Source: From Ref. [8]. Reproduced with permission of Elsevier.

Aside from V^0, the vacancy defect can also exist in the negatively charged form, designated as V^- or ND1 [24]. The V^- center shows a sharp ZPL at 3.149 eV (394 nm), which appears only when a substantial amount of nitrogen is present in the diamond lattice. In type Ia diamonds, the ND1 center has a peak intensity comparable to that of GR1 but can barely be detectable in type IIa diamonds (Figure 3.4a and b). There is no distinct photoluminescence band associated with ND1. It has been reported that intense UV illumination of the ND1 band results in a reduction of its absorption strength but concurrently increases the GR1 band intensity [31], presumably due to the release of the trapped electrons from the vacancies. However, heating of the samples in the dark can reverse these changes, suggesting the occurrence of some interesting ionization-recombination processes. More about this process is given later.

3.3.2 NV^0 and NV^-

Vacancies in diamond once produced by radiation damage are static at room temperature. They start to migrate when the irradiated diamond is heated above 500 °C. The activation energy barrier is 2.3 eV (or 53 kcal mol^{-1}) [28]. For type Ib diamond containing atomic nitrogen, while some of the vacancies are annihilated at the surface, the majority of them can form stable complexes (i.e. the nitrogen-vacancy (NV) centers) with nitrogen atoms in the diamond matrix. Figure 3.4c shows a representative UV-Vis absorption spectrum of a type Ib diamond after irradiation with 2-MeV electrons and subsequent annealing at 800 °C under vacuum. The signature absorption band of V^0 at 1.673 eV has disappeared nearly completely, replaced by a new feature emerging at 1.945 eV due to NV absorption. This large spectral change, a result of the combined irradiation and annealing treatments, shows up as a dramatic alteration in the diamond color from yellow to purple. Upon exposure to green light, these diamonds give off bright red emission vividly radiant in each and every way (Figure 3.6a). This red emission is possibly related to the extended red emission band detected in the interstellar medium as will be discussed in Chapter 14.

The NV center in diamond is a point defect comprising a substitutional nitrogen atom adjacent to a vacancy in a C_{3v} point group symmetry (Figure 3.3b). Similar to the

(a)   (b)

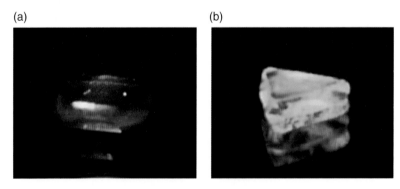

Figure 3.6 Photographs of fluorescent diamonds containing high-density ensembles of (a) NV and (b) H3 centers excited by green and blue light, respectively. *Source:* Adapted with permission from Refs. [9, 13]. (a) Reproduced with permission of American Chemical Society and (b) Reproduced with permission of Elsevier.

vacancies GR1 and ND1, a NV center can exist in one of the two states: NV^0 or NV^-. The neutral form NV^0 has a sharp ZPL at 575 nm (or 2.156 eV) [32], which is distinctively different from the ZPL at 637 nm (or 1.945 eV) of NV^- (Table 3.2) [33]. Both of the ZPLs are accompanied by broad phonon sidebands in the higher energy region of their absorption spectra (cf., Figure 3.4c for NV^- only). The corresponding emission bands peak approximately at 600 and 700 nm for NV^0 and NV^-, respectively (Figure 3.5), with the associated lifetimes of 11.6 ns [34] and 19 ns [35]. Compared to NV^-, NV^0 is significantly lower in concentration due to the transfer of electrons from N^0 to NV^0 during annealing [36, 37]:

$$NV^0 + N^0 \rightarrow NV^- + N^+. \tag{3.7}$$

In type Ib diamonds, since there exist about 100 ppm N^0 in the crystal lattice, the reaction in Eq. (3.7) can readily occur at high temperatures, thus reducing the concentration of NV^0 in the final products. For the NV^- center, its concentration (in cm^{-3}) can be estimated based on the integrated absorption coefficient (in cm^{-1}) of the ZPL at 80 K by [8]

$$A_{NV^-} = 1.4 \times 10^{-16} \left[NV^- \right]. \tag{3.8}$$

The charge transfer reaction as illustrated above can also be induced with electromagnetic radiation. This is recognized as *photochromism*, which describes a process where a chemical species can transform between two forms reversibly through the absorption of photons with each form having a distinct absorption spectrum [38]. For the NV centers in diamond, the relative concentrations of the two different forms, NV^0 and NV^-, may be changed by a shift of the Fermi level due to neutron irradiation [39] or laser illumination [40]. Figure 3.7 presents a schematic energy diagram of the photoinduced ionization and recombination of NV^- and NV^0 with respect to the energy band gap ($E_g = 5.5$ eV) of diamond. Detailed spectroscopic investigations have shown that the ionization processes (1 and 2 in Figure 3.7) begin with one-photon absorption by NV^-, followed by the excitation of the same center with the second photon, which promotes the electron from the excited state of NV^- to the conduction band, creating NV^0.

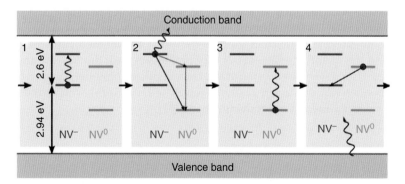

Figure 3.7 Schematic illustration of the photoinduced ionization and recombination of NV^- and NV^0. Processes 1 and 2 are associated with ionization and processes 3 and 4 are associated with recombination. *Source:* Reprinted with permission from Ref. [41].

For the charge recombination processes (3 and 4 in Figure 3.7), the NV^0 center is first photoexcited, followed by energy transfer and capture of an electron from the valence band to form NV^-. At the steady state, the NV^- population is close to 75% under continuous laser excitation in the 450–610 nm wavelength range [41].

3.3.3 H3 and N3

The nitrogen aggregates in type Ia (or natural) diamonds can take on several different forms in structures, which are identifiable based on their characteristic IR absorption bands [6, 11]. The smallest aggregate, known as the A center, consists of two adjacent lattice points occupied by two nitrogen atoms. They can combine with vacancies to form a stable nitrogen–vacancy complex, designated as H3 or N–V–N, with a C_{2v} point group symmetry (Figure 3.3c and Table 3.2). This color center has a ZPL at 503 nm, along with a broad phonon sideband around 470 nm (Figure 3.4d) [42, 43]. Similar to the cases of V^0 and NV^-, the concentration (in cm^{-3}) of this center can be estimated from the integrated absorption coefficient (in cm^{-1}) of its ZPL at 80 K by [8]

$$A_{H3} = 1.0 \times 10^{-16} [H3].$$ (3.9)

For diamonds containing high-density ensembles of H3 centers, bright green emission emerges when they are exposed to blue light around 470 nm (Figure 3.6b). The emission band peaks at 530 nm with a quantum efficiency of 0.95 and a lifetime of 16 ns (Table 3.2) [43]. The quantum efficiency remains high at temperatures exceeding 500 °C. Taking advantage of these remarkable photophysical properties, together with the high thermal conductivity and the non-hygroscopic characteristic of the material, Rand and coworkers [44, 45] explored the possibility of using type Ia diamonds containing H3 as a high-stability lasing medium for room-temperature color center lasers. They measured a gain coefficient of 0.201 cm^{-1} at 531 nm and an absorption cross-section in the range of $(1.4 - 2) \times 10^{-17}$ cm^2 $molecule^{-1}$ at 488 nm, well suited for laser development.

N3 is another color center frequently discovered in natural diamonds after irradiation and annealing treatments (Table 3.2) [42]. Often produced alongside H3, this structural defect consists of three substitutional nitrogen atoms surrounding a vacancy, designated as N3 + V, in a point group symmetry of C_{3v} (Figure 3.3d). It has a ZPL at 415 nm and shows a photoluminescence sideband around 450 nm. The lifetime of this blue emission is 41 ns and the quantum efficiency is 0.29 (Table 3.2) [46], too low for any viable design of a new laser. The blue emission of this center is most commonly observed in natural diamonds when exposed to UV photons (wavelength < 225 nm), which excite electrons from the valence band to the conduction band [47].

3.3.4 SiV⁻

Diamond is a highly dense material with eight carbon atoms squeezed in a 3.57-Å lattice cube (Figure 2.1). As a result, most natural diamonds contain only nitrogen as an impurity and, if any at all, the concentrations of other impurities are lower by several orders of magnitude in comparison. Silicon is a commonly found impurity in CVD diamonds (Section 2.3.2). A Si atom is about 1.5 times the size of a carbon atom. Because of the size difference, a Si can replace two neighboring carbons in the diamond lattice and situate in between two vacant sites, forming an optically active center [48]. In a D_{3d} point group symmetry, the negatively charged silicon–vacancy center (SiV⁻) shows a sharp ZPL at 738 nm (or 1.681 eV) with a weak phonon sideband around 770 nm in its photoluminescence spectra even at room temperature. The radiative decay lifetime of this center is 1.28 ns, which increases to 1.72 ns as the temperature is lowered from 298 to 4 K [49]. Bright photoluminescence, near-infrared wavelength, and weak vibronic couplings make this center appealing for quantum optics and quantum information applications [50].

3.4 The NV⁻ Center

Over the past six decades, the optical properties of vacancy-related color centers in natural and synthetic diamonds have been extensively studied using various spectroscopic techniques [3]. Table 3.2 summarizes the spectroscopic properties of V⁰, V⁻, NV⁰, NV⁻, H3, N3, and SiV⁻ centers in bulk diamonds. Of these seven color centers, NV⁻ deserves special attention with an in-depth discussion for its remarkable optical and magnetic properties. We will discuss in this section only some salient features of the center. Excellent reviews on its quantum optical and spectroscopic properties from both theoretical and experimental aspects can be found elsewhere [51–53].

The NV⁻ center consists of three carbons and one nitrogen surrounding an anionic vacancy in a tetrahedral configuration that belongs to a C_{3v} point group symmetry (cf., Table 3.4 for the character table). It is a six-electron system: 1 e⁻ from each of the three carbons, 2 e⁻ from N, and 1 e⁻ of the vacancy. *Linear combinations of atomic orbitals* (LCAO) suggest a ground electronic configuration of $a_1^2a_1^2e^2$ [54]. The two electrons in the e-orbital are unpaired, one in e_x and another one in e_y [55]. The unperturbed electronic energy structure of the center is $^3A_2$, a triplet ground state with the spin quantum numbers of $m_s = 0$ and ± 1 in its magnetic sublevels (Figure 3.8). The two $m_s = \pm 1$ sublevels are degenerate in a zero magnetic field, and they are energetically higher than the $m_s = 0$ sublevel by 2.87 GHz, which is caused by the crystal field splitting arising from the spin–spin interactions between the unpaired electrons at the e_x and e_y levels [56]. In the presence of a weak external magnetic field, the degeneracy of the $m_s = \pm 1$ sublevels at the ground state of NV⁻ is lifted via the *Zeeman effect*. Under the circumstance that the magnetic field is aligned with the NV's

Table 3.4 Character table of the C_{3v} point group.

C_{3v}	E	$2C_3$	$3\sigma_v$		
A_1	1	1	1	z	x^2+y^2, z^2
A_2	1	1	−1	R_z	
E	2	−1	0	$(x, y), (R_x, R_y)$	$(x^2 - y^2, xy), (xz, yz)$

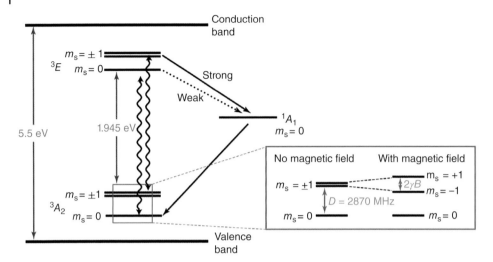

Figure 3.8 Electronic energy diagram of the NV⁻ center in diamond. *Source:* Adapted with permission from Ref. [53].

principal molecular axis passing through the C_3 rotation, the energy difference between these two levels is [53]

$$\Delta\omega = 2g_e\mu_B B, \tag{3.10}$$

where $g_e \sim 2.003$, $\mu_B = 14.00\,\text{MHz}\,\text{mT}^{-1}$ is the Bohr magneton, and B is the magnetic field strength (cf., Section 11.1 for further details).

In the optical region, the NV⁻ center shows a characteristic ZPL in both absorption and emission spectra at 637 nm. Uniaxial stress experiments confirmed that the ZPL originated from a transition of an "*A*" state to the "*E*" state of a trigonal center [32]. Since the spins must be conserved during optical transitions between the ground and excited states, one may designate the electronic transition as $^3A_2 \rightarrow {}^3E$ (Figure 3.8). Further studies found that the photoluminescence of the transition had a lifetime of 11.6 ns for a synthetic type Ib diamond [34] with a quantum yield close to 1 [57]. Moreover, the fluorescence intensity was exceptionally stable as the center acted like a pseudo-atom in an inert and highly thermally conductive matrix. Gruber et al. [58] first reported that the NV⁻ centers were so stable that even under intensive laser excitation ($>1 \times 10^6$ W cm^{-2}) over an extended period of time, they exhibited no sign of photobleaching. The high photostability allows facile detection of individual NV⁻ centers by fluorescence imaging, especially in biological systems, which constitutes one of the main themes of this book.

With the magneto-optical properties, the NV⁻ centers are amenable for analysis by *optically detected magnetic resonance* (ODMR), which is a double resonance technique that enables the transitions between spin sublevels to be detected by optical means. The technique is so powerful that it is capable of enhancing the sensitivity of spin resonance spectroscopy by several orders of magnitude. For the NV⁻ center in diamond, the origin of this effect lies in its unique electronic structure as discussed below.

Figure 3.9 ODMR spectra of a single NV⁻ center in bulk diamond in the presence of an increasing magnetic field. *Source:* Reprinted with permission from Ref. [60]. Reproduced with permission of Nature Publishing Group.

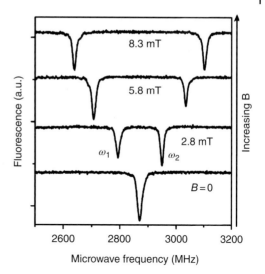

In the electronic energy diagram of NV⁻ as illustrated in Figure 3.8, there exists a metastable singlet state, ^1A. The *intersystem crossing* (ISC) from ^3E to this state is occasionally possible, depending on the sublevel of the excited state. Experimentally, it has been found that the ISC rate from the $m_s = \pm 1$ sublevel of the excited state to the singlet state is significantly higher than that from the $m_s = 0$ counterpart by non-radiative energy transfer processes [59]. The singlet state then decays non-radiatively to the $m_s = 0$ sublevel of the ground state. As a result, the optical pumping can preferentially populate the NV⁻ center in the $m_s = 0$ sublevel of the ground state, a process known as *optically induced spin polarization.* When a resonant microwave radiation of 2.87 GHz is applied, the electron spin will undergo the transition from $m_s = 0$ to ± 1 sublevels of the ground state. Such a transition will reduce the fluorescence intensity by about 30% after several cycles of excitation, because the populations in the $m_s = \pm 1$ sublevels of the excited state return to the ground state through non-radiative decays (Figure 3.8). That is to say, the information about the spins at the electronic ground states can be read out optically. Moreover, the spin states of the NV⁻ centers can be probed and manipulated individually by using continuous and/or pulsed microwave radiations for more advanced applications [53].

The ODMR detection of NV⁻ is so unique and significant that it is achievable at the single spin level even at room temperature [58], thanks to the excellent photostability of the center and the high quantum yield of the fluorescence emission. Furthermore, the Zeeman splitting of the ODMR bands of a single NV⁻ center can also be observed with high contrast (Figure 3.9) [60]. At present, the NV⁻ in diamond is the only known solid-state system that allows detection and manipulation of the spin states of a single localized electron at room temperature. In addition to that, the NV⁻ center has an exceptionally long coherence time and as such its spin states have been reported to remain coherent for longer than 1 ms in isotope-free bulk diamonds [61]. All these features combined together make diamonds hosting NV⁻ centers an extraordinary tool for ultra-sensitive measurements of temperature, magnetic field, electric field, and mechanical stress at the nanometer scale [52]. First discovered in the 1950s, and throughout the years well into the twenty-first century, the center has found remarkable applications in a wide range of research areas, including quantum information [62], quantum computing [63], nanoscale magnetometry [59, 64], high-precision temperature sensing [65–67], and super-resolution imaging [68].

Finally, the NV^- center is particularly appealing for use in biological imaging and tracking applications owing to its unmatchable physicochemical properties as summarized here [69–72]:

1) The NV^- centers can be produced as high-density ensembles in nanoscale diamonds, which are biologically inert and photostable materials.
2) The emission band of NV^- peaks at 685 nm where the light has a long penetration depth through tissue, and more than 70% of the emitted photons lie in the near-infrared window of bioimaging.
3) The fluorescence lifetime of NV^- is significantly longer than those of cell and tissue autofluorescence, allowing for time-gated fluorescence imaging both *in vitro* and *in vivo*.
4) The fluorescence intensity of NV^- can be modulated by applying either sinusoidal magnetic field or sinusoidal microwave radiation to achieve background-free detection in tissue imaging.

The combination of these outstanding features has attracted scientists to employ FNDs containing high-density ensembles of the NV^- centers as bioimaging contrast agents. We will continue the discussion on the biological aspects of the applications in the following chapters of the book.

References

1 Kittel, C. (2005). *Introduction to Solid State Physics*, 8e. Wiley.
2 Pohl, R.W. (1937). Electron conductivity and photochemical processes in alkali-halide crystals. *Proc Phys Soc (London)* 49 (supplement): 3–31.
3 Zaitsev, A.M. (2001). *Optical Properties of Diamond: A Data Handbook*. Springer-Verlag.
4 Hardy, J.R. and Smith, S.D. (1961). Two-phonon infra-red lattice absorption in diamond. *Phil Mag* 6: 1163–1172.
5 Dresselhaus, M.S., Jorio, A., and Saito, R. (2010). Characterizing graphene, graphite, and carbon nanotubes by Raman spectroscopy. *Annu Rev Condens Matter Phys* 1: 89–108.
6 Kaiser, W. and Bond, W.L. (1959). Nitrogen, a major impurity in common type I diamond. *Phys Rev* 115: 857–863.
7 Woods, G.S., Purser, G.C., Mtimkulu, A.S.S., and Collins, A.T. (1990). The nitrogen content of type Ia natural diamonds. *J Phys Chem Solids* 51: 1191–1197.
8 Davies, G. (1999). Current problems in diamond: towards a quantitative understanding. *Physica B* 273–274: 15–23.
9 Wee, T.L., Tzeng, Y.K., Han, C.C. et al. (2007). Two-photon excited fluorescence of nitrogen-vacancy centers in proton-irradiated type Ib diamond. *J Phys Chem A* 111: 9379–9386.
10 Woods, G.S., van Wyk, J.A., and Collins, A.T. (1990). The nitrogen-content of type-1b synthetic diamond. *Phil Mag B* 62: 589–595.

11 Kiflawi, I., Mayer, A.E., Spear, P.M. et al. (1994). Infrared-absorption by the single nitrogen and a defect centers in diamond. *Phil Mag B* 69: 1141–1147.

12 Lawson, S.C., Fisher, D., Hunt, D.C., and Newton, M. (1998). On the existence of positively charged single-substitutional nitrogen in diamond. *J Phys Condens Matter* 10: 6171–6180.

13 Wee, T.L., Mau, Y.W., Fang, C.Y. et al. (2009). Preparation and characterization of green fluorescent nanodiamonds for biological applications. *Diam Relat Mater* 18: 567–573.

14 Boyd, S.R., Kiflawi, I., and Woods, G.S. (1994). The relationship between infrared absorption and the a defect concentration in diamond. *Phil Mag B* 69: 1149–1153.

15 Liang, Z.Z., Jia, X., Ma, H.A. et al. (2005). Synthesis of HPHT diamond containing high concentrations of nitrogen impurities using NaN_3 as dopant in metal-carbon system. *Diam Relat Mater* 14: 1932–1935.

16 Su, L.J., Fang, C.Y., Chang, Y.T. et al. (2013). Creation of high density ensembles of nitrogen-vacancy centers in nitrogen-rich type Ib nanodiamonds. *Nanotechnology* 24: 315702.

17 Ando, T., Ishii, M., Kamo, M., and Sato, Y. (1993). Diffuse reflectance infrared Fourier-transform study of the plasma hydrogenation of diamond surfaces. *J Chem Soc Faraday Trans* 89: 1383–1386.

18 Griffiths, P.R. and de Haseth, J.A. (2006). *Fourier Transform Infrared Spectrometry*, 2e. Wiley.

19 Shames, A.I., Osipov, V.Y., Bogdanov, K.V. et al. (2017). Does progressive nitrogen doping intensify negatively charged nitrogen vacancy emission from e-beam-irradiated Ib type high-pressure-high-temperature diamonds? *J Phys Chem C* 121: 5232–5240.

20 The Nobel Prize in Physics 1903. *Nobelprize.org*. Nobel Media AB 2014. https://www.nobelprize.org/nobel_prizes/physics/laureates/1903 (accessed 16 April 2018).

21 Overton, T.W. and Shigley, J.W. (2008). A history of diamond treatments. *Gems Gemology* 44: 32–55.

22 Royal Society of Chemistry (2017). Radium. http://www.rsc.org/periodic-table/element/88/radium (accessed 16 April 2018).

23 Campbell, B., Choudhury, W., Mainwood, A. et al. (2002). Lattice damage caused by the irradiation of diamond. *Nucl Instrum Meth A* 476: 680–685.

24 Clark, C.D., Ditchburn, R.W., and Dyer, H.B. (1956). The absorption spectra of natural and irradiated diamonds. *Proc R Soc Lond A* 234: 363–381.

25 Clark, C.D., Ditchburn, R.W., and Dyer, H.B. (1956). The absorption spectra of irradiated diamonds after heat treatment. *Proc R Soc Lond A* 237: 75–89.

26 Walker, J. (1979). Optical absorption and luminescence in diamond. *Rep Prog Phys* 42: 1605–1659.

27 Hui, Y.Y. and Chang, H.C. (2014). Recent developments and applications of nanodiamonds as versatile bioimaging agents. *J Chin Chem Soc* 61: 67–76.

28 Davies, G., Lawson, S.C., Collins, A.T. et al. (1992). Vacancy-related centers in diamond. *Phys Rev B* 46: 13157–13170.

29 Clark, C.D. and Norris, C.A. (1971). Photoluminescence associated with the 1.673, 1.944 and 2.498 eV centres in diamond. *J Phys C Solid State Phys* 4: 2223–2229.

30 Davies, G., Thomaz, M.F., Nazare, M.H. et al. (1987). The radiative decay time of luminescence from the vacancy in diamond. *J Phys C Solid State Phys* 20: L13–L17.

31 Dyer, H.B. and du Preez, L. (1965). Irradiation damage in type I diamond. *J Chem Phys* 42: 1898–1906.

32 Davies, G. (1979). Dynamic Jahn-Teller distortions at trigonal optical centres in diamond. *J Phys C* 12: 2551–2566.

33 Davies, G. and Hamer, M.F. (1976). Optical studies of 1.945 eV vibronic band in diamond. *Proc R Soc A* 348: 285–298.

34 Collins, A.T., Thomaz, M.F., and Jorge, M.I.B. (1983). Luminescence decay time of the 1.945 eV center in type 1b diamond. *J Phys C Solid State* 16: 2177–2181.

35 Liaugaudas, G., Davies, G., Suhling, K. et al. (2012). Luminescence lifetimes of neutral nitrogen-vacancy centres in synthetic diamond containing nitrogen. *J Phys Condens Matter* 24: 435503.

36 Collins, A.T. (2002). The Fermi level in diamond. *J Phys Condens Matter* 14: 3743–3750.

37 Collins, A.T., Connor, A., Ly, C.H. et al. (2005). High-temperature annealing of optical centers in type-1 diamond. *J Appl Phys* 97: 083517.

38 Dürr, H. (2003). General introduction. In: *Molecules and Systems: Photochromism* (ed. H. Bouas-Laurent), 1–14. Elsevier.

39 Mita, Y. (1996). Change of absorption spectra in type-1b diamond with heavy neutron irradiation. *Phys Rev B* 53: 11360–11364.

40 Iakoubovskii, K., Adriaenssens, G.J., and Nesladek, M. (2000). Photochromism of vacancy-related centres in diamond. *J Phys Condens Matter* 12: 189–199.

41 Aslam, N., Waldherr, G., Neumann, P. et al. (2013). Photo-induced ionization dynamics of the nitrogen vacancy defect in diamond investigated by single-shot charge state detection. *New J Phys* 15: 013064.

42 Clark, C.D. and Norris, C.A. (1970). The polarization of luminescence associated with the 4150 and 5032 Å centres in diamond. *J Phys C Solid State Phys* 3: 651–658.

43 Crossfield, M.D., Davies, G., Collins, A.T., and Lightowlers, E.C. (1974). The role of defect interactions in reducing the decay time of H3 luminescence in diamond. *J Phys C Solid State Phys* 7: 1909–1917.

44 Rand, S.C. and DeShazer, L.G. (1985). Visible color-center laser in diamond. *Opt Lett* 10: 481–483.

45 Roberts, W.T., Rand, S.C., and Redmond, S. (2005). Measuring two key parameters of H3 color centers in diamond. *NASA Tech Briefs* November: 24–25.

46 Thomaz, M.F. and Davies, G. (1978). The decay time of N3 luminescence in natural diamond. *Proc R Soc A* 362: 405–419.

47 Lu, H.C., Lin, M.Y., Chou, S.L. et al. (2012). Identification of nitrogen defects in diamond with photoluminescence excited in the 160–240 nm region. *Anal Chem* 84: 9596–9600.

48 Feng, T. and Schwartz, B.D. (1993). Characteristics and origin of the 1.681 eV luminescence centre in chemical-vapor-deposited diamond films. *J Appl Phys* 73: 1415–1425.

49 Rogers, L.J., Jahnke, K.D., Teraji, T. et al. (2014). Multiple intrinsically identical single-photon emitters in the solid state. *Nat Commun* 5: 4739.

50 Aharonovich, I., Castelletto, S., Simpson, D.A. et al. (2011). Diamond-based single-photon emitters. *Rep Prog Phys* 74: 076501.

51 Jelezko, F. and Wrachtrup, J. (2006). Single defect centers in diamond: a review. *Phys Stat Sol (a)* 203: 3207–3225.

52 Doherty, M.W., Manson, N.B., Delaney, P. et al. (2013). The nitrogen-vacancy colour centre in diamond. *Phys Rep* 528: 1–45.

53 Schirhagl, R., Chang, K., Loretz, M., and Degen, C.L. (2014). Nitrogen-vacancy centers in diamond: nanoscale sensors for physics and biology. *Annu Rev Phys Chem* 65: 83–105.

54 Lenef, A. and Rand, S.C. (1996). Electronic structure of the N-V center in diamond: theory. *Phys Rev B* 53: 13441.

55 van Oort, E., Manson, N.B., and Glasbeek, M. (1988). Optically detected spin coherence of the diamond N-V centre in its triplet ground state. *J Phys C Solid State Phys* 21: 4385–4391.

56 Loubser, J. and van Wyk, J.A. (1978). Electron spin resonance in the study of diamond. *Rep Prog Phys* 41: 1201–1248.

57 Rand, S.C. (1994). Diamond lasers. In: *Properties and Growth of Diamond*, EMIS Datareviews Series, vol. 9 (ed. G. Davies), 235–240. Institute of Electrical Engineers.

58 Gruber, A., Drabenstedt, A., Tietz, C. et al. (1997). Scanning confocal optical microscopy and magnetic resonance on single defect centers. *Science* 276: 2012–2014.

59 Tetienne, J.-P., Rondin, L., Spinicelli, P. et al. (2012). Magnetic-field-dependent photodynamics of single NV defects in diamond: an application to qualitative all-optical magnetic imaging. *New J Phys* 14: 103033.

60 Balasubramanian, G., Chan, I.Y., Kolesov, R. et al. (2008). Nanoscale imaging magnetometry with diamond spins under ambient conditions. *Nature* 455: 648–651.

61 Balasubramanian, G., Neumann, P., Twitchen, D. et al. (2009). Ultralong spin coherence time in isotopically engineered diamond. *Nat Mater* 8: 383–387.

62 Dutt, M.V.G., Childress, L., Jiang, L. et al. (2007). Quantum register based on individual electronic and nuclear spin qubits in diamond. *Science* 316: 1312–1316.

63 Weber, J.R., Koehl, W.F., Varley, J.B. et al. (2010). Quantum computing with defects. *Proc Natl Acad Sci USA* 107: 8513–8518.

64 Taylor, J.M., Cappellaro, P., Childress, L. et al. (2008). High-sensitivity diamond magnetometer with nanoscale resolution. *Nat Phys* 4: 810–816.

65 Toyli, D.M., de las Casas, C.F., Christle, D.J. et al. (2013). Fluorescence thermometry enhanced by the quantum coherence of single spins in diamond. *Proc Natl Acad Sci USA* 110: 8417–8421.

66 Kucsko, G., Maurer, P.C., Yao, N.Y. et al. (2013). Nanometer-scale thermometry in a living cell. *Nature* 500: 54–58.

67 Neumann, P., Jakobi, I., Dolde, F. et al. (2013). High-precision nanoscale temperature sensing using single defects in diamond. *Nano Lett* 13: 2738–2742.

68 Rittweger, E., Han, K.Y., Irvine, S.E. et al. (2009). STED microscopy reveals crystal colour centers with nanometric resolution. *Nat Photon* 3: 144–147.

69 Vaijayanthimala, V. and Chang, H.C. (2009). Functionalized fluorescent nanodiamonds for biomedical applications. *Nanomedicine* 4: 47–55.

70 Hui, Y.Y., Cheng, C.L., and Chang, H.C. (2010). Nanodiamonds for optical bioimaging. *J Phys D Appl Phys* 43: 374021.

71 Hsiao, W.W.W., Hui, Y.Y., Tsai, P.C., and Chang, H.C. (2016). Fluorescent nanodiamond: a versatile tool for long-term cell tracking, super-resolution imaging, and nanoscale temperature sensing. *Acc Chem Res* 49: 400–407.

72 Hui, Y.Y., Hsiao, W.W.W., Haziza, S. et al. (2017). Single particle tracking of fluorescent nanodiamonds in cells and organisms. *Curr Opin Solid State Mater Sci* 21: 35–42.

73 Lu, H.C., Peng, Y.C., Chou, S.L. et al. (2017). Far-UV excited luminescence of nitrogen-vacancy centers: evidence for diamonds in space. *Angew Chem Int Ed* 56: 14469–14472.

4

Surface Chemistry of Nanodiamonds

We discuss surface modification chemistry of *nanodiamonds* (NDs) in this chapter. The chemistry of a substance is ultimately determined by its molecular structure and physical conditions surrounding it. A quick look over the three allotropes of nanocarbon material reveals some subtle differences between the sp^2 endcaps of *fullerenes*, the sp^2 sidewalls of *carbon nanotubes* (CNTs), the sp^2 sheets of *graphenes*, and their dangling carbon bonds along the edges (Figure 1.1–1.3). Both CNT and graphene have mixed chemical reactions as discussed in Section 1.2 and, depending where on a CNT or graphene, different chemical reactions may occur at different locations. Now with the ND's sp^3 saturated carbon bonds, it is the most stable bond configuration of carbon and therefore is expected to be chemically inert. Why would anyone care for the chemistry of such an inert material? Well, it is precisely because of the inertness (chemical stability) and other trademark properties (Table 2.1) that make ND an excellent nanomaterial for a diversified array of applications, ranging from laboratories to industries and from cosmetics to lubrications, including as a supporting medium for biomolecules. Similar to CNTs and graphenes, all chemical activities if/when occurs with NDs must happen on their surface. Fortunately, for a given amount, ND is equipped with a large surface area and a deep loading capacity perfectly for a nano-sized carrier. But, the question is: How could it be possible for a tiny ND to carry a large load of cargo? We need to look into this matter of ND's surface area a bit closer before any further discussion of chemistry.

Due to their small size, nanoparticles have a large specific surface area or a large surface-area-to-volume ratio. The specific surface area is defined as the surface area of a material per unit mass (in $m^2\,g^{-1}$). Consider a cubic ND particle with a length of l. Its specific surface area increases linearly as l is reduced, because the surface area scales with the length in the second power as $6l^2$, and the mass in the third power as ρl^3, where ρ is the density. For ND cubes of $l = 2.86\,nm$ (or eight unit cells in one dimension) as an example (Figure 4.1), 1 g of these particles has a total surface area of $600\,m^2$, which is about 1.5 times as large as that of a basketball court ($15\,m \times 28\,m = 420\,m^2$). Out of the 4096 carbon atoms in each cube, nearly 20% of them are located on the surface. Clearly, the smaller the particle size, the more important role the surface atoms play in determining the chemistry. Therefore, prior to any practical use of NDs, one must understand their surface chemistry and their interactions with different target molecules in specific environments. In light of many potential applications of the carbon-based nanoparticles in electronics, optics, and medicine, studies on the surface chemistry of NDs have been actively and extensively conducted over the past two decades [1]. This

Fluorescent Nanodiamonds, First Edition. Huan-Cheng Chang, Wesley Wei-Wen Hsiao and Meng-Chih Su.
© 2019 John Wiley & Sons Ltd. Published 2019 by John Wiley & Sons Ltd.

chapter provides a condensed but comprehensive overview of these research studies and developments of NDs for biological applications.

4.1 Functionalization

As a carbon-based nanoparticle, ND is inherently biocompatible. It is more suitable for applications in life science research than other nanoparticles such as gold, silver, and silica nanobeads. However, in order to make any use of them, NDs must be chemically modified and/or functionalized so that they are susceptible for ensuing interactions with the target molecules. Only atoms on the surface of NDs are subjected to such chemical modifications. This is because a bare ND's surface always contains many dangling bonds, i.e. unsatisfied valence bonds associated with carbon atoms (Figure 4.1). These surface atoms, which are chemically unstable, can either bind to each other to form double bonds or with other atoms (such as H or O atoms) to fill their valence shells. The chemical reactions involved here essentially follow the basic principles of organic chemistry but are largely hindered by steric effects and thus less effective than small organic molecules in solution. The reactions are slow if the particles are large and have a monocrystalline structure.

NDs, either produced by the *high-pressure high-temperature* (HPHT), *chemical vapor deposition* (CVD), or detonation method, are always contaminated with residual chemical compounds from the manufacturing processes, leaving sp^2 or graphitic carbon atoms on the surface. To remove these components, methods have been developed to treat the as-formed ND materials in strong oxidative acids such as H_2SO_4/HNO_3 mixtures [2] or ozone [3] at elevated temperatures. Consequently, these particles are derivatized with a variety of oxygen-containing functional groups, including carboxylic (–COOH) and carbonyl (–C=O) groups as well as different alcohol (primary, secondary, and tertiary) and ether groups, etc. Fourier-transform infrared (FTIR) spectroscopy, Raman spectroscopy, X-ray photoelectron spectroscopy, thermal desorption mass spectrometry, and thermogravimetric analysis are common tools used to characterize the chemical compositions and surface terminations of these surface-modified NDs [4]. Through FTIR, researchers have been able to distinguish different types of functional groups and adsorbates on the ND surface based on their signature vibrational frequencies. Furthermore, these analytical techniques can also detect subtle changes in the chemical composition before and after surface modification. For instance, the FTIR spectra of surface-oxidized NDs often exhibit prominent absorption bands at 1700–1800 cm^{-1}, ascribable to the —C=O stretches

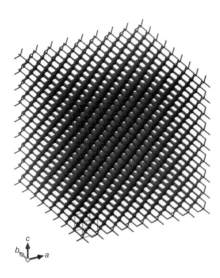

Figure 4.1 An idealized cubic ND with a perfect crystal structure and a length of 2.85 nm (or eight unit cells in one dimension).

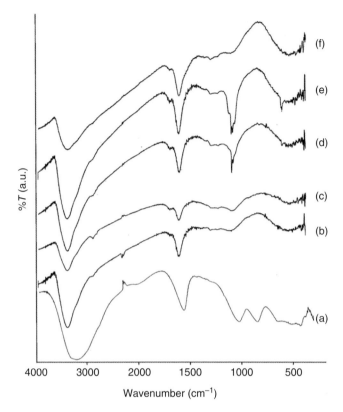

Figure 4.2 IR transmission spectra of DNDs subjected to (a) LiAlH$_4$ reduction, (b) borane reduction, (c) no treatment, (d) oxidation with HClO$_4$, (e) oxidation with HNO$_3$ and H$_2$SO$_4$ (1 : 1, v/v), and (f) reaction with ozone under ultraviolet irradiation. *Source:* Reprinted with permission from Ref. [5].

of carbonyl, ketone, aldehyde, carboxylic acid, ester, and other oxygen-containing groups (Figure 4.2). The spectra can be substantially simplified if a homogeneous layer of hydroxyl groups is formed on the surface by borane reduction [5].

High-temperature gas treatments also serve as a means to modify the ND surface. Specifically, oxidation in air at temperatures higher than 500 °C has been applied to control the sp^2/sp^3 carbon ratios and the surface chemistry of *detonation nanodiamond* (DND) powders [6]. Treatment in hydrogen plasma at high temperatures can reduce —C=O to —C—OH and further to —C—H groups under favorable conditions. Heating NDs in NH$_3$ can lead to the production of a variety of nitrogen-containing groups including —NH$_2$, —C≡N, and moieties containing —C=N. Similarly, heating NDs in Cl$_2$ produces acylchlorides and in F$_2$ produces the —C—F groups. Annealing NDs in N$_2$, Ar, or a vacuum at high temperatures completely eliminates all functional groups on the surface, converting sub-10-nm NDs such as DNDs to graphitic carbon nano-onions. Krueger and coworkers [7] have presented a full description of all possible techniques for the chemical treatment and functionalization of ND surfaces. Figure 4.3 provides an overview of some common strategies for the surface modifications of NDs [8]. It illustrates how wet chemistry combined with high-temperature gas treatments may attach various functional groups to the ND surface. An in-depth review of the preparation of

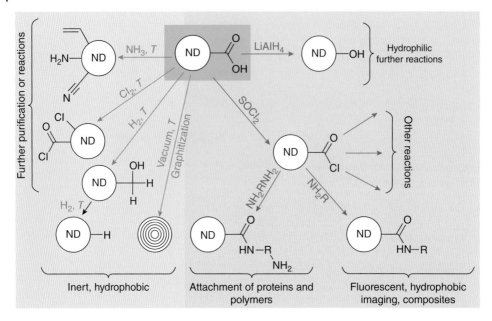

Figure 4.3 Overview of the commonly used methods for chemical modification and functionalization of ND surfaces. *Source:* Reprinted with permission from Ref. [8]. Reproduced with permission of Nature Publishing Group.

hydrogenated NDs and the characterization of their surface properties has been given by Arnault and Girard [9].

Another useful method for the surface modification of NDs is by radical reactions. Benzoyl peroxide is one of the radical initiators to induce the reactions. Experiments of Tsubota et al. [10] demonstrated the feasibility of modifying hydrogenated diamond surface with various carboxylic acids using the peroxide radicals under mild conditions. The process could be greatly facilitated by ultrasonication and microwave irradiation, which initiated the radical copolymerization of surface-graphitized NDs with molecules containing various functional groups, including –COOH, –NH$_2$, or aliphatic moieties [11]. The development of these techniques, together with the chemical modification schemes as illustrated in Figure 4.3, gives researchers access to further conjugation of NDs with various bioactive ligands or biomolecules for diverse biotechnological and biomedical applications.

Now is a good time to ask a simple and yet critical question: How can we effectively control the surface affinity, namely, hydrophilic or hydrophobic surface? Conventional hydrogenation and fluorination are two obvious choices to prepare hydrophobic NDs. However, they both require high-temperature gas treatments. Alternatively, wet chemistry offers a more versatile approach. A notable example in this direction is that of Hui et al. [12] who first treated acidified NDs with BH$_3$ in tetrahydrofuran to homogenize their surfaces with hydroxyl groups through reductive reactions, and then terminated the hydroxylated ND surface with octadecyltrimethoxysilane through silanization (Figure 4.4a). As shown in the photograph of Figure 4.4b, the NDs were

(a)

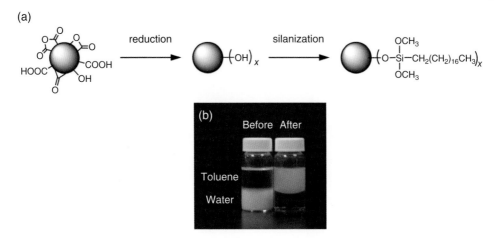

(b)

Figure 4.4 (a) Synthesis of hydrophobic HPHT-NDs by reduction and silanization of acid-treated NDs. (b) Photograph of 40-nm HPHT-NDs suspended in toluene/water before and after the reduction and silanization treatments. *Source:* Adapted with permission from Ref. [12].

originally hydrophilic, staying preferably in water (the lower layer), but became highly hydrophobic and all suspended in the organic phase (toluene in this case) after surface modification. These hydrophobic particles can further be encapsulated in lipids to form liposomal NDs (cf., Section 4.3).

A most useful functional group derivatized on the ND surface is –COOH, which is capable of forming direct covalent conjugation with biomolecules as indicated in Figure 4.3. *Zeta potential analysis*, which measures the electrokinetic potential of a colloidal dispersion [13], serves as a convenient tool to identify the presence of –COOH groups on the surface. The potential describes the stability of a colloid and its tendency toward agglomeration. The colloid is considered stable when the zeta potential is lower than –30 mV or higher than +30 mV. It has been reported that NDs can form colloids with zeta potentials ranging from –40 to +40 mV, depending on their surface terminations [14–16]. For HPHT-NDs treated with strong acid mixtures such as H_2SO_4/HNO_3 (3 : 1, v/v) as an example, the colloids can have a zeta potential of –36 mV at pH ≥ 5 (Figure 4.5a) and exhibit high dispersibility in water [17]. The zero-point potential occurs at pH 2.9, slightly smaller than $pK_a = 3.75$ of HCOOH, meaning that their surface is negatively charged at neutral pH and in physiological medium (pH 7.4).

The zeta potential analysis described above provides only a qualitative measurement for the status of surface charge. What exactly is the amount of –COOH groups on the ND surface? To address this issue, Nguyen et al. [17] have adopted a conductometric back-titration method that measures the electrolytic conductivity as a means to monitor the progress of a chemical reaction in solution [18]. Direct titration of the sample solution with NaOH is not suitable here because the surface carboxylic acid groups on the NDs are low in both acidity and density. Shown in Figure 4.5b is a typical backward titration curve obtained by first reacting the 100-nm HPHT-NDs with an excessive amount of 0.1 N NaOH, then titrating back with standardized 0.1 N HCl. The curve consists of three regions, of which the first region corresponds to the neutralization of excess OH⁻,

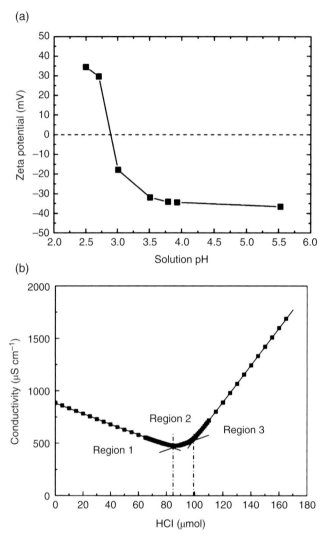

Figure 4.5 (a) Zeta potentials of oxidative-acid-treated HPHT-NDs (nominal size ~100 nm) as a function of solution pH. (b) Conductometric titration of oxidative-acid-treated 100-nm HPHT-NDs. In this titration, an excessive amount of NaOH was first added into the ND suspension and then neutralized with 0.1 N HCl. *Source:* Adapted with permission from Ref. [17]. Reproduced with permission of Elsevier.

the second one corresponds to the titration of the surface −COOH, and the third corresponds to the increase of solution H_3O^+ due to the addition of excess titrant, 0.1 N HCl. The conductivity is lowest when the solution OH^- ions are completely neutralized. With this method, the researchers determined a quantity of $100\,\mu mol\,g^{-1}$ for the carboxylate groups on the surface of the 100-nm HPHT-ND. However, due to the high polydispersity and irregular shape of the ND particles, only a rough estimate of approximately 7% could be obtained for the fraction of surface carbon atoms in carboxyl form. A higher percentage is anticipated for DNDs which are polycrystalline and contain more defects (and thus more sites to form −COOH groups) on their surface.

4.2 Bioconjugation

4.2.1 Noncovalent Conjugation

A ND's surface may contain a rich variety of functional groups, depending on how the nanomaterial is synthesized and subsequently chemically treated. The chemical treatment can change the properties of the particles from inert to highly reactive or even highly toxic in cells and living organisms. Conjugation of NDs with biomolecules is one of such chemical treatments and can be achieved by either physical adsorption through noncovalent interactions or covalent linkage with surface functional groups. A convenient way to characterize the outcome of the conjugation and the stability of the nanoparticle bioconjugates in solution is *dynamic light scattering* (DLS).

DLS is a technique measuring the temporal fluctuation of the intensities of scattered light from molecules or small particles undergoing Brownian motion in solution by using lasers as the light sources [19]. It provides information about the hydrodynamic sizes and size distribution profiles of the particles before and after surface modification. The first-order result of the DLS measurement is the size-dependent intensity distribution. This distribution, however, does not reflect directly the hydrodynamic diameters (d_h) of the particles, because the intensity is proportional to d_h^6, instead of d_h. Thus, particles of larger size dominate the distribution in observation. It is possible to convert the intensity distributions to volume and number distributions according to the Mie scattering theory [20] if the particles are spherical, sizes are homogeneous, and their optical properties are known. For NDs, due to their irregular shape and large variation in size, the volume and number distributions derived from their intensity distributions are best used only for comparative purposes. Nonetheless, the comparison still sheds significant insight into the result of surface modification.

An array of organic and biological molecules have been conjugated onto NDs either covalently or noncovalently. These include chemotherapeutic drugs [21–24], carbohydrates [25, 26], peptides [14, 27], proteins [17, 28], small interfering RNA [29, 30], and DNA [31, 32] (cf., Chapter 13 for further details). When attaching biomolecules such as proteins to NDs, it is crucial to confirm that their functionalities are conserved. Nguyen et al. [17] addressed this concern in a study of lysozyme adsorption and immobilization on the surface of 100-nm HPHT-NDs. They detected the hydrolytic activity of the enzyme after physical attachment and found that their activity was significantly lower than that of free lysozymes in aqueous solution. The relative activity was only about 15% at the surface coverage of 10%. The activity, however, could be boosted as the ND surface was blocked with supplementary proteins such as cytochrome *c* to create a more "crowded" environment. The tactic effectively increased the activity of ND-bound lysozymes from 60 to 70%. Such a surface effect is expected to be found in other enzymes as well.

A question often asked is: What is the maximum loading capacity of proteins on a ND particle? The answer can be derived from a measurement of the adsorption isotherms of the protein of interest on NDs. Specifically, Kong et al. [33] investigated the high affinity capture of proteins (including cytochrome *c*, myoglobin, and serum albumin) by the acid-treated HPHT-NDs. They determined the loading capacity based on the change of protein concentration before and after adding diamond powders into the solution by ultraviolet-visible (UV-Vis) spectroscopy (Figure 4.6a). It was found that the loading maximized at the isoelectric points of the individual proteins, as a result of the competition

(a)

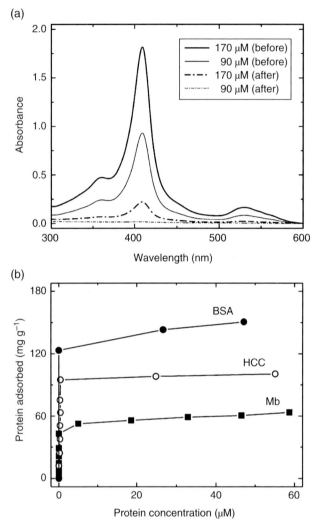

(b)

Figure 4.6 (a) UV-Vis absorption spectra of horse cytochrome *c* solutions before (solid curves) and after (dash dot curves) exposure to HPHT-NDs at two different protein concentrations. (b) Adsorption isotherms of horse heart cytochrome *c* (HCC), horse heart myoglobin (Mb), and bovine serum albumin (BSA) on 100-nm HPHT-NDs at pH 10.5, 6.9, and 4.7, respectively. *Source:* Reprinted with permission from Refs. [2] and [33]. Reproduced with permission of American Chemical Society.

between protein–protein and protein–ND interactions. The protein loading capacity varied from 60 to 150 mg g^{-1}, depending on the sizes and molecular weights (12–66 kDa) of the adsorbed proteins. For a 100-nm ND that weighs approximately 2 fg particle^{-1}, this loading capacity suggests that more than 1000 protein molecules can be attached to the surface per particle. Another significant finding of the same study is that the acid-treated HPHT-NDs have an exceptionally high affinity for all proteins investigated. The feature is clearly shown in the isotherms where the 100-nm HPHT-ND surface is readily saturated with protein molecules at the protein concentration less than 5 μM

(Figure 4.6b). A combination of electrostatic forces, hydrogen bonding, and hydrophobic interactions between adsorbents (NDs) and adsorbates (proteins) may be the origin of this high affinity [33]. The affinity, however, significantly diminishes as the particle size decreases. For DNDs of approximately 5 nm in diameter, both myoglobin and serum albumin exhibit a Langmuir-type adsorption behavior with the surface saturation occurring only at the protein concentration greater than 100 μm [34]. The particles form 1 : 1 complexes with these two types of proteins at saturation.

Given excellent chemical stability, small size, ease of purification, and high affinity for proteins, the acid-treated HPHT-NDs have been proposed as a solid-phase extraction device for bioanalytical applications [35]. Chang and coworkers [36] demonstrated the utility of 100-nm NDs as a tool to facilitate proteomic analysis by showing that the particles were able to capture proteins from complex medium or highly diluted solution in minutes. The affinity was so high that these protein–ND complexes could sustain repeated washing with deionized water without much loss. Moreover, after separation by centrifugation or filtering, the captured protein molecules could be analyzed immediately by gel electrophoresis or mass spectrometry. A platform called *SPEED* (solid-phase extraction and elution on diamond) was developed for proteomics research by the group [36, 37]. A distinct advantage of the SPEED platform is that it facilitates purification and concentration of intact proteins and their enzymatic digests for ensuing *sodium dodecyl sulfate-polyacrylamide gel electrophoresis* (SDS-PAGE) or *matrix-assisted laser desorption/ionization mass spectrometry* (MALDI-MS) analysis without prior removal of the ND adsorbent (Figure 4.7). Moreover, one-pot workflow involving the reduction of disulfide bonds, protection of free cysteine residues, and proteolytic digestion of the

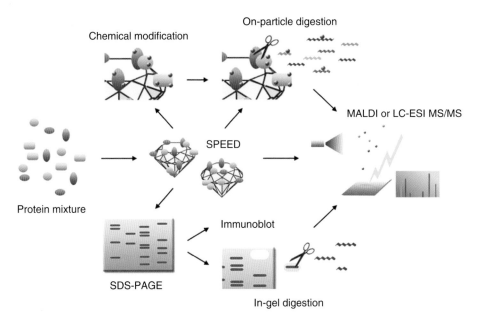

Figure 4.7 Scope of typical applications of the SPEED platform to proteome analysis. ESI, electrospray ionization; LC, liquid chromatography; MALDI, matrix-assisted laser desorption/ionization; MS, mass spectrometry; PAGE, polyacrylamide gel electrophoresis; and SDS, sodium dodecyl sulfate. *Source:* Reprinted with permission from Ref. [35]. Reproduced with permission of John Wiley & Sons.

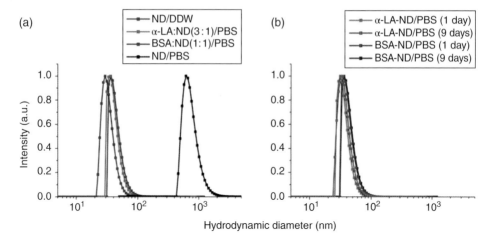

Figure 4.8 (a) Size distributions of HPHT-NDs before and after noncovalent conjugation with BSA and α-LA in DDW and PBS, measured by DLS. The mean diameters of the particles with size distributions from left to right are 30.0, 35.1, 41.9, and 752.7 nm, respectively. (b) Stability tests of the colloidal suspensions of BSA- and α-LA-conjugated HPHT-NDs in PBS at room temperature for nine days. The mean diameters of the particles with size distributions from left to right are 35.0, 37.0, 41.9, and 43.2 nm, respectively. *Source:* Adapted with permission from Ref. [43]. Reproduced with permission of John Wiley & Sons.

adsorbed proteins can be directly carried out on the particles. The platform, in combination with two-phase separation techniques, is further applicable to extract membrane proteins in detergent micelles for shotgun proteomic analysis [38, 39]. It is potentially useful as a nucleating agent to facilitate the crystallization of low abundant proteins under unfavorable growth conditions too [40].

In applying NDs for biological research, a problem often encountered is the low colloidal stability due to facile aggregation of the nanoparticles in physiological medium. The colloidal stability of surface-unmodified NDs is not static and, in fact, it drops substantially at high ionic strengths in solution, a behavior commonly found in many nanoparticles [41, 42]. This poses a serious problem since biological buffers such as phosphate-buffered saline (PBS) and cultivation media all contain high concentrations of salts. Fortunately, difficulties of this kind can be easily overcome by noncovalent conjugation of NDs with protein molecules like albumin [28, 43]. Figure 4.8a shows an example of the DLS measurements for the hydrodynamic sizes of 30-nm HPHT-NDs noncovalently conjugated with α-lactalbumin (α-LA) and bovine serum albumin (BSA) at different weight ratios in distilled deionized water (DDW) and PBS [43]. The nanoparticle bioconjugates maintain their good dispersibility over weeks without noticeable agglomeration and precipitation in the buffers (Figure 4.8b). Colloidally stable α-LA-coated NDs as small as 18 nm can also be prepared with this simple method.

4.2.2 Covalent Conjugation

Despite the simple and straightforward approach that noncovalent conjugation may provide, there is a concern for the long-term stability of the nanoparticle bioconjugates in physiological medium. Covalent conjugation is a more desirable approach. However, direct and chemical bonding of NDs with nanometer-sized biomolecules such as proteins is nontrivial due to the high steric hindrance of the reactions as discussed in the previous section. To reduce the steric constraints and retain the adsorbate's activity, a number of methods have been developed to insert spacers between NDs and the biomolecules to be conjugated. The step is particularly crucial for the immobilization of enzymes whose active sites may be sterically hindered after attachment to the ND surface. Additional functions of the spacers are that they help suppress nonspecific interactions and prevent protein conformational changes caused by strong biomolecule–ND interactions.

One of the most commonly used spacers is polyethylene glycol (PEG), which has a high biocompatibility but a low degree of nonspecific interactions with biomolecules [44]. PEG molecules derivatized with amino, carboxyl, and other functional groups are all available commercially. They can be covalently conjugated with carboxylated NDs via the amino groups on their termini by carbodiimide chemistry as follows:

Moreover, these PEG spacers are available in various discrete lengths and also provide additional functional groups for carbodiimide or other crosslinking with drugs such as doxorubicin [24], bioactive ligands such as folic acid [45], or proteins such as streptavidin [28]. Furthermore, they can be noncovalently conjugated with a cationic–hydrophobic block to form (PEG-*b*-poly(2-(dimethylamino)ethyl methacrylate-co-butyl methacrylate)) copolymers on ND surface to avoid aggregation of the nanoparticles in biological buffers [46].

Poly-lysine is the first polymer covalently grafted on NDs [47]. The grafting terminates the ND surface with amino groups, which are also useful ligands for conjugation with peptides, proteins, and other biomolecules. The method is facile and more effective than high-temperature gas treatments involving NH_3. It allows the formation of multiple amide bonds between surface carboxyl groups and free lysine residues in the polymers, generating a stable overlayer with a high density of amino groups on the surface. The same concept is applicable to other polyelectrolytes such as polyethylenimine and polyallylamine or organosilanes for surface functionalization with amino groups [14]. Further research in the field has also found that atom-transfer radical-polymerization is an effective method of attaching polymer brushes onto the ND surface terminated with an initiator [48]. Additionally, coating of NDs with hyperbranched polyglycerol by ring-opening polymerization can significantly improve their colloidal stability in DDW and PBS [49]. Subsequent reaction of polyglycerol-conjugated NDs with succinic anhydride and covalent conjugation with antibiotics (such as ampicillin) allows selective targeting of cytokine receptors on cell membrane [50].

Apart from homopolymers, copolymers such as that consisting of 95% *N*-(2-hydroxypropyl)methacrylamide and 5% propargylacrylamide or 3-(azidopropyl)methacrylamide) have been also developed as biocompatible protein-resistant coatings on NDs after silica coating as previously described [51]. These polymeric molecules are hydrophilic and highly flexible, allowing bioorthogonal attachment of various molecules by *click chemistry,*

In this reaction, an azide forms a five-membered heterogeneous ring with an alkyne group, known as the *catalyzed Cu(I)–azide–alkyne cycloaddition* [52]. Because of its gentle nature and high specificity, the chemistry is gaining popularity for the bioconjugation of NDs terminated first with azides by carbodiimide chemistry or other methods [53, 54].

Many potent drugs, such as those can potentially treat cancers, are bundled with delivery challenges. For instance, some of the drugs are insoluble in polar protic solvents (such as water) but are soluble in nonpolar solvents (such as tetrahydrofuran) that are harmful to the body. Through surface modification in conjunction with drug loading on NDs, researchers have created new delivery methods to solve this problem. One of such examples is the covalent linkage of DNDs with paclitaxel, a chemotherapy drug, for cancer therapy [23]. The benefits of this approach are many as NDs are inherently biocompatible (Chapter 5) and have the ability to carry a significant amount of drugs. While some studies have already used ND surfaces to conjugate with drugs via chemical bonding [22–24], the majority of the research studies are focusing on physical adsorption procedures [55, 56]. Detailed discussion of the therapeutic applications of NDs can be found in Chapter 13.

4.3   Encapsulation

In nanotechnology, encapsulation is a process that one nanomaterial is enclosed in another material, either a pure element or a compound. The encapsulation can be carried out covalently and/or noncovalently, and the thickness of the overlayers is typically in the range of 1–10 nm. The overlayers may be organic (such as lipids), inorganic (such as silica and metals), or a combination of both. Crosslinking often occurs between the atoms or molecules in the overlayers to form a network on the nanomaterial's surface and thus stabilize the encapsulation. In this section, we introduce only two types of encapsulation that have been successfully applied to NDs: Lipid layers and silica shells. Further discussions of the silica-encapsulated NDs and other hybrid ND materials are given in Chapter 12.

4.3.1   Lipid Layers

Liposomization of pharmaceuticals has been a promising technique in drug delivery since its invention in the 1970s [57]. A well-developed method for liposome preparation in the field is the thin-film hydration, which involves dissolving powdered lipids and lipid-soluble drugs in organic solvent, followed by deposition of a thin film of lipids on the surface of a round bottom flask by evaporating the organic phase, and finally hydration of the lipid film in aqueous medium. Ho and coworkers [58] applied the technique to synthesize self-assembled ND–lipid hybrid nanoparticles that allowed for

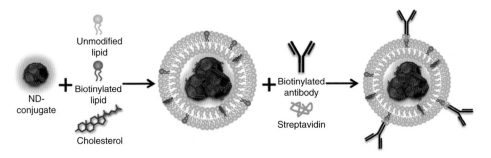

Figure 4.9 Encapsulation of NDs in liposome by rehydration of lipid thin films containing cholesterol and biotinylated lipids in concentrated ND solution. The molecule in red is epirubicin, a chemotherapy drug. *Source:* Reprinted with permission from Ref. [58]. Reproduced with permission of John Wiley & Sons.

potent interactions between NDs and small drug molecules. To achieve cell-specific targeting, the researchers synthesized ND-based nanohybrids containing cholesterol and biotinylated lipids, which could selectively bind to streptavidin and then with biotinylated antibodies of various specificities (Figure 4.9). Hui et al. [12] reported a similar procedure to encapsulate HPHT-NDs within cationic cholesterol-based lipids after surface reduction and silanization of the particles, as illustrated in Figure 4.4a.

Recently, a simple and effective method to encapsulate NDs in biofunctionalized lipid layers has been developed by Hsieh et al. [59]. The method takes advantage of the Ouzo effect [60], which involves the addition of a mixture of hydrophobic solute and water-miscible solvent into water to form stable microdroplets. The hydrophobic solute used in this work is a lipid layer consisting of egg phosphatidylcholine, cholesterol, and PEGylated 1,2-distearoyl-sn-glycero-3-phosphoethanolamine, which are dissolved in tetrahydrofuran (i.e. the water-miscible solvent) and then added to water containing surface-oxidized NDs to form emulsions. Subsequent evaporation of tetrahydrofuran in a vacuum allows the lipid layer to coat on NDs. The method enables not only robust coating but also the synthesis of NDs with desired functional groups such as biotin. The particles exhibit exceptionally high dispersibility in PBS and cell medium, well suited for biolabeling applications. Further stabilization of the surface coating can be established with photo-crosslinked lipids [61]. The effective encapsulation of NDs (size of 30–100 nm) in liposomes opens a promising new avenue to conjugate the particles with bioactive ligands or proteins on the lipid layer for specific cell labeling, targeting, and imaging with both light microscopy and electron microscopy (cf., Section 10.4).

4.3.2 Silica Shells

Along with liposomization, researchers have also encapsulated NDs in silica shells [62–68]. The encapsulation provides a novel platform for subsequent chemical treatment based on the silica chemistry. For example, the silica-encapsulated ND surface may contain a variety of free silanol groups that allow conjugation of the biomolecules of interest with the encapsulated nanoparticles. The approach enhances not only the colloidal stability but also the functionality of NDs by using core-shell rational designs with the benefit of synthetic versatility [69]. Figure 4.10a and b show, respectively, the *transmission electron microscopy* (TEM) images of HPHT-NDs before and after silica coating [65].

(a)

(b)

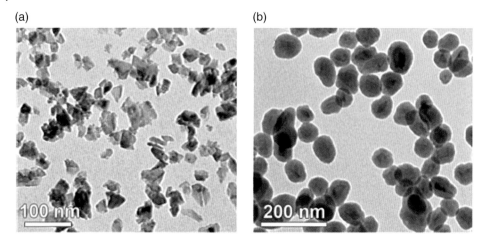

Figure 4.10 TEM images of (a) as-received and (b) silica-coated HPHT-ND particles. *Source:* Reprinted with permission from Ref. [65]. Reproduced with permission of John Wiley & Sons.

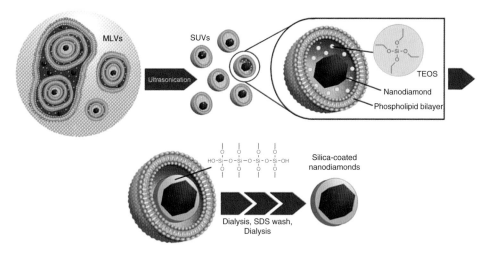

Figure 4.11 Synthesis of silica-coated NDs by liposome-based encapsulation. MLVs, multilamellar vesicles; SUVs, unilamellar vesicles; TEOS, tetraethyl orthosilicate; and SDS, sodium dodecyl sulfate. *Source:* Adapted with permission from Ref. [64]. Reproduced with permission of American Chemical Society.

The shell thickness can be as thin as 10 nm if the coating is properly prepared [67]. An added benefit of the shell coating is that it normalizes the irregular shape of the prickly ND particles [65, 68], yielding egg-like spheroids. Moreover, the shell serves as a multifunctional interface for further conjugation with bioactive molecules like biotin for subsequent conjugation with streptavidin, antibodies, and other protein molecules.

An example of the synthesis of core-shell ND-silica particles through liposome-based encapsulation is given in Figure 4.11. In this protocol developed by Bumb et al. [64], NDs in a solution of tetraethyl orthosilicate (TEOS) were first trapped in multilamellar vesicles (MLVs) that ranged in size from 500 to 10 000 nm. Ultrasonication broke the MLVs into small unilamellar vesicles (SUVs) with a nominal diameter of approximately 100 nm. TEOS was then converted into silica, catalyzed by triethylamine. After removal of free TEOS and triethylamine by dialysis, a sodium dodecyl sulfate wash ruptured the liposomes to free the coated NDs. The final products consisted of stabilized and monodisperse silica-encapsulated NDs. Although the surface of these silica-encapsulated NDs presented mainly free silanol groups, attachment of biomolecules to the nanoparticles could be readily achieved by replacing these silanol groups with amino groups. The same strategy has been applied to grafting copolymers on the silica shell for further conjugation with fluorescent probes and targeting peptides via click chemistry [67]. With this combined approach, it is possible to selectively attach bioactive ligands to NDs of various sizes and concurrently improve the colloidal stability of these nanoparticles in biological buffers.

References

1 Krueger, A. (2014). The chemistry of nanodiamond. In: *Nanodiamonds* (ed. O.A. Williams), 49–88. Royal Society of Chemistry.

2 Huang, L.C.L. and Chang, H.C. (2004). Adsorption and immobilization of cytochrome c on nanodiamonds. *Langmuir* 20: 5879–5884.

3 Shenderova, O., Koscheev, A., Zaripov, N. et al. (2011). Surface chemistry and properties of ozone-purified detonation nanodiamonds. *J Phys Chem C* 115: 9827–9837.

4 Sotoma, S., Akagi, K., Hosokawa, S. et al. (2015). Comprehensive and quantitative analysis for controlling the physical/chemical states and particle properties of nanodiamonds for biological applications. *RSC Adv* 5: 13818–13827.

5 Krueger, A., Liang, Y.J., Jarre, G., and Stegk, J. (2006). Surface functionalisation of detonation diamond suitable for biological applications. *J Mater Chem* 16: 2322–2328.

6 Osswald, S., Yushin, G., Mochalin, V. et al. (2006). Control of sp^2/sp^3 carbon ratio and surface chemistry of nanodiamond powders by selective oxidation in air. *J Am Chem Soc* 128: 11635–11642.

7 Krueger, A. and Lang, D. (2012). Functionality is key: recent progress in the surface modification of nanodiamond. *Adv Funct Mater* 22: 890–906.

8 Mochalin, V.N., Shenderova, O., Ho, D., and Gogotsi, Y. (2012). The properties and applications of nanodiamonds. *Nat Nanotechnol* 7: 11–23.

9 Arnault, J.C. and Girard, H.A. (2017). Hydrogenated nanodiamonds: synthesis and surface properties. *Curr Opin Solid State Mater Sci* 21: 10–16.

10 Tsubota, T., Tanii, S., Ida, S. et al. (2004). Chemical modification of diamond surface with various carboxylic acids by radical reaction in liquid phase. *Diam Relat Mater* 13: 1093–1097.

11 Chang, I.P., Hwang, K.C., Ho, J.A. et al. (2010). Facile surface functionalization of nanodiamonds. *Langmuir* 26: 3685–3689.

12 Hui, Y.Y., Zhang, B.L., Chang, Y.C. et al. (2010). Two-photon fluorescence correlation spectroscopy of lipid-encapsulated fluorescent nanodiamonds in living cells. *Opt Express* 18: 5896–5905.

13 Moore, W.J. (1963). *Physical Chemistry*. New York: Longmans Green.

14 Vial, S., Mansuy, C., Sagan, S. et al. (2008). Peptide-grafted nanodiamonds: preparation, cytotoxicity and uptake in cells. *ChemBioChem* 9: 2113–2119.

15 Williams, O.A., Hees, J., Dieker, C. et al. (2010). Size-dependent reactivity of diamond nanoparticles. *ACS Nano* 4: 4824–4830.

16 Kaur, R. and Badea, I. (2013). Nanodiamonds as novel nanomaterials for biomedical applications: drug delivery and imaging systems. *Int J Nanomed* 8: 203–220.

17 Nguyen, T.T.B., Chang, H.C., and Wu, V.W.K. (2007). Adsorption and hydrolytic activity of lysozyme on diamond nanocrystallites. *Diam Relat Mater* 16: 872–876.

18 Vogel, A.I. (1989). *Vogel's Textbook of Quantitative Chemical Analysis*. New York: Wiley.

19 Berne, B.J. and Pecora, R. (2000). *Dynamic Light Scattering: With Applications to Chemistry, Biology, and Physics*. Dover Publications.

20 Bohren, C.F. and Huffman, D.R. (1983). *Absorption and Scattering of Light by Small Particles*. Wiley.

21 Huang, H., Pierstorff, E., Osawa, E., and Ho, D. (2007). Active nanodiamond hydrogels for chemotherapeutic delivery. *Nano Lett* 7: 3305–3314.

22 Li, J., Zhu, Y., Li, W.X. et al. (2010). Nanodiamonds as intracellular transporters of chemotherapeutic drug. *Biomaterials* 31: 8410–8418.

23 Liu, K.K., Zheng, W.W., Wang, C.C. et al. (2010). Covalent linkage of nanodiamond-paclitaxel for drug delivery and cancer therapy. *Nanotechnology* 21: 315106.

24 Wang, D.X., Tong, Y.L., Li, Y.Q. et al. (2013). PEGylated nanodiamond for chemotherapeutic drug delivery. *Diam Relat Mater* 36: 26–34.

25 Hartmann, M., Betz, P., Sun, Y.C. et al. (2012). Saccharide-modified nanodiamond conjugates for the efficient detection and removal of pathogenic bacteria. *Chem-Eur J* 18: 6485–6492.

26 Barras, A., Martin, F.A., Bande, O. et al. (2013). Glycan-functionalized diamond nanoparticles as potent *E. coli* anti-adhesives. *Nanoscale* 5: 12678–12678.

27 Shimkunas, R.A., Robinson, E., Lam, R. et al. (2009). Nanodiamond-insulin complexes as pH-dependent protein delivery vehicles. *Biomaterials* 30: 5720–5728.

28 Chang, B.M., Lin, H.H., Su, L.J. et al. (2013). Highly fluorescent nanodiamonds protein-functionalized for cell labeling and targeting. *Adv Funct Mater* 23: 5737–5745.

29 Chen, M., Zhang, X.Q., Man, H.B. et al. (2010). Nanodiamond vectors functionalized with polyethylenimine for siRNA delivery. *J Phys Chem Lett* 1: 3167–3171.

30 Alhaddad, A., Adam, M.P., Botsoa, J. et al. (2011). Nanodiamond as a vector for siRNA delivery to *Ewing sarcoma* cells. *Small* 7: 3087–3095.

31 Zhang, X.Q., Chen, M., Lam, R. et al. (2009). Polymer-functionalized nanodiamond platforms as vehicles for gene delivery. *ACS Nano* 3: 2609–2616.

32 Petrakova, V., Benson, V., Buncek, M. et al. (2016). Imaging of transfection and intracellular release of intact, non-labeled DNA using fluorescent nanodiamonds. *Nanoscale* 8: 12002–12012.

33 Kong, X.L., Huang, L.C.L., Hsu, C.M. et al. (2005). High-affinity capture of proteins by diamond nanoparticles for mass spectrometric analysis. *Anal Chem* 77: 259–265.

34 Lin, C.L., Lin, C.H., Chang, H.C., and Su, M.C. (2015). Protein attachment on nanodiamonds. *J Phys Chem A* 119: 7704–7711.

35 Wu, C.C., Han, C.C., and Chang, H.C. (2010). Applications of surface-functionalized diamond nanoparticles for mass-spectrometry-based proteomics. *J Chin Chem Soc-Taip* 57: 583–594.

36 Chen, W.H., Lee, S.C., Sabu, S. et al. (2006). Solid-phase extraction and elution on diamond (SPEED): a fast and general platform for proteome analysis with mass spectrometry. *Anal Chem* 78: 4228–4234.

37 Sabu, S., Yang, F.C., Wang, Y.S. et al. (2007). Peptide analysis: solid phase extraction-elution on diamond (SPEED) combined with atmospheric pressure MALDI-FTICR mass spectrometry. *Anal Biochem* 367: 190–200.

38 Pham, M.D., Yu, S.S.F., Han, C.C., and Chan, S.I. (2013). Improved mass spectrometric analysis of membrane proteins based on rapid and versatile sample preparation on nanodiamond particles. *Anal Chem* 85: 6748–6755.

39 Pham, M.D., Wen, T.C., Li, H.C. et al. (2016). Streamlined membrane proteome preparation for shotgun proteomics analysis with Triton X-100 cloud point extraction and nanodiamond solid phase extraction. *Materials* 9: 385.

40 Chen, Y.W., Lee, C.H., Wang, Y.L. et al. (2017). Nanodiamonds as nucleating agents for protein crystallization. *Langmuir* 33: 6521–6527.

41 Lim, J.K., Majetich, S.A., and Tilton, R.D. (2009). Stabilization of superparamagnetic iron oxide core-gold shell nanoparticles in high ionic strength media. *Langmuir* 25: 13384–13393.

42 Zhang, W. (2014). Nanoparticle aggregation: principles and modeling. *Adv Exp Med Biol* 811: 19–43.

43 Tzeng, Y.K., Faklaris, O., Chang, B.M. et al. (2011). Superresolution imaging of albumin-conjugated fluorescent nanodiamonds in cells by stimulated emission depletion. *Angew Chem Int Ed* 50: 2262–2265.

44 Zhang, X.Y., Fu, C.K., Feng, L. et al. (2012). PEGylation and polyPEGylation of nanodiamond. *Polymer* 53: 3178–3184.

45 Zhang, B.L., Li, Y.Q., Fang, C.Y. et al. (2009). Receptor-mediated cellular uptake of folate-conjugated fluorescent nanodiamonds: a combined ensemble and single-particle study. *Small* 5: 2716–2721.

46 Lee, J.W., Lee, S., Jang, S. et al. (2013). Preparation of non-aggregated fluorescent nanodiamonds (FNDs) by non-covalent coating with a block copolymer and proteins for enhancement of intracellular uptake. *Mol Biosyst* 9: 1004–1011.

47 Fu, C.C., Lee, H.Y., Chen, K. et al. (2007). Characterization and application of single fluorescent nanodiamonds as cellular biomarkers. *Proc Natl Acad Sci USA* 104: 727–732.

48 Dahoumane, S.A., Nguyen, M.N., Thorel, A. et al. (2009). Protein-functionalized hairy diamond nanoparticles. *Langmuir* 25: 9633–9638.

49 Zhao, L., Takimoto, T., Ito, M. et al. (2011). Chromatographic separation of highly soluble diamond nanoparticles prepared by polyglycerol grafting. *Angew Chem Int Ed* 50: 1388–1392.

50 Sotoma, S., Iimura, J., Igarashi, R. et al. (2016). Selective labeling of proteins on living cell membranes using fluorescent nanodiamond probes. *Nanomaterials* 6: 56.

51 Rehor, I., Mackova, H., Filippov, S.K. et al. (2014). Fluorescent nanodiamonds with bioorthogonally reactive protein-resistant polymeric coatings. *ChemPlusChem* 79: 21–24.

52 Kolb, H.C., Finn, M.G., and Sharpless, K.B. (2001). Click chemistry: diverse chemical function from a few good reactions. *Angew Chem Int Ed* 40: 2004–2021.

53 Barras, A., Szunerits, S., Marcon, L. et al. (2010). Functionalization of diamond nanoparticles using "click" chemistry. *Langmuir* 26: 13168–13172.

54 Meinhardt, T., Lang, D., Dill, H., and Krueger, A. (2011). Pushing the functionality of diamond nanoparticles to new horizons – orthogonally functionalized nanodiamond using click chemistry. *Adv Funct Mater* 21: 494–500.

55 Chen, M., Pierstorff, E.D., Lam, R. et al. (2009). Nanodiamond-mediated delivery of water-insoluble therapeutics. *ACS Nano* 3: 2016–2022.

56 Chow, E.K., Zhang, X.Q., Chen, M. et al. (2011). Nanodiamond therapeutic delivery agents mediate enhanced chemoresistant tumor treatment. *Sci Transl Med* 3: 73ra21.

57 Gregoriadis, G. and Ryman, B.E. (1971). Liposomes as carriers of enzymes or drugs: a new approach to the treatment of storage diseases. *Biochem J* 124: 58.

58 Moore, L., Chow, E.K., Osawa, E. et al. (2013). Diamond-lipid hybrids enhance chemotherapeutic tolerance and mediate tumor regression. *Adv Mater* 25: 3532–3541.

59 Hsieh, F.J., Chen, Y.W., Hui, Y.Y. et al. (2018). Correlative light-electron microscopy of lipid-encapsulated fluorescent nanodiamonds for nanometric localization of cell surface antigens. *Anal Chem* 90: 1566–1571.

60 Vitale, S.A. and Katz, J.L. (2003). Liquid droplet dispersions formed by homogeneous liquid-liquid nucleation: "the ouzo effect". *Langmuir* 19: 4105–4110.

61 Sotoma, S., Hsieh, F.J., Chen, Y.W. et al. (2018). Highly stable lipid-encapsulation of fluorescent nanodiamonds for bioimaging applications. *Chem Commun* 54: 1000–1003.

62 von Haartman, E., Jiang, H., Khomich, A.A. et al. (2013). Core-shell designs of photoluminescent nanodiamonds with porous silica coatings for bioimaging and drug delivery I: fabrication. *J Mater Chem B* 1: 2358–2366.

63 Prabhakar, N., Nareoja, T., von Haartman, E. et al. (2013). Core-shell designs of photoluminescent nanodiamonds with porous silica coatings for bioimaging and drug delivery II: application. *Nanoscale* 5: 3713–3722.

64 Bumb, A., Sarkar, S.K., Billington, N. et al. (2013). Silica encapsulation of fluorescent nanodiamonds for colloidal stability and facile surface functionalization. *J Am Chem Soc* 135: 7815–7818.

65 Rehor, I., Slegerova, J., Kucka, J. et al. (2014). Fluorescent nanodiamonds embedded in biocompatible translucent shells. *Small* 10: 1106–1115.

66 Rehor, I., Lee, K.L., Chen, K. et al. (2015). Plasmonic nanodiamonds: targeted core-shell type nanoparticles for cancer cell thermoablation. *Adv Healthc Mater* 4: 460–468.

67 Slegerova, J., Hajek, M., Rehor, I. et al. (2015). Designing the nanobiointerface of fluorescent nanodiamonds: highly selective targeting of glioma cancer cells. *Nanoscale* 7: 415–420.

68 Chu, Z., Zhang, S., Zhang, B. et al. (2014). Unambiguous observation of shape effects on cellular fate of nanoparticles. *Sci Rep* 4: 4495.

69 Guerrero-Martínez, A., Pérez-Juste, J., and Liz-Marzán, L.M. (2010). Silica-coated nanomaterials: recent progress on silica coating of nanoparticles and related nanomaterials. *Adv Mater* 22: 1182–1195.

5

Biocompatibility of Nanodiamonds

"A diamond is forever," De Beers' diamond campaign launched on the *New York Times*, September 1948, is the same slogan still used today [1]. Funny as it may sound, if asked, scientists working on *nanodiamonds* (NDs) would say the same. Scientifically, the slogan vividly depicts the exceptionally high physical and chemical stability of the gemstone under ambient conditions. However, some may be wondering: Stable, yes, but is it safe to use? How to tell if it is safe for humans? Can we measure safety in a quantitative way and, if so, how? These are some questions that we try to answer in this chapter.

Diamond, categorized as an inorganic material, is considered both chemically inert and biologically compatible because it is composed of pure sp^3-hybridized carbon atoms, except those on the surface. According to *International Union of Pure and Applied Chemistry* (IUPAC), the term "biological compatibility" or "biocompatibility" in short is defined in a general context as "the ability to be in contact with a living system without producing an adverse effect" or, in the context of medical therapy, "the ability of a material to perform with an appropriate host response in a specific application" [2]. The exceptionally low chemical reactivity of diamond is well in line with the first definition. This characteristic, together with the facts that diamond can be synthesized by chemical vapor deposition methods for coating of biomedical devices and the surface of diamond can be readily derivatized with various functional groups for bioconjugation, has earned diamond the name "Biomaterial of the 21st century" [3].

Focusing on nanoscale diamonds, we discuss in this chapter the biocompatibility studies of NDs and their comparisons with other members in the nanocarbon family (Section 1.2). How NDs can be used to perform specific medical applications will be the subject of discussion for Chapter 13. Here, we begin with a brief review of some representative protocols used in research laboratories to assess the safety of nanoparticles in cells and organisms.

5.1 Biocompatibility Testing

The purpose of testing for biocompatibility is to determine whether or not a substance or material when introduced to a living host would harm or even kill the host. The events can occur at the cellular (*cyto*) or animal levels. To examine the damages to cells, for example, we may want to test "cytotoxicity" by measuring the proportion of cells suffered from the introduction of a certain substance under study to the sample.

Fluorescent Nanodiamonds, First Edition. Huan-Cheng Chang, Wesley Wei-Wen Hsiao and Meng-Chih Su.
© 2019 John Wiley & Sons Ltd. Published 2019 by John Wiley & Sons Ltd.

A quantitative analysis of this kind will provide a fair comparison for the biocompatibility of different materials in a study. Several well-established assays are discussed here, some being more specific than the others, and all together serve as a general tool to evaluate how safe a material is to use in a biological system.

5.1.1 Cytotoxicity

Cultured cells grown under controlled conditions are simple living systems routinely used for assessing the biocompatibility of a material under study *in vitro* [4]. The cells can be put through various testing procedures with the chemicals of interest in laboratories, allowing for detailed evaluation of the cytotoxicity or irritancy potential of the material being tested. They also provide an excellent platform for any chemicals and materials prior to *in vivo* studies. Immortal cell lines are the most commonly used cells for testing. In contrast to normal cells, which have a limited lifespan in culture, these cell lines are derived from multicellular organisms that have undergone mutations to become "immortal" [5]. They can continue to divide and grow for a prolonged period of time in culture. The constant supply of almost identical cells makes immortal cell lines an important tool for research in biochemistry and cell biology.

A major pathway by which a cell dies is *necrosis* [6], which is a form of cell damage so severe that it eventually leads to a loss of plasma membrane integrity. Some organic dyes such as trypan blue, propidium iodide, and ethidium homodimer are impermeant to healthy cells but can enter membrane-compromised cells [7]. They are capable of distinguishing live cells from dead cells. Trypan blue (Figure 5.1a), for example, is negatively charged at neutral pH and thus does not interact with the membrane of live cells. When stained by this molecule, dead cells show a distinctive blue color and can be quantified colorimetrically at 607 nm. Propidium iodide and ethidium homodimer (Figure 5.1b and c), on the other hand, are positively charged fluorescent probes at neutral pH. Both dye molecules can bind to nucleic acids by intercalating between base pairs, yielding nuclear-localized red fluorescence (emission maximum at ~620 nm) with an enhanced intensity by more than one order of magnitude. The technique is suitable for the identification and quantification of dead or damaged cells by fluorescence microscopy, flow cytometry, and fluorometry.

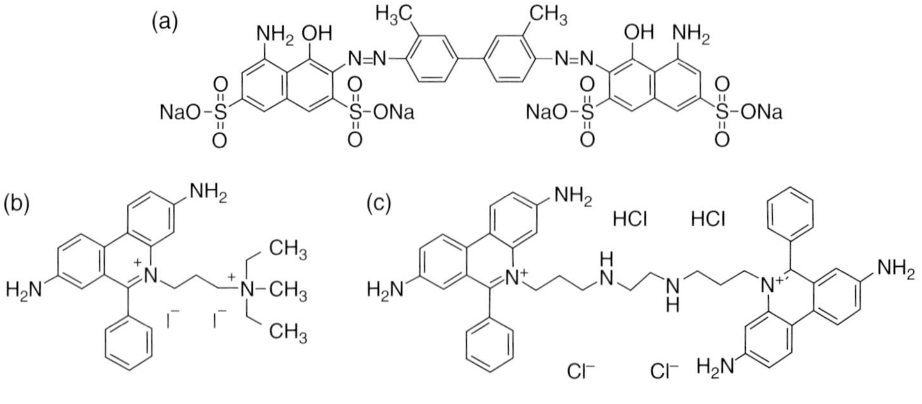

Figure 5.1 Molecular structures of (a) trypan blue, (b) propidium iodide, and (c) ethidium homodimer.

Aside from using nucleic acid stains, the cytotoxicity testing of a nanomaterial can also be conducted by measuring *reactive oxygen species* (ROS) [6]. The oxygen-containing species, such as peroxide, superoxide, hydroxyl radical, or singlet oxygen, are routine byproducts of metabolism in biological systems. The ROS level can significantly increase in the presence of environmental stresses (e.g. ultraviolet or heat exposure) that may cause cellular damage to lipids, proteins, and nucleic acids, leading to a myriad of pathological disorders and diseases [8]. A commonly used probe to detect intracellular ROS production is 2′,7′-dichlorodihydrofluorescein diacetate, a membrane-permeant molecule. Upon cleavage of the acetate groups by esterases and subsequent oxidation in cells, the nonfluorescent probe is converted to the highly fluorescent 2′,7′-dichlorofluorescein that can be detected at 514 nm [9].

Apoptosis is another pathway of cell death. It is genetically regulated and occurs in multicellular organisms [6]. Cells undergoing the final stages of apoptosis often display death-signaling molecules such as phosphatidylserine on their cell surface for phagocytic recognition [10]. The apoptotic cells can be recognized with a Ca^{2+}-dependent phospholipid-binding protein called annexin V, which has a high affinity for phosphatidylserine [11]. The assay using fluorescence-tagged annexin V as the probe can be combined with propidium iodide to distinguish viable cells from apoptotic cells and necrotic cells in a population.

The second method useful for the identification and quantification of apoptotic/necrotic cells is the *terminal deoxynucleotidyl transferase mediated dUTP nick end labeling (TUNEL) assay* [12]. The method detects DNA fragmentation by labeling the terminal end of nucleic acids with modified dUTPs using terminal deoxynucleotidyl transferase. The modified dUTPs can be later fluorescence-labeled and probed, allowing for ensuing analysis by both fluorescence microscopy and flow cytometry.

A good indicator of cell health is the cell viability or the cell proliferation rate. The cell viability, defined as the number of healthy cells in a sample, can be measured by several different ways. A commonly used method is the *MTT assay*, which measures the activity of mitochondrial reductase that reduces the tetrazolium salt (e.g., 3-[4,5-dimethylthiazol-2-yl]-2,5-diphenyltetrazolium bromide, MTT) to formazan (Figure 5.2). After dissolving the reduction end product in acidified isopropyl alcohol [13], the resulting purple solution is measured spectrophotomerically for the absorbance at 570 nm. A more recent advance in the field to measure the cell proliferation is to monitor the impedance of cells cultured on growth-compatible microplates with prepatterned gold electrodes at the bottom surface [14]. Measurement of the impedance changes over time provides high-temporal resolution readouts of the cell growth as well as the attachment characteristics.

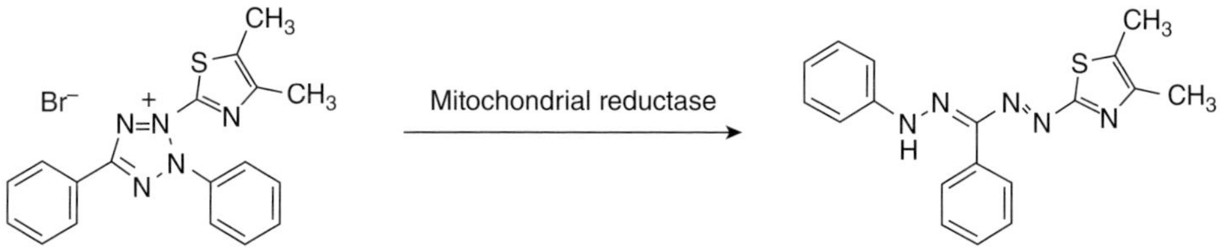

Figure 5.2 The MTT assay involving the reduction of 3-[4,5-dimethylthiazol-2-yl]-2,5-diphenyltetrazolium bromide (left) to 1-(4,5-dimethylthiazol-2-yl)-3,5-diphenylformazan (right) by mitochondrial reductase.

5.1.2 Genotoxicity

Genotoxicity is a destructive effect caused by chemical agents that damage the genetic makeup within a cell. The damage can occur in either somatic or germline cells, leading to mutations and possibly cancers. One of the most common tests for genotoxicity is single-cell gel electrophoresis, known as the *comet assay* [15]. The technique involves lysing cells using detergent and high-concentration salt to form nucleoids containing supercoiled loops of DNA linked to the nuclear matrix. The lysed cells are then analyzed by gel electrophoresis, yielding comet-like structures for the individual cells. The comet heads are contributed mainly by undamaged DNA strands, whereas the tails represent DNA fragments that are drawn faster than the intact DNA towards the positively charged electrode. The number of DNA breaks is finally estimated by measuring the fluorescence intensity of the comet tail relative to that of the head after staining and visualization under a microscope.

The *micronucleus assay* is another useful tool for screening potentially genotoxic compounds [16]. Micronuclei are small membrane-bound DNA fragments formed during the metaphase-to-anaphase transition of cell division. They could originate from acentric chromosome fragments or whole chromosomes that are unable to migrate with the rest of the chromosomes during the anaphase. A positive result from the assay, i.e. more micronuclei found in the treatment groups than in the control groups, is an indication that the tested substance induces chromosomal damage.

5.1.3 Hemocompatibility

Materials used in medical therapy are often in direct contact with blood (*hemo*) and, therefore, they must be assessed for hemocompatibility to establish their safety. A complete list of the assays to evaluate the blood compatibility of nanoparticles can be found in the literature [17]. Here, we focus only on three major issues pertaining to the hemocompatibility, including (i) disruption of blood cells (*hemolysis*), (ii) activation of the coagulation pathways (*thrombogenicity*), and (iii) increase of the cytokine levels in blood (*inflammation*).

Human erythrocytes, or *red blood cells* (RBCs), are typical samples used to assess the hemolytic activity of a material [18]. When RBCs are in direct or indirect contact with the materials being tested, hemolysis may be caused by either physical interactions or chemical reactions of the cells with toxins, metal ions, or other compounds. As a result, the erythrocytes are destroyed and the hemoglobin contained within now released. After centrifuging the blood sample, one can easily separate the freed hemoglobin (in supernatant) from the lysed erythrocytes (in precipitate). The heme group in hemoglobin is known to have a strong absorption band, peaking at 410 nm, which is responsible for the deep red-color in blood. Spectrophotometric measurement of the heme absorption allows quantification for the concentrations of hemoglobin in the supernatant. With water-treated erythrocytes as the positive control and saline-treated RBCs as the negative control, one can calculate the hemolysis percentage from the absorbance (A) of hemoglobin at the wavelength (λ) of the measurement as

$$\% \text{ hemolysis} = \frac{\text{Sample } A(\lambda) - \text{Negative control } A(\lambda)}{\text{Positive control } A(\lambda) - \text{Negative control } A(\lambda)} \times 100. \qquad (5.1)$$

A high percentage of hemolysis signals a high-risk factor of using the nanomaterial under testing.

Activated partial thromboplastin time (APTT) is a medical test commonly used to evaluate the function of a blood clotting system [19]. The test measures the overall speed at which human blood clots by means of the *intrinsic* coagulation pathway. It is typically conducted with an APTT kit consisting of phospholipid, a surface activator (e.g. kaolin), and $CaCl_2$. The standard protocol calls for an extraction of plasma from a blood sample. An excess of calcium chloride (in a phospholipid suspension) is then mixed with the plasma and the surface activator is added to activate the intrinsic pathway of coagulation. Finally, the time that the sample takes to clot is optically measured. The APPT test along with the *prothrombin time* (PT) measurement [19], which evaluates the *extrinsic* pathway of blood coagulation, allows for detailed characterization of the thrombogenic activity of a nanomaterial.

Cytokines are small glycoproteins (~20 kDa) responsible for cell signaling and cell-to-cell communications [6]. They are produced by a broad range of cells including immune cells, endothelial cells, fibroblasts, and various stromal cells. Cytokines are primarily involved in host responses to disease such as infection and inflammation. *Interleukins* are a class of cytokines acting between leukocytes and other cell types [20]. There are 17 common families of interleukins and some of them are inflammatory mediators including IL-1, IL-4, and IL-6. Evaluation of the cytokine levels in the blood of an animal model (such as a mouse) provides important information for its immune response to the blood-contacting material. To measure the overall cytokine response elicited, an *enzyme-linked immunosorbent assay* (ELISA) [21] is typically used to quantify the number of signal protein molecules in serum with antibodies for colorimetric detection. The antibodies can be either covalently linked to an enzyme (such as horseradish peroxidase) or detected by a secondary antibody linked to the same enzyme through covalent conjugation. An elevating amount of cytokine in the blood sample measured by ELISA is an indication for a higher level of immune response to the material under testing, suggesting a high risk of using it in humans.

5.2 *In Vitro* Studies

High-pressure high-temperature nanodiamonds (HPHT-NDs) and *detonation nanodiamonds* (DNDs) are two major types of diamond nanoparticles applied to biological research. DNDs hold great potentials for use as drug delivery devices because of their small size and, therefore, large specific surface area (Chapter 4). HPHT-NDs, on the other hand, may serve as excellent medical contrast agents due to their imaging capability with built-in color centers (Chapter 3). We start our discussion with HPHT-NDs as a step stone toward *fluorescent nanodiamonds* (FNDs).

5.2.1 HPHT-ND

Yu et al. [22] was the first group to examine the cytotoxicity of surface-oxidized HPHT-NDs in immortal cell lines. The HPHT-NDs used in the study contained high-density ensembles of NV centers as fluorophores and thus were called FNDs. Upon feeding human embryonic kidney cells with FNDs (~100 nm in diameter), the

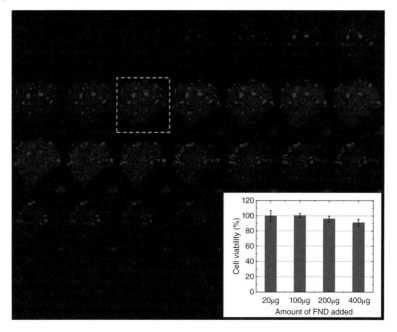

Figure 5.3 Confocal fluorescence images of a single 293T human kidney cell after FND uptake. The cross-sectional image in each three-dimensional scan (as indicated by the yellow dashed square) has a vertical thickness of 0.25 μm and an area of 42×42 μm^2. The bright red spots correspond to FNDs. Inset: Cytotoxicity tests with the 293T cells and the MTT reduction assay. *Source:* Reprinted with permission from Ref. [22]. Reproduced with permission of American Chemical Society.

researchers found that the particles were avidly taken up by the cells under serum-free conditions (Figure 5.3). The presence of FNDs in cells was confirmed by detecting the red fluorescence emission from the NV centers. MTT assays showed no noticeable toxicity of the particles at the concentration as high as $100\,\mu g\,ml^{-1}$ (inset in Figure 5.3). Subsequent studies of the cell viability for HeLa cells (a cervical cancer cell line) with MTT assays [23] as well as the cell proliferation (Figure 5.4a) and genotoxicity (Figure 5.4b) of human fibroblasts with impedance sensing and single-cell gel electrophoresis assays all indicated that the 100-nm FNDs neither impaired cell growth nor caused DNA damage [24].

Paget et al. [25] have recently made a thorough and systematic investigation for the *in vitro* biocompatibility of HPHT-NDs. Their work was aimed at a rigorous risk assessment for NDs in human health. The research team examined the cytotoxicity and genotoxicity of two sets of carboxylated HPHT-NDs with nominal diameters of 20 and 100 nm. Six human cell lines were chosen as representatives of potential target organs: HepG2 and Hep3B (liver), Caki-1 and HEK293 (kidney), HT29 (intestine), and A549 (lung). The cytotoxicity was assessed by impedance sensing for cell proliferation and flow cytometric analysis for dead cells. The genotoxicity was measured according to the distribution of the number of γ-H2Ax foci per nucleus, another highly sensitive technique for the study of DNA double-strand breaks [26]. Their results indicated that the HPHT-NDs could effectively enter the cells but did not cause any significant cytotoxic or genotoxic effects on the six cell lines even when the dosage went up as high as

(a)

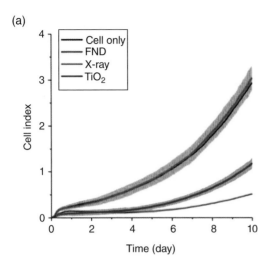

(b)

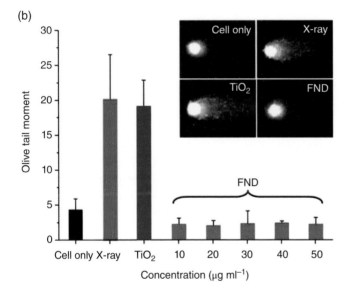

Figure 5.4 (a) Cell proliferation and (b) comet assays of human fibroblasts after treatments with X-ray, TiO$_2$, and FNDs. The X-ray and TiO$_2$ treatments served as positive controls. *Source:* Adapted with permission from Ref. [24]. Reproduced with permission of John Wiley & Sons.

250 µg ml^{-1}. Further studies using Chinese hamster ovary cells confirmed that the surface modification of these particles with biomolecules such as peptides [27] did not alter their biocompatibility either.

While HPHT-NDs have been found to be nontoxic to a wide range of cell lines grown in culture, how they may influence the functions of primary cells remains a concern. Compared to immortal cell lines, *primary cells* are the more biologically relevant *in vitro* model. They are isolated directly from tissues and have a finite lifespan and limited expansion capacity. Additionally, they have normal cell morphology and possess many important characteristics originally present *in vivo*. Two types of primary cells have been employed to address this issue: Mouse lung stem/progenitor cells (LSCs) and mouse embryonal primary neurons. Wu et al. [28] reported that the labeling of LSCs with 100-nm FNDs did not eliminate the cells' abilities of self-renewal and differentiation into type I and type II pneumocytes. Huang et al. [29] reported that the FND labeling did not cause any noticeable toxicity in primary neurons derived from either central or peripheral nervous systems. However, a decrease of the neurite length in both types of the cells was found, which was attributed to the spatial hindrance by the FND particles in advancing axonal growth cones.

In a separate study, FNDs have also been tested for their effects on the differentiation of embryonal carcinoma stem (ECS) cells, e.g. mouse P19 and human NT2/D1 ECS cells, into neuronal cells [30]. It was found that the 100-nm FNDs could be effectively internalized by the ECS cells, but their physical presence in the cytoplasm did not significantly alter the cells' morphology and growth ability. Moreover, the FNDs caused no noticeable changes in the protein expression of the stem cell marker, stage-specific embryonic antigen-1 [31], and induced no cytotoxicity (including apoptosis) during the neuronal differentiation. In the differentiated neuronal cells, the FNDs did not reduce the cell viability or affect the expression of the neuron-specific marker, β-III-tubulin [32], presumably due to the exceptional chemical inertness of the nanoparticles. Altogether, the results have highlighted the great potential of HPHT-NDs for nanomedicine and their practical use as negative controls in nanotoxicology studies for both immortal and primary cell lines.

5.2.2 DND

DNDs are synthesized by detonation of an oxygen-deficient explosive mixture in a closed chamber under extremely high temperature and pressure, achieved at the front of the detonation wave in several microseconds (Section 2.3.3). The primary particles of DNDs are small, approximately 5 nm in diameter, but contain a significant amount of impurities including N, O, and H from the reactants. The typical content of C atoms in the particles is in the range of 90–99% by weight, depending on the manufacturing processes [33]. Unlike HPHT-NDs, DNDs are predominantly polycrystalline in structure and their surface is always covered with a shell of graphitic carbon atoms. Elemental analysis showed that more than 1% metal impurities (including Fe, Cu, Cr, Ti, and several others) are incorporated into the samples [34]. These impurities appear to come from the interactions of the explosion wave with the reaction chamber walls and the instrumentation corrosion during purification. The high impurity content raises considerable concerns about the biocompatibility of this nanocarbon material.

A large number of experiments have been carried out to study the *in vitro* toxicity of DNDs [35–44]. Schrand et al. [36, 37] conducted the first experiments to test the differential biocompatibility of different carbon nanoparticles. Using two different cell lines (neuroblastoma cells and rat alveolar macrophage) and the MTT assays, they found that DND had a greater biocompatibility than carbon black (CB), multi-walled carbon nanotubes (MWCNTs), and single-walled carbon nanotubes (SWCNTs) (Figure 5.5a and b). The biocompatibility trend followed the order of DND > CB > MWCNT > SWCNT for both cell lines. A later study using HeLa cells treated with MWCNTs, graphene oxides (GOs), and NDs reached a similar

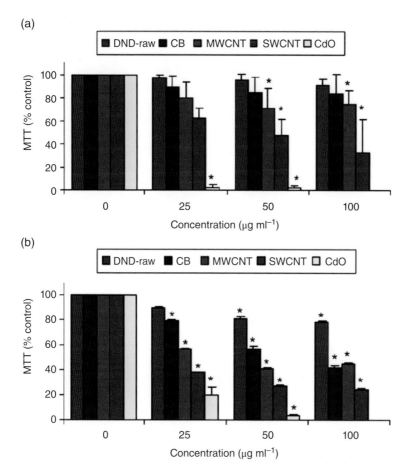

Figure 5.5 Cytotoxicity measurements after 24-h incubation of various nanocarbons in (a) neuroblastoma cells and (b) macrophages. GdO nanoparticles served as the positive control. *Source:* Adapted with permission from Ref. [36]. Reproduced with permission of Elsevier.

conclusion, with a decreasing biocompatibility of DND > MWCNT ≈ GO [38]. The low toxicity of DND makes it a more desirable candidate than other carbon nanoparticles as a drug delivery vehicle for biomedical applications, including personalized medicine [45] (cf., Chapter 13 for details).

The high biocompatibility and low cytotoxicity of DNDs were well founded as per the studies of Schrand et al. [35–37] on the subject. The research team did not observe any disruption of mitochondrial membrane permeability, morphological alterations, or viability changes when exposing the cells to 5–100 μg ml⁻¹ DNDs. Moreover, the DNDs were discovered to neither induce ROS generation nor cause oxidative stress, which could have led to membrane dysfunction, protein degradation, or DNA damage. They also confirmed the lack of change in the expression level of genes that served as the indicators of inflammation and protection against apoptosis in macrophages and neuroblastoma cells when incubated with DNDs. Their results are in accord with other cytotoxicity and genotoxicity tests, showing that the carbon-based nanoparticles are well tolerated by multiple cell types at both functional and gene expression levels [38–41].

While several experiments have provided evidences for the innate biocompatibility of DNDs, some studies refute this finding, arguing that DNDs can induce both geno-toxicity and cytotoxic responses under certain conditions [42–44]. Researchers making such an argument found that the toxicity of DNDs varied, depending on the dosage and surface chemistry of the particles, the type of cell lines used for the assessments, as well as the composition of the treatment medium. Concerns about the toxicity typically arose from the small size of DNDs and their ability to enter cells and localize in critical organelles [42]. Also, the oxidative stress induced by elevation of ROS after the DND treatment could result in DNA damage [43]. Furthermore, the graphitic surface content is a possible determinant of the bioactivity of these carbonaceous nanoparticles [44]. A comprehensive account of the cellular response to FNDs, DNDs, and other functionalized NDs can be found in the work by Moore et al. [46].

5.3 *Ex Vivo* Studies

Different from *in vitro* studies with cultured cells under controlled environment, *ex vivo* biocompatibility evaluations employ tissues isolated from organisms to explore the potential impacts of the tested materials on human health. When tested in animal models such as mice and rats, the majority of nanoparticles are administered through the bloodstream. Therefore, understanding the particles' blood compatibility is vitally

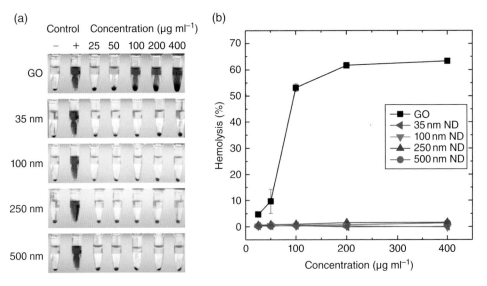

Figure 5.6 Hemolysis studies of GOs and oxidized NDs of different size with human RBCs. (a) Photographs of human RBCs treated with GOs and NDs of four different sizes at the concentration range of 25–400 µg ml^{-1}. (b) Hemolysis percentages measured at the concentration range of 25–400 µg ml^{-1} for GOs and four different ND samples (35, 100, 250 and 500 nm in diameter) incubated with RBCs at 25 °C for two hours. GOs served as the positive control. *Source:* Adapted with permission from Ref. [47]. Reproduced with permission of Nature Publishing Group.

important [16]. Figure 5.6a displays a typical result of the testing for the hemolytic activity of surface-oxidized HPHT-NDs of various sizes (35–500 nm) using human RBCs as the samples [47]. The red color of the solution in panel (a) is due to the release of hemoglobin from damaged RBCs and the red pellets at the bottom of the tubes are intact RBCs precipitated by centrifugation. In this particular experiment, phosphate-buffered saline and distilled deionized water served as the negative (−) and positive (+) controls, respectively. The results showed that irrespective of the particle size, all HPHT-NDs caused no significant RBC destruction at the concentration as high as 400 mg ml^{-1}. In contrast, the membrane of the RBCs incubated with GOs was substantially damaged even at a dosage lower than 25 mg ml^{-1} (Figure 5.6b). Therefore, the influence of HPHT-NDs on the oxygenation states and microrheological properties of RBCs was negligible [48].

Following the hemolysis study, Li et al. [47] examined the thrombogenicity of surface-oxidized HPHT-NDs with human RBCs. In contrast to GOs which exhibited a significant anticoagulant activity, the oxidized NDs (35–500 nm) were completely inert, showing neither thrombogenic potential nor anticoagulant activity at the concentration of up to 400 mg ml^{-1}. Such a remarkable hemocompatibility was attributed to the exceptionally high affinity of the nanoparticles for proteins (Section 4.2.1). A protein corona immediately formed on the surface when the HPHT-NDs were in contact with serum. It was the so-called *protein corona effect* [49] that prevented the RBC membrane from being damaged and the blood clotting from occurring. As to DNDs, controversies exist. Mona et al. [50] reported that there was no delay in time when the coagulation was initiated through the intrinsic pathway in the APPT tests, whereas Kumari et al. [51] claimed that these particles without careful purification could activate blood platelets and induce thromboembolism.

5.4   *In Vivo* Studies

A number of model organisms have been employed to assess the *in vivo* biocompatibility of NDs, including *Caenorhabditis elegans* (*C. elegans*) [52], zebrafish embryos [53, 54], *Xenopus* embryos [55], mice [56–59], rats [59, 60], rabbits [61], and monkeys [60]. Covering a wide range of species from the nematode to the primate, these studies used both feeding and microinjection methods to introduce NDs into the living organisms.

Caenorhabditis elegans is a free living soil nematode with simple and well-defined anatomy. The 1-mm long adult hermaphrodite consists of an invariable number of 959 cells, which are organized to form complex tissues including intestine, muscle, hypodermis, gonad, and nerve systems [62]. The genome of *C. elegans* has been completely sequenced [63], making it feasible to study biological processes at the molecular level. It is an ideal model organism to assess the biocompatibility of nanoparticles owing to its short life cycles (about three days), easy handling, and high sensitivity to various types of stresses. Mohan et al. [52] was the first group to introduce surface-oxidized and protein-conjugated FNDs into the nematodes. Through feeding and optical imaging, they found that the particles could be readily taken up by the worms and accumulated in lumen. Toxicity assessments, performed by using longevity, reproductive potential, and the ROS level as physiological indicators, showed that the particles were nontoxic and did not cause any detectable stress to the worms.

Extended from their *C. elegans* work, Chang et al. [53] microinjected FNDs after coating with bovine serum albumin (BSA) into the yolk cells of zebrafish embryos at the one-cell stage. By fluorescence imaging, they found that the FND particles could be incorporated into the dividing cells in the blastomeres through cytoplasmic streaming [54]. The FND-labeled larvae were able to develop into whole fishes without any apparent morphological anomalies during their embryogenesis, indicating that the HPHT-ND particles did not cause any deleterious effects on the development of the vertebrate organism. Parallel to the FND studies above, Marcon et al. [55] used DND to assess the *in vivo* toxicity of the nanomaterials with different surface modifications (including –OH, –NH$_2$, or –COOH) in *Xenopus* embryos. They reported that microinjection of DND-COOH into early-stage embryos could cause significant embryotoxicity and teratogenicity, despite the fact that DND-NH$_2$ and DND-OH were only slightly toxic.

In the studies using murine models to test the biocompatibility of HPHT-NDs, Li et al. [47] detected no significant elevation of the inflammatory cytokine levels of interleukin-1β (IL-1β) and interleukin-6 (IL-6) after intravenous injection of the particles into mice through tail veins [47]. Similarly, no visible toxicity and side effects (e.g. stress response) were reported by Vaijayanthimala, et al. [59] for rats subjected to intraperitoneal injection of FNDs over a three-month period with a total quantity of up to 75 mg kg^{-1} body weight. The differences in fodder consumption, body weight, and organ index between the control and FND-treated animals were insignificant. Histopathological examination of tissues revealed that the injected FND particles were engulfed by macrophages, but there was no observable inflammation, necrosis, or tissue reaction surrounding these carbon-laden macrophages (Figure 5.7).

A question often asked is: Where would these nanoparticles end up? Information concerning the biodistribution and fate of the injected nanoparticles is exceedingly important in assessing the *in vivo* biocompatibility. Yuan et al. [56] carried out experiments pertaining to this study by using HPHT-NDs labeled with ^{125}I radioisotopes. They found that the particles with a size of 50 nm predominantly accumulated in livers of the mice after intravenous injection, followed by lungs as the target organs. About 37% of the initially injected particles were entrapped in livers and 6% in lungs after 0.5 hour post-dose (Figure 5.8). High-resolution transmission electron microscopy and Raman spectroscopy of digested organ solutions confirmed the long-term entrapment of the BSA-coated HPHT-NDs in livers and lungs. However, no mice showed any symptoms of abnormality, such as weight loss, lethargy, anorexia, vomiting, and diarrhea during the course of the treatments.

Studies have also been made for the *in vivo* biocompatibility of DNDs. Zhang et al. [57] found that intratracheal instillation of DNDs did not lead to any differences in body weight or abnormal pathologies of treated mice. However, it did have a negative effect on the lungs, liver, kidneys, and hematological systems. Of these organs, the lungs suffered the worst toxicological effects, becoming inflamed and tissue damaged, as a result of the high uptake and long retention time of the nanoparticles in the lung tissue. In contrast to their findings, the results of Yuan et al. [58] showed no pulmonary toxicity in mice after intratracheal instillation of DNDs. The discrepancies between these two results could be associated with the dosage and sources of the DNDs as well as how their surface might have been modified, as discussed in Section 5.2.2.

Rabbits are one of the earliest animal models used for *in vivo* nanotoxicity studies. Puzyr et al. [61] gave a high dose (125 mg) of DNDs to the rabbits through intravenous

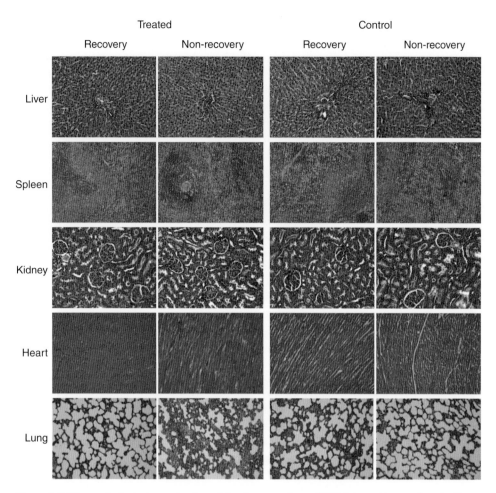

Figure 5.7 Histopathological examination of the tissue sections of FND-injected (treated) and saline-injected (control) rats with and without recovery. The results show no specific pathological changes in both FND-treated and control groups (magnification 200×). *Source:* Reprinted with permission from Ref. [59]. Reproduced with permission of Elsevier.

Figure 5.8 Biodistribution of ^{125}I-labeled HPHT-NDs in mice after intravenous injection with a dose of 20 mg kg^{-1} body weight for 30 minutes. *Source:* Adapted with permission from Ref. [56]. Reproduced with permission of Elsevier.

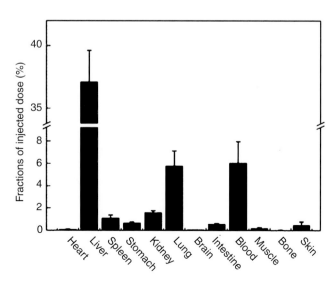

administration and yet did not cause any deaths. Both the RBC count and the hemoglobin level of the rabbits remained stable for 15 min after administration. Later, in a longer time period (48 hours), the levels of biochemical molecules such as total bilirubin, triglyceride, low-density lipoprotein, etc. changed to an extent that was statistically significant. Three months later, the rabbits showed no signs of inflammation, suggesting a long-term biocompatibility of the nanoparticles.

Despite the discrepancies mentioned above, a broad spectrum of biocompatibility studies has indicated that DNDs cause no adverse effects on cells and organisms. Aiming for clinical translation of DNDs, Ho and coworkers [60] have recently conducted a comprehensive assessment for the safety of the nanomaterial with both small and large animal preclinical models: rats and monkeys. They performed the studies in two cohorts that lasted for two weeks (rats) or six months (monkeys) based on histological, serum, urine, and body weight analysis. Their results showed that DNDs were well tolerated at the clinically relevant doses of $6.75–13.5 \, \mathrm{mg \, kg^{-1}}$. However, in order for in-human validation, more detailed biodistribution and pharmacokinetics analysis of the injected nanomaterial is needed. FNDs are expected to make significant contributions to such research studies and developments as discussed in Chapter 9.

References

1 Martin, G. (1998). The meaning and origin of the expression: a diamond is forever. http://www.phrases.org.uk/meanings/a-diamond-is-forever.html (accessed 16 April 2018).

2 Vert, M., Doi, Y., Hellwich, K.H. et al. (2012). Terminology for biorelated polymers and applications (IUPAC recommendations 2012). *Pure Appl Chem* 84: 377–410.

3 Dion, I., Baquey, C., and Monties, J.R. (1993). Diamond – the biomaterial of the 21st century. *Int J Artif Organs* 16: 623–627.

4 Hodgson, E., Leblanc, G.A., Meyer, S.A., and Smart, R.C. (2010). Introduction to biochemical and molecular methods in toxicology. In: *A Textbook of Modern Toxicology*, 4e (ed. E. Hodgson), 15–28. New York: Wiley.

5 Masters, J.R.W. (2000). Human cancer cell lines: fact and fantasy. *Nat Rev Mol Cell Biol* 1: 233–236.

6 Alberts, B., Johnson, A., Lewis, J. et al. (2014). *Molecular Biology of the Cell*, 6e. Garland Science.

7 Johnson, I. and Spence, M. (2010). *Molecular Probes Handbook, a Guide to Fluorescent Probes and Labeling Technologies*, 11e. Springer, Sec. 8.1.

8 Schieber, M. and Chandel, N.S. (2014). ROS function in redox signaling and oxidative stress. *Curr Biol* 24: R453–R462.

9 Johnson, I. and Spence, M. (2010). *Molecular Probes Handbook, a Guide to Fluorescent Probes and Labeling Technologies*, 11e. Springer, Sec. 18.2.

10 Li, M.O., Sarkisian, M.R., Mehal, W.Z. et al. (2003). Phosphatidylserine receptor is required for clearance of apoptotic cells. *Science* 302: 1560–1563.

11 Johnson, I. and Spence, M. (2010). *Molecular Probes Handbook, a Guide to Fluorescent Probes and Labeling Technologies*, 11e. Springer, Sec. 15.5.

12 Loo, D.T. (2002). TUNEL assay. An overview of techniques. *Methods Mol Biol* 203: 21–30.

13 van Meerloo, J., Kaspers, G.J., and Cloos, J. (2011). Cell sensitivity assays: the MTT assay. *Methods Mol Biol* 731: 237–245.

14 Ke, N., Wang, X., Xu, X., and Abassi, Y.A. (2011). The xCELLigence system for real-time and label-free monitoring of cell viability. *Methods Mol Biol* 740: 33–43.

15 Speit, G. and Hartmann, A. (2006). The comet assay: a sensitive genotoxicity test for the detection of DNA damage and repair. *Methods Mol Biol* 314: 275–286.

16 Fenech, M. (2008). The micronucleus assay determination of chromosomal level DNA damage. *Methods Mol Biol* 410: 185–216.

17 Evani, S.J. and Ramasubramanian, A.K. (2011). Hemocompatibility of nanoparticles. In: *Nanobiomaterials Handbook* (ed. B. Sitharaman). CRC Press Chapter 31.

18 Neun, B.W. and Dobrovolskaia, M.A. (2011). Method for analysis of nanoparticle hemolytic properties *in vitro*. *Methods Mol Biol* 697: 215–224.

19 Neun, B.W. and Dobrovolskaia, M.A. (2011). Method for *in vitro* analysis of nanoparticle thrombogenic properties. *Methods Mol Biol* 697: 225–235.

20 Meager, A. (2004). Cytokines: interleukins. In: *Encyclopedia of Molecular Cell Biology and Molecular Medicine*, 2e (ed. R.A. Meyers), 115–151. New York: Wiley.

21 Crowther, J.R. (1995). ELISA. Theory and practice. *Methods Mol Biol* 42: 1–218.

22 Yu, S.J., Kang, M.W., Chang, H.C. et al. (2005). Bright fluorescent nanodiamonds: no photobleaching and low cytotoxicity. *J Am Chem Soc* 127: 17604–17605.

23 Vaijayanthimala, V., Tzeng, Y.K., Chang, H.C., and Li, C.L. (2009). The biocompatibility of fluorescent nanodiamonds and their mechanism of cellular uptake. *Nanotechnology* 20: 425103.

24 Lin, H.H., Lee, H.W., Lin, R.J. et al. (2015). Tracking and finding slow-proliferating/quiescent cancer stem cells with fluorescent nanodiamonds. *Small* 11: 4394–4402.

25 Paget, V., Sergent, J.A., Grall, R. et al. (2014). Carboxylated nanodiamonds are neither cytotoxic nor genotoxic on liver, kidney, intestine and lung human cell lines. *Nanotoxicology* 8: 46–56.

26 Smart, D.J., Ahmedi, K.P., Harvey, J.S., and Lynch, A.M. (2011). Genotoxicity screening via the gammaH2AX by flow assay. *Mutat Res* 715: 25–31.

27 Vial, S., Mansuy, C., Sagan, S. et al. (2008). Peptide-grafted nanodiamonds: preparation, cytotoxicity and uptake in cells. *ChemBioChem* 9: 2113–2119.

28 Wu, T.J., Tzeng, Y.K., Chang, W.W. et al. (2013). Tracking the engraftment and regenerative capabilities of transplanted lung stem cells using fluorescent nanodiamonds. *Nat Nanotechnol* 8: 682–689.

29 Huang, Y.A., Kao, C.W., Liu, K.K. et al. (2014). The effect of fluorescent nanodiamonds on neuronal survival and morphogenesis. *Sci Rep* 4: 6919.

30 Hsu, T.C., Liu, K.K., Chang, H.C. et al. (2014). Labeling of neuronal differentiation and neuron cells with biocompatible fluorescent nanodiamonds. *Sci Rep* 4: 5004.

31 Solter, D. and Knowles, B.B. (1978). Monoclonal antibody defining a stage-specific mouse embryonic antigen (SSEA-1). *Proc Natl Acad Sci USA* 75: 5565–5569.

32 Memberg, S.P. and Hall, A.K. (1995). Dividing neuron precursors express neuron-specific tubulin. *J Neurobiol* 27: 26–43.

33 Schrand, A.M., Hens, S.A.C., and Shenderova, O.A. (2009). Nanodiamond particles: properties and perspectives for bioapplications. *Crit Rev Solid State Mater Sci* 34: 18–74.

34 Volkov, D., Proskurnin, M., and Korobov, M. (2014). Elemental analysis of nanodiamonds by inductively-coupled plasma atomic emission spectroscopy. *Carbon* 74: 1–13.

35 Schrand, A.M., Huang, H., Carlson, C. et al. (2006). Are diamond nanoparticles cytotoxic? *J Phys Chem B* 111: 2–7.

36 Schrand, A.M., Dai, L., Schlager, J.J. et al. (2007). Differential biocompatibility of carbon nanotubes and nanodiamonds. *Diam Relat Mater* 16: 2118–2123.

37 Schrand, A.M., Johnson, J., Dai, L. et al. (2009). Cytotoxicity and genotoxicity of carbon nanomaterials. In: *Safety of Nanoparticles* (ed. T.J. Webster), 159–187. Springer.

38 Liu, K.K., Cheng, C.L., Chang, C.C., and Chao, J.I. (2007). Biocompatible and detectable carboxylated nanodiamond on human cell. *Nanotechnology* 18: 325102.

39 Huang, H., Pierstorff, E., Osawa, E., and Ho, D. (2007). Active nanodiamond hydrogels for chemotherapeutic delivery. *Nano Lett* 7: 3305–3314.

40 Xing, Y., Xiong, W., Zhu, L. et al. (2011). DNA damage in embryonic stem cells caused by nanodiamonds. *ACS Nano* 5: 2376–2384.

41 Zhang, X.Y., Hu, W.B., Li, J. et al. (2012). A comparative study of cellular uptake and cytotoxicity of multi-walled carbon nanotubes, graphene oxide, and nanodiamond. *Toxicol Res* 1: 62–68.

42 Solarska, K., Gajewska, A., Bartosz, G., and Mitura, K. (2012). Induction of apoptosis in human endothelial cells by nanodiamond particles. *J Nanosci Nanotechnol* 12: 5117–5121.

43 Dworaka, N., Wnuk, M., Zebrowski, J. et al. (2013). Genotoxic and mutagenic activity of diamond nanoparticles in human peripheral lymphocytes *in vitro*. *Carbon* 68: 763–776.

44 Silbajoris, R., Linak, W., Shenderova, O. et al. (2015). Detonation nanodiamond toxicity in human airway epithelial cells is modulated by air oxidation. *Diam Relat Mater* 58: 16–23.

45 Ho, D., Wang, C.H., and Chow, E.K. (2015). Nanodiamonds: the intersection of nanotechnology, drug development, and personalized medicine. *Sci Adv* 1: e1500439.

46 Moore, L., Grobarova, V., Shen, H. et al. (2014). Comprehensive interrogation of the cellular response to fluorescent, detonation and functionalized nanodiamonds. *Nanoscale* 6: 11712–11721.

47 Li, H.C., Hsieh, F.J., Chen, C.P. et al. (2013). The hemocompatibility of oxidized diamond nanocrystals for biomedical applications. *Sci Rep* 3: 3044.

48 Lin, Y.C., Tsai, L.W., Perevedentseva, E. et al. (2012). The influence of nanodiamond on the oxygenation states and micro rheological properties of human red blood cells *in vitro*. *J Biomed Opt* 17: 101512.

49 Hamad-Schifferli, K. (2015). Exploiting the novel properties of protein coronas: emerging applications in nanomedicine. *Nanomedicine* 10: 1663–1674.

50 Mona, J., Kuo, C.J., Perevedentseva, E. et al. (2013). Adsorption of human blood plasma on nanodiamond and its influence on activated partial thromboplastin time. *Diam Relat Mater* 39: 73–77.

51 Kumari, S., Singh, M.K., Singh, S.K. et al. (2014). Nanodiamonds activate blood platelets and induce thromboembolism. *Nanomedicine (Lond)* 9: 427–440.

52 Mohan, N., Chen, C.S., Hsieh, H.H. et al. (2010). *In vivo* imaging and toxicity assessments of fluorescent nanodiamonds in *Caenorhabditis elegans*. *Nano Lett* 10: 3692–3699.

53 Mohan, N., Zhang, B., Chang, C.C. et al. (2011). Fluorescent nanodiamond – a novel nanomaterial for *in vivo* applications. *MRS Proc* 1362: 25–35.

54 Chang, C.C., Zhang, B., Li, C.Y. et al. (2012). Exploring cytoplasmic dynamics in zebrafish yolk cells by single particle tracking of fluorescent nanodiamonds. *Proc SPIE* 8272: 827205.

55 Marcon, L., Riquet, F., Vicogne, D. et al. (2010). Cellular and *in vivo* toxicity of functionalized nanodiamond in *Xenopus* embryos. *J Mater Chem* 20: 8064–8069.

56 Yuan, Y., Chen, Y., Liu, J.H. et al. (2009). Biodistribution and fate of nanodiamonds *in vivo*. *Diam Relat Mater* 18: 95–100.

57 Zhang, X., Yin, J., Kang, C. et al. (2010). Biodistribution and toxicity of nanodiamonds in mice after intratracheal instillation. *Toxicol Lett* 198: 237–243.

58 Yuan, Y., Wang, X., Jia, G. et al. (2010). Pulmonary toxicity and translocation of nanodiamonds in mice. *Diam Relat Mater* 19: 291–299.

59 Vaijayanthimala, V., Cheng, P.Y., Yeh, S.H. et al. (2012). The long-term stability and biocompatibility of fluorescent nanodiamond as an *in vivo* contrast agent. *Biomaterials* 33: 7794–7802.

60 Moore, L., Yang, J., Lan, T.T. et al. (2016). Biocompatibility assessment of detonation nanodiamond in non-human primates and rats using histological, hematologic, and urine analysis. *ACS Nano* 10: 7385–7400.

61 Puzyr, A.P., Baron, A.V., Purtov, K.V. et al. (2007). Nanodiamonds with novel properties: a biological study. *Diam Relat Mater* 16: 2124–2128.

62 Sulston, J.E., Schierenberg, E., White, J.G., and Thomson, J.N. (1983). The embryonic cell lineage of the nematode *Caenorhabditis elegans*. *Dev Biol* 100: 64–119.

63 *C. elegans* Sequencing Consortium (1998). Genome sequence of the nematode *C. elegans*: a platform for investigating biology. *Science* 282: 2012–2018.

Part II

Specific Topics

6

Producing Fluorescent Nanodiamonds

Natural diamonds in colors are commonly known as *fancies*, or fancy color diamonds, in gemstone industries. They are rare, beautiful, and some even carry impressive price tags in the jewelry market. By comparison, micro- and nanoscale diamond powders are low in price, with or without colors and fluorescent or not. These powders have been used as abrasives for grinding and polishing purposes since ancient time, mainly because of their extraordinary hardness. Little or no attention has been paid over the centuries to other properties of nanodiamonds such as their innate biocompatibility and light-emitting capability. The invention of *fluorescent nanodiamond* (FND) in 2005 has revolutionized the field, opening a new area of research and development with diamonds [1]. Experiments with FNDs in the last decade have demonstrated various promising applications of surface-functionalized FNDs in diversified fields, ranging from physics and chemistry to biology and medicine [2]. It is worthy of noting that as originated from the discovery of Radium by Marie Skłodowska Curie (Section 3.2), FNDs may very well be called *Madame Curie's gemstones*, valued appropriately as a scientist's best friend.

In this chapter, we focus our attention on the principles and practices of producing FNDs in a large-scale quantity, which is the first step towards broad applications of the nanomaterials in biology and nanoscale medicine (cf., Chapters 7–10). Additionally, we will explore the topics such as size reduction and spectroscopic characterization of FNDs at the ensemble level with various optical methods. Single particle detection is the subject of the next chapter.

As a reminder, the reader may wish to review some key concepts before proceeding to the details of the discussion in this chapter. These concepts include: (i) general characteristics of nanodiamonds (Section 2.2) and nitrogen impurities (Section 3.1), (ii) important spectral features of diamond color centers (Table 3.2), (iii) optical properties of NV^0, NV^-, and H3 (Section 3.3), and (iv) optically detected magnetic resonance (Section 3.4), which is one of the major players in the next few chapters.

6.1 Production

6.1.1 Theoretical Simulations

FNDs are diamond nanoparticles containing vacancy-related color centers as fluorophores. Particles containing H3 centers are called *green* FNDs and particles containing NV centers are called *red* FNDs, corresponding to their emission colors. Structural

Fluorescent Nanodiamonds, First Edition. Huan-Cheng Chang, Wesley Wei-Wen Hsiao and Meng-Chih Su.
© 2019 John Wiley & Sons Ltd. Published 2019 by John Wiley & Sons Ltd.

defects like vacancies always exist in the crystal matrices of both natural and synthetic diamonds, although the concentration is low (typically <1 ppm). Studies on thermally annealed HPHT-NDs have shown that the smallest nanodiamond particle able to host a stable NV center is approximately 8 nm in diameter [3]. This number corresponds to a NV concentration of 20 ppm, given that 1 ppm is equal to 1.76×10^{17} atoms cm^{-3} in diamond. In order to achieve such a high concentration, the NDs must undergo high-energy particle bombardment that knocks carbon atoms out of their normal bonding positions to create vacant sites in the crystal lattice. Many techniques serve the purpose well, including electron beam, neutron beam, and ion beam irradiations (Section 3.2). We discuss here only two most commonly used methods, namely, the bombardment with electrons and low-mass ions.

In producing FNDs, an important issue is how to create an appropriate amount of internal defects in the diamond matrix. For the two techniques of our interest, electron and ion irradiations, the optimal dose varies not only with the types of particles used in the bombardment but also with the energies they carry. Overdose leading to the graphitization of the irradiated diamond substrate should be avoided. Theoretical simulations thus play a critical role here in optimizing the procedures of making FNDs. Through the simulations, one can predict the outcomes of the bombardment with different charged particles at various energies before carrying out real experiments. The simulations are particularly valuable when using ions of different sources (Table 6.1) and, among anything else, they provide important post-collision information such as the depth of the ion penetration through the target as well as the vacancy density in the crystal.

To start with, diamond is known to be radiation-tolerated due to its "toughness." For electron irradiation, only particles with sufficient energies are able to create vacancies in the targeted diamond materials. Campbell et al. [4, 5] have studied in detail the mechanism of radiation damage by electrons. Two major processes are involved in the damage: (i) Rutherford scattering and (ii) knock-on atoms. Rutherford scattering is the elastic scattering of charged particles by Coulombic interactions [6]. The carbon atom in diamond can be displaced by the incident electron if, after collision, the atom has received an energy greater than what is known as the *displacement energy*, E_{dis}. The typical value of E_{dis} for diamond is 35–43 eV, which means that electrons with a kinetic energy of lower than 165–197 keV cannot cause radiation damage. A Monte

Table 6.1 Methods of creating vacancies in nanodiamonds by radiation damage.

Damaging agent	Energy range	References
e$^-$	10 MeV	Boudou et al. [19]
e$^-$	13.9 MeV	Dantelle et al. [48]
H$^+$	40 keV	Chang et al. [12]
H$^+$	3 MeV	Yu et al. [1]
H$^+$	15.5 MeV	Havlik et al. [49]
He$^+$	20 keV	Mahfouz et al. [50]
He$^+$	40 keV	Wee et al. [13]
H$^+$, He$^+$, Li$^+$, N$^+$	20–250 keV	Sotoma et al. [51]

Carlo program has been developed by the researchers to simulate the progress of incident electrons through diamond. In cases that the displaced carbon atom has a sufficient energy to displace further atoms (i.e. knock-on atoms), carbon–carbon collisions should be taken into account. A well-tested computer program, known as *Transport of Ions in Matter* (TRIM) [7] is available to accurately predict the distribution of the knock-on atoms and the trail of damage left by them. Figure 6.1a displays the damage profiles of a diamond, created by electron irradiation with energies of 1, 2, and 5 MeV. The three profiles all show a sharp cut-off (i.e. the penetration depth), as predicted by the theoretical simulations. The maximum penetration depth of the electrons increases nearly linearly with energy from 1.6, 3.0, to 7.9 mm at 1, 2, and

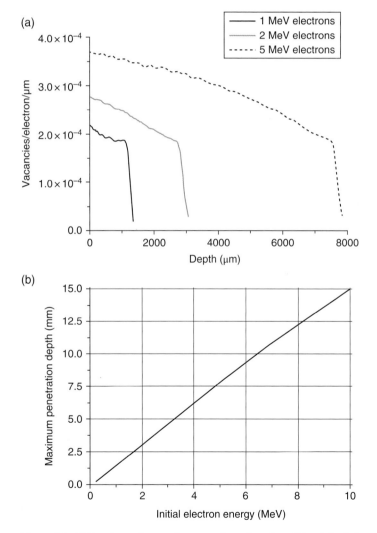

Figure 6.1 (a) Numbers of vacancies created as a function of depth by 1, 2, and 5 MeV electrons through diamond. (b) Maximum penetration depths of diamond damage as a function of electron energy. *Source:* Adapted with permission from Ref. [4].

5 MeV, respectively (Figure 6.1b). The corresponding total vacancy production was 0.23, 0.66, and 2.23 vacancies electron^{-1} [4, 5].

Ion irradiation is another way to create defects in the crystal structure of diamond. Similar to electron irradiation, the ion–matter collisions can be simulated with the computer program *Stopping and Range of Ions in Matter* (SRIM) [8], which includes the ion stopping and range tables with TRIM as the core program. The TRIM first calculates the impact and energy changes after ion–matter interactions based on the binary collision approximation [9]. The SR tables provide the stopping status and range when the ions of different energies interact with the matter. In the simulation, the types of ions (such as the simplest ion, H^+) and their energies (such as 10 eV to 2 GeV) are first chosen as the input parameters. Target materials, sample thicknesses, incident angles, and combinations of different materials (if applicable) are also entered in the program.

TRIM is a Monte Carlo-based simulation tool. When high-energy charged particles enter the target material, the simulation tracks the trajectories of the incident particles colliding with the target material. Neglecting the impact caused by neighboring atoms, TRIM calculates the distance between two collisions and other associated parameters by random sampling. The parameters recorded during the processes include the final position, energy loss, and secondary particles, with all corresponding expectation errors analyzed statistically. The incident particles are displayed in a three-dimensional distribution plot, effectively showing the depth of penetration. Reported in the final results are the kinetic energy produced during the collision process, the damage to the target, ionization, phonon generation, defect condition, and sputtering. The program is now widely used by researchers in the field to simulate the ion distributions of irradiation before any real experiments.

For the illustration purpose, the SRIM simulations are performed here using protons and helium ions as the damaging agents [10–15]. The values of the parameters used to carry out the simulations for both ions are all built in the software. The simulation starts with low incident energy, which is raised gradually. The selected target material is carbon, with its well-known density of 3.52 g cm^{-3} and E_{dis} of 37.5 eV for diamond. The target thickness is set as 100 μm for both ions with the energy of 0–3 MeV. The calculation time and accuracy depend on the numbers of sampling, which is typically set at 10 000. Figure 6.2a and b display, respectively, the number of vacancies produced by bombardment with these two types of ions and the corresponding maximum penetration depths. At 2 MeV, the proton has a total vacancy production yield of approximately 10 vacancies ion^{-1}, which is 14 times that (i.e. ~0.7 vacancies electron^{-1}) of 2-MeV electrons, estimated from the integrated area in Figure 6.1a. However, its maximum penetration depth is only 24 μm, which is about 1/120 that of the 2-MeV electrons. Notably, the ion distribution of the 3-MeV protons (as well as other ions) exhibits a maximum at the region where they stop in the crystal (Figure 6.3a). The peak, known as the *spread-out Bragg peak* in proton beam therapy [16], becomes sharper and more pronounced as the beam energy increases.

6.1.2 Electron/Ion Irradiation

Now, on the experimental side, creating vacancies in diamond can be achieved by bombarding the material with high-energy (typically 2 MeV) electrons from devices like the Van de Graaff accelerator [17]. However, the lack of easy accessibility to these research-type

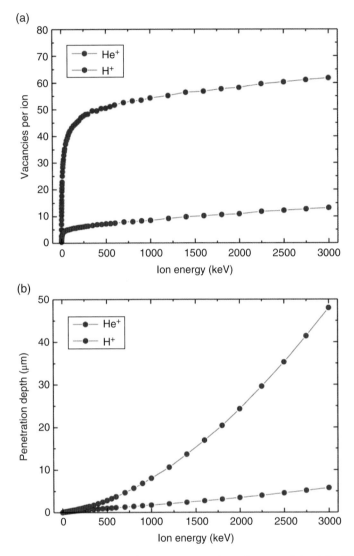

(a)

(b)

Figure 6.2 (a) Numbers of vacancies created as a function of depth by H$^+$ and He$^+$ ions through diamond. (b) Maximum penetration depths of diamond damage as a function of the energy of H$^+$ and He$^+$ ions. A displacement energy of 37.5 eV was used in the simulations. *Source:* Adapted with permission from Ref. [13]. Reproduced with permission of Elsevier.

accelerators makes it difficult to produce color centers in diamond on a routine basis. Another way of doing this is to use Rhodotron, a new commercial type of continuous-wave electron accelerator that has found broad industrial applications such as sterilization of medical devices. The accelerator produces a powerful electron beam (30–200 kW) in the energy range of 1–20 MeV [18]. Boudou et al. [19, 20] first reported the use of the Rhodotron technique for high-yield fabrication of FNDs. They fabricated *fluorescent microdiamonds* (FMDs) by bombarding submillimeter-sized diamond powders with 10-MeV electrons from the accelerator operating at a beam power of 80 kW (e.g. 10 MeV and 8 mA), followed

(a)

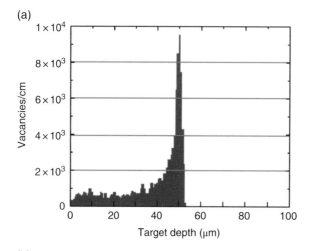

(b)

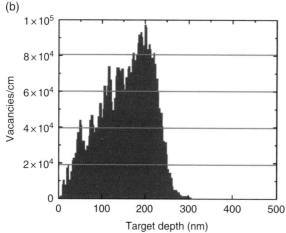

Figure 6.3 Spatial distribution of vacancies produced in diamond as a function of target depth predicted by SRIM Monte Carlo simulations. The proton beams used for the irradiation are (a) 3 MeV and (b) 40 keV, respectively. The numbers of damage events used in both simulations are 9999. Note that the vacancies are more uniformly distributed along the target depth in (b). *Source:* Adapted with permission from Ref. [11]. Reproduced with permission of American Chemical Society.

by annealing at 800 °C. The size of the FMDs was then reduced to the sub-100-nm range by high-energy air-jet milling and ball milling. Due to the long penetration depth (up to 15 mm in Figure 6.1b) of the electrons, the FMDs (and similarly FNDs) could be readily produced on the gram scale. Figure 6.4a and b show the color change of HPHT microdiamonds before and after the electron irradiation and annealing treatment.

While the irradiation with 10-MeV electrons allows for scale-up production of FNDs, they are not really suitable for daily operations because of the elaborate setup for necessary operations. Moreover, optimizing the irradiation conditions to produce the brightest FNDs is quite difficult and costly. The ion irradiation, alternatively, provides a more feasible approach since its radiation damage threshold is substantially lowered due to the much larger mass of ions compared with the electron. To maximize the

(a) (b)

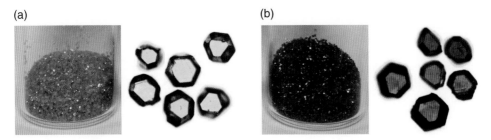

Figure 6.4 Photographs of HPHT microdiamonds before (a) and after (b) electron irradiation and thermal annealing. A bright-field transmission image of the corresponding particles (~400 μm in diameter) is shown on the right of each panel.

penetration depth without having to use a particle accelerator, Chang and coworkers [12, 13] have employed 40-keV protons or helium ion beams that can be readily installed and operated in any ordinary laboratory. According to the SRIM simulations (Figures 6.2 and 6.3b), each 40-keV H^+ ion will penetrate diamond by 230 nm with four vacancies created along the way. If the 40-keV He^+ ions are used instead, the penetration depth would decrease to 200 nm but the number of vacancies produced per ion can increase markedly to 40. Although the helium ion has a shallow penetration depth, it is able to penetrate through NDs of 100 nm or smaller in size without problems. It has the advantages of reducing the needed ion dose and reaction time, a feature of particular importance to achieve high vacancy densities in NDs.

Figure 6.5 shows a schematic diagram and a photograph of the experimental setup used to create vacancies in NDs. In this experiment, He^+ ions are generated by radio-frequency discharge of pure He to form a positive ion source and accelerated to 40 keV via a high-voltage acceleration tube. The typical current of the unfocused ion beam without mass discrimination is in the 10 μA range. To cope with the small penetration depth problem, 100-nm ND powders are first spread thin on the surface of a long copper tape before being placed in the ion implantation chamber. With the tape rotating in the chamber, all ND particles can be exposed to the ion bombardment over several cycles of treatment. Using this high-flux ion beam facility and subsequent annealing, radiation-damaged NDs in tens of milligram quantities can be produced on a routine basis [12].

6.1.3 Size Reduction

As discussed in Section 3.3, radiation-damaged NDs must go through annealing to promote migration of vacancies in the diamond crystal lattice to form vacancy-related color centers. Typically, this is performed in a vacuum at 800 °C or above for two hours. However, the annealing together with irradiation by electrons or ions inevitably results in graphitization on the diamond surface, leading to quenching of the FND fluorescence [21]. The effect becomes a serious concern for 100 nm or smaller FNDs that have large surface-area-to-volume ratios. A way to overcome this problem is to oxidize the freshly prepared FNDs (i.e. right after irradiation and annealing) in air at 450 °C for one hour to remove graphitic surface structures [22], followed by washing in a concentrated H_2SO_4:HNO_3 mixture at 100 °C to remove metal and other impurities (Figure 6.6) [23]. After separation by centrifugation and extensive rinsing in distilled deionized water, the

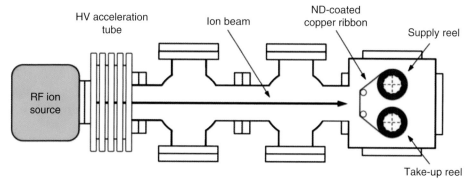

Figure 6.5 Experimental setup used to create vacancies in NDs with a medium-energy ion beam. *Source:* Reprinted with permission from Refs. [2, 12]. Reproduced with permission of American Chemical Society.

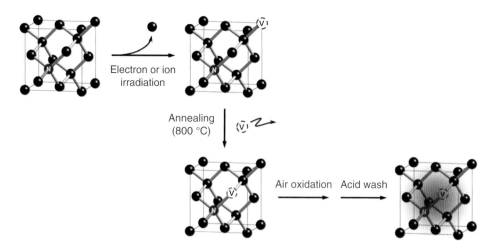

Figure 6.6 Procedures of FND production. The method is applicable for all color centers, not limited to NV as illustrated herein.

purified FND powders can be either suspended in water or dried in air prior to use. The final size of the nanoparticles is amenable to characterization by *dynamic light scattering* (DLS), *transmission electron microscopy* (TEM), or *atomic force microscopy* (AFM). It is worth noting that since FNDs are prepared under such harsh conditions, the shelf lifetime of the final product is extremely long (more than one year) at room temperature.

The FNDs produced by radiation damage of HPHT-NDs usually have a broad size distribution (Figure 4.8a). They vary not only in size but also in shape from batch to batch (Figure 4.10a). For any given commercial HPHT-ND powders with a median size of 35 nm, it is typical that one-fifth of the particles are smaller than 20 nm. These particles can be separated and collected by differential centrifugation after optimization of the experimental conditions. The yield can be further increased by combining air oxidation and the differential centrifugation techniques. For example, Mohan et al. [24] reported a procedure to produce sub-20-nm FNDs based on the above two-step approach. They first extracted sub-20-nm FNDs from the 50-nm ensemble by differential centrifugation and then reduced the size of larger particles in the sample by air oxidation at 500 °C for one to two hours. Iteration of these two steps allowed the production of smaller FNDs without damaging their crystal structures. Ultimately, the method can be applied to produce FNDs down to 1 nm in size with high yields at the cost of an extended processing time [25, 26].

With the air-oxidation method, Gaebel et al. [3] characterized in detail the size reduction and its effects on the nitrogen-vacancy (NV) centers in unirradiated HPHT-NDs using a combined AFM and confocal fluorescence microscope (Figure 6.7a). They found that the average height reduction rate of the individual nanocrystals, as measured by AFM, was $10\,nm\,h^{-1}$ at 600 °C, $4\,nm\,h^{-1}$ at 550 °C, and less than $1\,nm\,h^{-1}$ at 500 °C by air oxidation at atmospheric pressure. The oxidation effectively removed graphitic materials from the surface, thereby reducing the background fluorescence signals. By conducting fluorescence imaging for the individual particles (Figure 6.7b and c), they observed the annihilation of NV centers on the surface, which provided important insight into the behavior of these color centers in small diamond crystallites.

While air oxidation is a feasible approach to decrease the particle size, the yield is low and the process is time-consuming and laborious. Another way of reducing the particle size is by high-pressure crushing. Su et al. [27] developed such a method to produce smaller FND particles but with low damage to the crystal structures. They first mixed 30-nm FNDs with NaF powders at a weight ratio of 1 : 10 and pressed them to form a pellet under a pressure of 10 tons by a hydraulic oil press. The pellet was then annealed at 720 °C in a vacuum for two hours, after which it was dissolved in hot water to remove NaF. The method is much gentler than ball milling and introduces almost no loss in the yield of smaller FNDs with contamination-free quality. The yield is typically 5% for particles of less than 20 nm in diameter.

6.2 Characterization

6.2.1 Fluorescence Intensity

The final FNDs, produced by irradiation, annealing, oxidation, and acid wash treatments as described above, are now ready for fluorescence testing. Figure 6.8 shows a schematic layout of the optical setup used to acquire the fluorescence spectrum of FNDs suspended

(a)

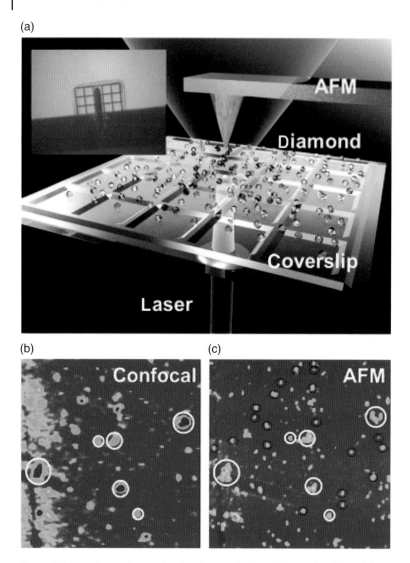

Figure 6.7 Experimental setup for the characterization of air-oxidized NDs. (a) An artistic view of the confocal beam incident from the bottom and through the glass coverslip combined with the AFM tip probing the sample from above. The inset is a photograph of the sample from directly above. (b, c) A confocal fluorescence intensity map (b) and an AFM height map (c) of the sample. The scan area is 50×50 μm. *Source:* Reprinted with permission from Ref. [3]. Reproduced with permission of Elsevier.

in water. The need for such a special arrangement is because diamond has a high refractive index ($n = 2.41$ in the visible region), which causes strong light scattering at the diamond–water interfaces. The scattering can seriously distort the observed spectra if the fluorescence is collected and detected with a CCD-based spectrophotometer in a perpendicular geometry. Displayed in Figure 6.9 are typical fluorescence spectra of green and red FNDs excited by 473- and 532-nm laser light, respectively. The fluorescence intensity of each sample provides a relative measure for the density of vacancy-related color centers in the

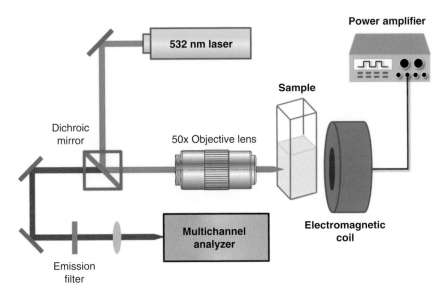

Figure 6.8 Optical layout of the experimental setup for fluorescence measurements of red FNDs suspended in water. A round electromagnet supplies a time-varying magnetic field of 0–50 mT, as controlled by a power amplifier. By changing the laser and optics, the setup can be applied to detect FNDs containing other color centers such as H3.

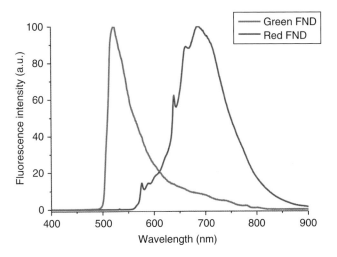

Figure 6.9 Fluorescence spectra of H3 and NV centers in green and red FNDs, respectively, suspended in water. The sizes of both particles are approximately 100 nm in diameter and the FND concentration is 1 mg ml^{-1} each.

diamond matrix. The results of the measurements serve as an important guide to the production of brighter FNDs through adjusting the doses of electron or ion irradiation.

The optical properties of FNDs can be further characterized by depositing them on a glass slide for fluorescence imaging. Bright-field imaging of the red FNDs produced by 3-MeV proton irradiation of HPHT-NDs showed that these particles formed

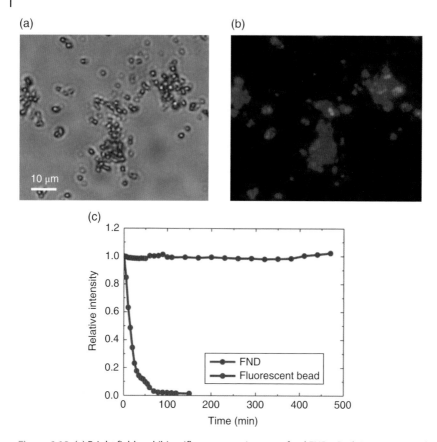

Figure 6.10 (a) Bright field and (b) epifluorescence images of red FNDs. Both images were obtained with a 40× objective. (c) Photostability tests of FND (red) and fluorescent polystyrene beads (blue) excited under the same conditions. The excitation was made with light of 510–560 nm in wavelength. *Source:* Reprinted with permission from Ref. [1]. Reproduced with permission of American Chemical Society.

aggregates on the glass slide when dried in air (Figure 6.10a). Excitation of the FNDs with the yellow lines from a 100 W mercury vapor lamp produced intense red emission (Figure 6.10b). The fluorescence intensity was nearly two orders of magnitude higher than that of the samples without irradiation but with annealing under the same conditions. No sign of photobleaching was observed for the FND even after eight hours of continuous excitation with the Hg lamp (Figure 6.10c). In contrast, the 0.1-µm red fluorescent polystyrene nanospheres containing approximately 10^4 dye equivalents photobleached within 0.5 hour when excited by the same lamp light [1]. The high photostability was similarly observed for green FNDs excited with the mercury lamp light at 450–490 nm [13].

6.2.2 Electron Spin Resonance

A question often asked is: What is the density of NV or H3 centers in the FNDs? According to the SRIM Monte Carlo simulations, a 3-MeV proton will produce 12 vacancies when penetrating diamond (Figure 6.2). The penetration depth is approximately $50\,\mu m$. If a dose of 1×10^{16} ions$\,cm^{-2}$ is used for irradiation, the density of the vacancies created in the crystals is 2.4×10^{19} centers$\,cm^{-3}$ (equivalent to 140 ppm, where 1 ppm $= 1.76 \times 10^{17}$ carbon atoms$\,cm^{-3}$ for diamond), corresponding to the production of 1.2×10^4 vacancies per 100-nm particle. This amount of vacancies is of the same magnitude as that of nitrogen atoms in each 100-nm HPHT-ND. Assuming that the efficiency of the NV$^-$ production is 10%, similar to that found for bulk diamonds [28], the number of NV$^-$ centers in the individual 100-nm FNDs, made of HPHT-NDs, is on the order of 10^3 or a density of approximately 10 ppm [1, 28]. The number agrees in the ballpark with that reported by Wee et al. [11] who bombarded type Ib bulk diamonds with a 3-MeV proton beam from a tandem accelerator and determined a NV$^-$ density in the range of 25 ppm.

A direct measurement for the number of NV$^-$ centers can be accomplished with *electron spin resonance* (ESR) spectroscopy [29]. Shames et al. [30] applied the technique to track the multistage process of the fabrication of FNDs produced by high-energy electron irradiation, annealing, and subsequent milling of type Ib diamond microcrystals (Figure 6.4). Their results indicated that the irradiation with 10-MeV electrons and subsequent annealing of the irradiated microdiamonds at 800 °C could produce triplet magnetic centers, identified as NV$^-$, with a density up to 5.4×10^{17} spins$\,g^{-1}$ (or 1.9×10^{18} spins$\,cm^{-3}$ or 11 ppm). However, after progressive milling of the FMDs down to a submicron scale, the relative abundance of NV$^-$ in the final product was reduced to 3.6×10^{17} spins$\,g^{-1}$ (or 7.2 ppm), corresponding to 660 NV$^-$ centers in each 100-nm FND particle. Although increasing the nitrogen content in the type Ib microdiamonds could facilitate the production of NV$^-$ centers, it also introduced structural imperfections, which were responsible for the appearance of additional nonradiative recombination centers in the ESR spectra and fluorescence quenching [31].

6.2.3 Fluorescence Lifetime

Fluorescence lifetime is another significant photophysical property of FNDs. For the NV$^-$ centers in bulk type Ib diamond, the lifetime of the $^3E \rightarrow {}^3A_2$ transition (Figure 3.8) is 11.6 ns as well-documented in the literature [32]. However, the lifetime can be significantly altered if the crystal size becomes smaller than the wavelength of the excitation light. It is known that for a molecular fluorophore in solution, its radiative decay rate is proportional to the square of the refractive index (n^2) of the environment due to the change in polarizability [33]. Although such a refractive index is not well defined for FNDs in water ($n = 1.33$), it is clearly smaller than that of bulk diamond ($n = 2.41$). So, one can expect a decrease of the radiative decay rate or an increase of the fluorescence lifetime for the color centers in FNDs.

Figure 6.11a displays typical fluorescence decay time traces of two red FND samples (100 and 35 nm in size) in water. Unlike the single-exponential decay observed for the NV$^-$ centers in bulk diamond, both the fluorescence time traces should be fitted with two decay constants:

$$I(t) = a_1 \exp\left(-\frac{t}{\tau_1}\right) + a_2 \exp\left(-\frac{t}{\tau_2}\right). \tag{6.1}$$

The observation of such a characteristic double-exponential decay is presumably due to the presence of a high-density ensemble ($\sim$10 ppm) of NV$^-$ centers in each

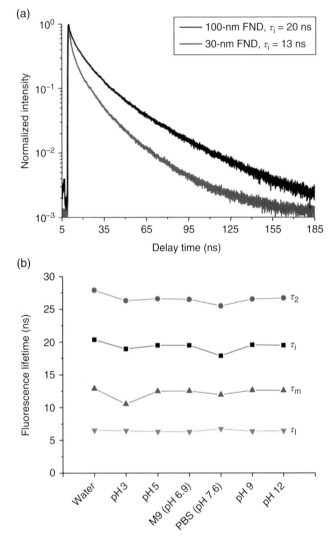

Figure 6.11 (a) Comparison of the fluorescence lifetimes of FNDs with different sizes (100 and 30 nm in diameter). The concentration of both FND suspensions is 1 mg ml^{-1}. (b) Variation of the fluorescence lifetimes of 100-nm FNDs in water, biological buffer, and aqueous solutions with different pHs. The amplitude-weighted mean lifetime (τ_m) and the intensity-weighted mean lifetime (τ_i) are calculated from τ_1 and τ_2 according to Eqs. (6.2) and (6.3) in text. *Source:* Adapted with permission from Refs. [27, 38].

FND particle that complicates the energy relaxation dynamics. The amplitude-weighted mean lifetime (τ_m) is

$$\tau_m = \frac{a_1\tau_1 + a_2\tau_2}{a_1 + a_2} \tag{6.2}$$

and the intensity-weighted mean lifetime (τ_i) is

$$\tau_i = \frac{a_1 \tau_1^2 + a_2 \tau_2^2}{a_1 \tau_1 + a_2 \tau_2} \tag{6.3}$$

Both lifetimes vary with the sample size; for example, the value of τ_i decreases from 20 ns of 100-nm FNDs to 13 ns of 30-nm FNDs [27]. The shortening of the observed lifetime (τ_{obs}) could be attributed to fluorescence quenching by surface defects [34], as the fluorescence quantum yield (Q_F) is related to the intrinsic radiative rate constant (k_r) and the nonradiative rate constant (k_{nr}) by

$$Q_F = \frac{k_r}{k_r + k_{nr}} = k_r \tau_{obs}. \tag{6.4}$$

Such a surface effect is expected to play a more significant role as the particle size decreases. The effect is particularly pronounced for DNDs, which are approximately 5 nm in diameter with a graphitic shell coated on the surface [35, 36]. As a result, DNDs are only weakly fluorescent and the emission is not photostable [37].

Described in the previous section, the density of NV^- centers in the 100-nm FNDs may be as high as 10 ppm. Assuming that these centers are uniformly distributed in the diamond lattice, the nearest neighbors are separated by roughly 10 nm. While most of the centers are embedded deeply in the crystal matrix, some of them are located within a 5-nm thickness of the outer surface shell, where their fluorescence properties can be significantly influenced by the external environment. Figure 6.11b shows how the fluorescence lifetimes of 100-nm FNDs are varied when they are suspended in water, biological buffer, and aqueous solutions with different pH values. All the fluorescence time traces exhibit double-exponential decay characteristic with a fast component of $\tau_1 \sim 6$ ns and a slow component of $\tau_2 \sim 26$ ns. Even under the extreme conditions of pH 3 and 12, the lifetime is virtually unchanged, revealing the exceptionally high chemical stability of the nanomaterial and the hosted color centers [38]. The result suggests that all the surface chemistry described in Chapter 4 will have only a minor effect on the fluorescence properties of the NV^- centers as long as the size of FNDs is maintained in the neighborhood of 100 nm. The same principle applies for the H3 centers in green FNDs of comparable size.

6.2.4 Magnetically Modulated Fluorescence

We now turn to the magnetic properties of NV^- centers in FNDs. As explained in Section 3.4, the NV^- center is unique in that it has two unpaired spins, exhibiting a magnetic resonance at 2.87 GHz ($\Delta m_s = 0 \rightarrow \pm 1$) in the ground electronic state. Interestingly, the spin states of NV^- can be optically polarized after several cycles of electronic excitation [39]. In the presence of an external magnetic field, not in alignment with the NV symmetry axis, the fluorescence intensity would decrease because the magnetic field lifts the degeneracy of the $m_s = \pm 1$ sublevels and mixes them with the $m_s = 0$ sublevel (Figure 3.8). For an ensemble of NV^- centers in bulk diamond, a change of the fluorescence intensity by more than 10% can result [40, 41]. Exploiting this unique magneto-optical property, Chapman and Plakhotnik [42] were able to achieve background-free imaging of FNDs on a glass slide by switching an external magnetic field on and off, followed by subtraction of the signals from these two measurements for contrast enhancement. The method is particularly suitable for detecting NV^- in high backgrounds. The magnetic modulation, though less common than optical modulation, has the advantage of being more applicable to samples with strong light scattering and complex chemical compositions [43].

The feasibility of modulating the fluorescence intensity of NV^- has also been tested for FNDs in solution (cf., Figure 6.8 for the experimental setup). Figure 6.12a displays a typical fluorescence spectrum of 100-nm FNDs suspended in water (concentration of $1\,mg\,ml^{-1}$) and excited by a 532-nm laser. The fluorescence intensity showed a maximum at 687 nm with a modulation depth of approximately 5% when the particles were exposed to a time-varying magnetic field with a strength of $B = 20\,mT$ at an on/off switching frequency of $f = 2\,Hz$ (Figure 6.12b). However, as the FND concentration was lowered to $1\,\mu g\,ml^{-1}$ or less, the NV^- emission could not be seen because the Raman signals of water dominated the observed spectrum. Luckily, the Raman signals were not affected by the magnetic modulation, thereby allowing removal of the background signals by *fast Fourier transform* (FFT) of the time traces using a computer program for frequency demodulation at each wavelength [44]. As shown in Figure 6.12c, the fluorescence signal of a highly diluted FND solution (concentration $\sim 0.1\,\mu g\,ml^{-1}$) was originally overwhelmed by the water Raman peaks; the background level, however, could be effectively reduced by more than two orders of magnitude with the magnetic modulation technique.

In a recent study, our group has also investigated how the modulation depth of the NV^- emission depends on the magnetic field strength. Instead of obtaining the whole emission spectra, we collected the total fluorescence signals at 650–750 nm using a bandpass filter and an avalanche photodiode as the detector. Figure 6.12d shows the modulation depth of the signals as a function of the magnetic field strength varying sinusoidally at 2 and 10 Hz. The depth increased with B and gradually reached a plateau ($\sim 15\%$) when $B \geq 50\,mT$. With the instrument operating at $B = 40\,mT$, $f = 10\,Hz$, and a data acquisition time of 10 seconds, a detection limit as low as $60\,ng\,ml^{-1}$ was readily achieved (Figure 6.12e). Assuming a spherical shape for the FND particles, the lowest detectable concentration was estimated to be 50 fM or 3×10^7 particles ml^{-1}.

Finally, it is worth noting that background-free detection by optically modulated fluorescence can also be achieved with FNDs. Using a green laser to drive the electronic transition of the NV^- centers in bulk diamond, Geiselmann et al. [45] reported an optical modulation of more than 80% and a time response of faster than 100 ns, controlled by a near-infrared gating laser, which promoted the population in the excited state (^3E) to an unknown dark band. Compared with magnetically modulated fluorescence, the method is advantageous in having a larger modulation depth and a faster modulation rate. It is compatible with optical lock-in detection fluorescence microscopy [46, 47], a technique developed to enhance the image contrast of fluorescent proteins or fluorescently labeled biomolecules in living cells by intensity modulation. Both the optical and magnetic modulation methods are ideally suited for high sensitivity detection of FNDs in extremely dilute solutions as well as high selectivity imaging of FNDs in complex biological environments as discussed in Chapter 9.

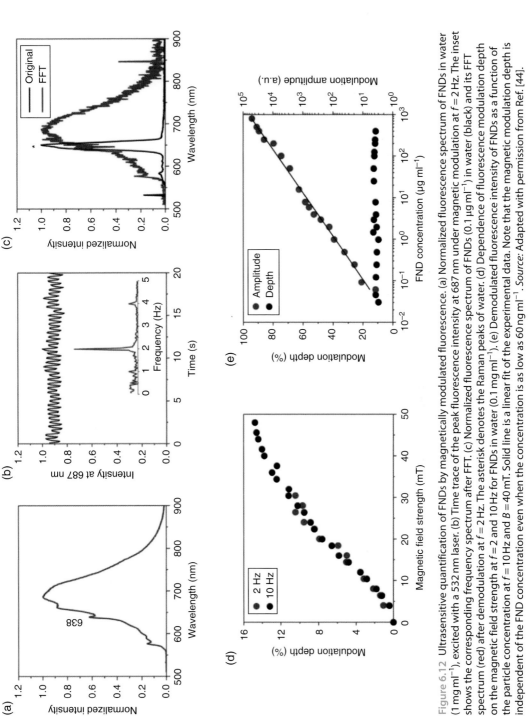

Figure 6.12 Ultrasensitive quantification of FNDs by magnetically modulated fluorescence. (a) Normalized fluorescence spectrum of FNDs in water (1 mg ml⁻¹), excited with a 532 nm laser. (b) Time trace of the peak fluorescence intensity at 687 nm under magnetic modulation at $f = 2$ Hz. The inset shows the corresponding frequency spectrum after FFT. (c) Normalized fluorescence spectrum of FNDs (0.1 µg ml⁻¹) in water (black) and its FFT spectrum (red) after demodulation at $f = 2$ Hz. The asterisk denotes the Raman peaks of water. (d) Dependence of fluorescence modulation depth on the magnetic field strength at $f = 2$ and 10 Hz for FNDs in water (0.1 mg ml⁻¹). (e) Demodulated fluorescence intensity of FNDs as a function of the particle concentration at $f = 10$ Hz and $B = 40$ mT. Solid line is a linear fit of the experimental data. Note that the magnetic modulation depth is independent of the FND concentration even when the concentration is as low as 60 ng ml⁻¹. *Source:* Adapted with permission from Ref. [44].

References

1 Yu, S.J., Kang, M.W., Chang, H.C. et al. (2005). Bright fluorescent nanodiamonds: no photobleaching and low cytotoxicity. *J Am Chem Soc* 127: 17604–17605.

2 Hsiao, W.W.W., Hui, Y.Y., Tsai, P.C., and Chang, H.C. (2016). Fluorescent nanodiamond: a versatile tool for long-term cell tracking, super-resolution imaging, and nanoscale temperature sensing. *Acc Chem Res* 49: 400–407.

3 Gaebel, T., Bradac, C., Chen, J. et al. (2011). Size-reduction of nanodiamonds via air oxidation. *Diam Relat Mater* 21: 28–32.

4 Campbell, B. and Mainwood, A. (2000). Radiation damage of diamond by electron and gamma irradiation. *Physica Stat Sol A* 181: 99–107.

5 Campbell, B., Choudhury, W., Mainwood, A. et al. (2002). Lattice damage caused by the irradiation of diamond. *Nucl Instrum Meth A* 476: 680–685.

6 Halliday, D., Walker, J., and Resnick, R. (2010). *Fundamentals of Physics*, 5e. New York: Wiley.

7 Biersack, J.P. and Haggmark, H.G. (1980). A Monte Carlo computer program for the transport of energetic ions in amorphous targets. *Nucl Instr Meth* 174: 257–269.

8 Ziegler, J.F., Biersack, J.P., and Littmark, U. (1985). The stopping and range of ions in solids. *Pergamon*. www.srim.org (accessed 16 April 2018).

9 Robinson, M.T. and Torrens, I.M. (1974). Computer simulation of atomic-displacement cascades in solids in the binary-collision approximation. *Phys Rev B* 9: 5008–5024.

10 Waldermann, F.C., Olivero, P., Nunn, J. et al. (2007). Creating diamond color centers for quantum optical applications. *Diam Relat Mater* 16: 1887–1895.

11 Wee, T.L., Tzeng, Y.K., Han, C.C. et al. (2007). Two-photon excited fluorescence of nitrogen-vacancy centers in proton-irradiated type Ib diamond. *J Phys Chem A* 111: 9379–9386.

12 Chang, Y.R., Lee, H.Y., Chen, K. et al. (2008). Mass production and dynamic imaging of fluorescent nanodiamonds. *Nat Nanotechnol* 3: 284–288.

13 Wee, T.L., Mau, Y.W., Fang, C.Y. et al. (2009). Preparation and characterization of green fluorescent nanodiamonds for biological applications. *Diam Relat Mater* 18: 567–573.

14 Acosta, V.M., Bauch, E., Ledbetter, M.P. et al. (2009). Diamonds with a high density of nitrogen-vacancy centers for magnetometry applications. *Phys Rev B* 80: 115202.

15 Botsoa, J., Sauvage, T., Adam, M.P. et al. (2011). Optimal conditions for NV- center formation in type-1b diamond studied using photoluminescence and positron annihilation spectroscopies. *Phys Rev B* 84: 125209.

16 Levin, W.P., Kooy, H., Loeffler, J.S., and DeLaney, T.F. (2005). Proton beam therapy. *Br J Cancer* 93: 849–854.

17 Davies, G., Lawson, S.C., Collins, A.T. et al. (1992). Vacancy-related centers in diamond. *Phys Rev B* 46: 13157–13170.

18 Jongen, Y., Abs, M., Capdevila, J.M. et al. (1994). The Rhodotron, a new high-energy, high-power, CW electron accelerator. *Nucl Instrum Meth B* 89: 60–64.

19 Boudou, J.P., Curmi, P.A., Jelezko, F. et al. (2009). High yield fabrication of fluorescent nanodiamonds. *Nanotechnology* 20: 235602.

20 Boudou, J.P., Tisler, J., Reuter, R. et al. (2013). Fluorescent nanodiamonds derived from HPHT with a size of less than 10 nm. *Diam Relat Mat* 37: 80–86.

21 Smith, B.R., Gruber, D., and Plakhotnik, T. (2010). The effects of surface oxidation on luminescence of nanodiamonds. *Diam Relat Mater* 19: 314–318.

22 Osswald, S., Yushin, G., Mochalin, V. et al. (2006). Control of sp^2/sp^3 carbon ratio and surface chemistry of nanodiamond powders by selective oxidation in air. *J Am Chem Soc* 128: 11635–11642.

23 Huang, L.C.L. and Chang, H.C. (2004). Adsorption and immobilization of cytochrome *c* on nanodiamonds. *Langmuir* 20: 5879–5884.

24 Mohan, N., Tzeng, Y.K., Yang, L. et al. (2010). Sub-20-nm fluorescent nanodiamonds as photostable biolabels and fluorescence resonance energy transfer donors. *Adv Mater* 22: 842–847.

25 Stehlik, S., Varga, M., Ledinsky, M. et al. (2015). Size and purity control of HPHT nanodiamonds down to 1 nm. *J Phys Chem C* 119: 27708–27720.

26 Stehlik, S., Varga, M., Ledinsky, M. et al. (2016). High-yield fabrication and properties of 1.4 nm nanodiamonds with narrow size distribution. *Sci Rep* 6: 38419.

27 Su, L.J., Fang, C.Y., Chang, Y.T. et al. (2013). Creation of high density ensembles of nitrogen-vacancy centers in nitrogen-rich type Ib nanodiamonds. *Nanotechnology* 24: 315702.

28 Pezzagna, S., Naydenov, B., Jelezko, F. et al. (2010). Creation efficiency of nitrogen-vacancy centres in diamond. *New J Phys* 12: 065017.

29 Atherton, N.M. (1993). *Principles of Electron Spin Resonance*. PTR Prentice Hall, Ellis Horwood.

30 Shames, A.I., Yu Osipov, V., Boudou, J.P. et al. (2015). Magnetic resonance tracking of fluorescent nanodiamond fabrication. *J Phys D: Appl Phys* 155302: 48.

31 Shames, A.I., Osipov, V.Y., Bogdanov, K.V. et al. (2017). Does progressive nitrogen doping intensify negatively charged nitrogen vacancy emission from e-beam-irradiated Ib type high-pressure-high-temperature diamonds? *J Phys Chem C* 121: 5232–5240.

32 Collins, A.T., Thomaz, M.F., and Jorge, M.I.B. (1983). Luminescence decay time of the 1.945 eV center in type 1b diamond. *J Phys C Solid State* 16: 2177–2181.

33 Strickler, S.J. and Berg, R.A. (1962). Relationship between absorption intensity and fluorescence lifetime of molecules. *J Chem Phys* 37: 814–822.

34 Chen, L.H., Lim, T.S., and Chang, H.C. (2012). Measuring the number of $(N-V)^-$ centers in single fluorescent nanodiamonds in the presence of quenching effects. *J Opt Soc Am B* 29: 2309–2313.

35 Bradac, C., Gaebel, T., Naidoo, N. et al. (2010). Observation and control of blinking nitrogen-vacancy centres in discrete nanodiamonds. *Nat Nanotechnol* 5: 345–349.

36 Smith, B.R., Inglis, D.W., Sandnes, B. et al. (2009). Five-nanometer diamond with luminescent nitrogen-vacancy defect centers. *Small* 5: 1649–1653.

37 Vlasov, I.I., Shenderova, O., Turner, S. et al. (2010). Nitrogen and luminescent nitrogen-vacancy defects in detonation nanodiamond. *Small* 6: 687–694.

38 Kuo, Y., Hsu, T.Y., Wu, Y.C. et al. (2013). Fluorescence lifetime imaging microscopy of nanodiamonds *in vivo*. *Proc SPIE* 8635: 863503.

39 Gruber, A., Drabenstedt, A., Tietz, C. et al. (1997). Scanning confocal optical microscopy and magnetic resonance on single defect centers. *Science* 276: 2012–2014.

40 Lai, N.D., Zheng, D., Jelezko, F. et al. (2009). Influence of a static magnetic field on the photoluminescence of an ensemble of nitrogen-vacancy color centers in a diamond single-crystal. *Appl Phys Lett* 95: 133101.

41 Rondin, L., Tetienne, J.P., Hingant, T. et al. (2014). Magnetometry with nitrogen-vacancy defects in diamond. *Rep Prog Phys* 77: 056503.

42 Chapman, R. and Plakhotnik, T. (2013). Background-free imaging of luminescent nanodiamonds using external magnetic field for contrast enhancement. *Opt Lett* 38: 1847–1849.

43 Yang, N. and Cohen, A.E. (2010). Optical imaging through scattering media via magnetically modulated fluorescence. *Opt Express* 18: 25461–25467.

44 Su, L.J., Wu, M.H., Hui, Y.Y. et al. (2017). Fluorescent nanodiamonds enable quantitative tracking of human mesenchymal stem cells in miniature pigs. *Sci Rep* 7: 45607.

45 Geiselmann, M., Marty, R., García de Abajo, F.J., and Quidant, R. (2013). Fast optical modulation of the fluorescence from a single nitrogen-vacancy centre. *Nat Phys* 9: 785–789.

46 Marriott, G., Mao, S., Sakata, T. et al. (2008). Optical lock-in detection imaging microscopy for contrast-enhanced imaging in living cells. *Proc Natl Acad Sci USA* 105: 17789–17794.

47 Hsiang, J.C., Jablonski, A.E., and Dickson, R.M. (2014). Optically modulated fluorescence bioimaging: visualizing obscured fluorophores in high background. *Acc Chem Res* 47: 1545–1554.

48 Dantelle, G., Slablab, A., Rondin, L. et al. (2010). Efficient production of NV colour centres in nanodiamonds using high-energy electron irradiation. *J Lumin* 130: 1655–1658.

49 Havlik, J., Petrakova, V., Rehor, I. et al. (2013). Boosting nanodiamond fluorescence: towards development of brighter probes. *Nanoscale* 5: 3208–3211.

50 Mahfouz, R., Floyd, D.L., Peng, W. et al. (2013). Size-controlled fluorescent nanodiamonds: a facile method of fabrication and color-center counting. *Nanoscale* 5: 11776–11782.

51 Sotoma, S., Yoshinari, Y., Igarashi, R. et al. (2014). Effective production of fluorescent nanodiamonds containing negatively-charged nitrogen-vacancy centers by ion irradiation. *Diam Relat Mater* 49: 33–38.

7

Single Particle Detection and Tracking

Single-molecule spectroscopy is a technique that investigates the photophysical properties of individual molecules [1]. It overcomes the limitations of ensemble-averaged measurements to allow the examination of chemical and biological systems with unprecedented details. The first single-molecule measurement by fluorescence detection in the condensed phase was conducted in 1990 by Orrit and Bernard [2] for pentacene in a p-terphenyl crystal at temperatures lower than 4 K. Seven years later, Wrächtrup and coworkers [3] took advantage of the excellent photostability and the high fluorescence brightness of NV$^-$ centers to study the individual defects in bulk diamond. Moreover, the team went a step further and obtained the first spin resonance spectra of single NV$^-$ centers at room temperature [3]. The studies have aroused considerable interest of using these single spins as room-temperature quantum sensors for both physics and biology since then [4].

As described in the previous chapter, red *fluorescent nanodiamonds* (FNDs) are nanoscale diamonds containing an ensemble of NV$^-$ centers as built-in fluorophores. More than 10 NV$^-$ centers can be hosted in a FND particle of 30 nm in diameter or larger. Therefore, each of them can be readily detected by confocal fluorescence microscopy, a standard tool used in biological research [5]. This outstanding feature, together with the capability of emitting far-red emission, makes it possible to use red FNDs as photostable markers to follow intra- and inter-cellular communications by *single particle tracking* (SPT) [6].

In a biological system, complicated and interconnected as such, the ability to continuously track and monitor the transport of target agents through intracellular medium at the single particle level is without any doubt a heavenly dream coming true for biophysicists and biochemists. With hard work by scientists all over the world, FND is now quickly closing the gap to turn this dream into reality. This chapter focuses on single particle detection and tracking of FNDs, with more to follow on practical cell labeling and imaging techniques in the next chapter.

7.1 Single Particle Detection

7.1.1 Photostability

The detection of single red FNDs was first demonstrated by Fu et al. [7] in 2007. Particles with a size down to 35 nm in diameter could be individually visualized by using a confocal fluorescence microscope optimized to detect NV$^-$ centers. Figure 7.1a shows

Fluorescent Nanodiamonds, First Edition. Huan-Cheng Chang, Wesley Wei-Wen Hsiao and Meng-Chih Su.
© 2019 John Wiley & Sons Ltd. Published 2019 by John Wiley & Sons Ltd.

(a)

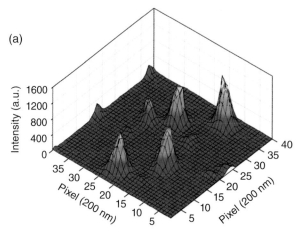

Figure 7.1 Fluorescence images and spectra of single 35-nm FNDs. Each pixel corresponds to 200 nm. (a) Confocal scanning image of 35-nm FNDs dispersed on a coverglass slide. (b) Fluorescence spectra of three different 35-nm FND particles. *Source:* Adapted with permission from Ref. [7].

(b)

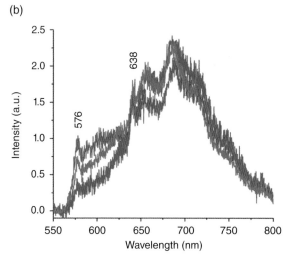

a representative fluorescence image of these particles dispersed on a glass substrate. The full width at half maximum of each peak in the image is 2–3 pixels, corresponding to a physical distance of 400–600 nm, which is close to the diffraction limit of optical microscopy, strongly suggesting that they are derived from single isolated FND particles. The suggestion is supported by their dispersed fluorescence spectra (Figure 7.1b), in which each 35-nm FND exhibits a distinct spectrum in the wavelength range of 550–800 nm, a signature of single particle detection.

Arguably being the most profound feature, the enduring photostability of FNDs distinguishes themselves from other molecular fluorophores by a large margin. Figure 7.2a shows the time traces for the fluorescence intensities of a single 35-nm FND and a single 100-nm FND spin-coated on a glass coverslip. Excellent photostability was observed for both particles. Under excitation with 532-nm light at a power density of $8 \times 10^3 \, \mathrm{W \, cm^{-2}}$, the fluorescence intensities of the FNDs investigated stayed essentially the same over a time period of 300 seconds. No sign of photoblinking was detected within the time resolution of 1 ms (Figure 7.2b). In contrast, single dye molecules such as Alexa Flour 546

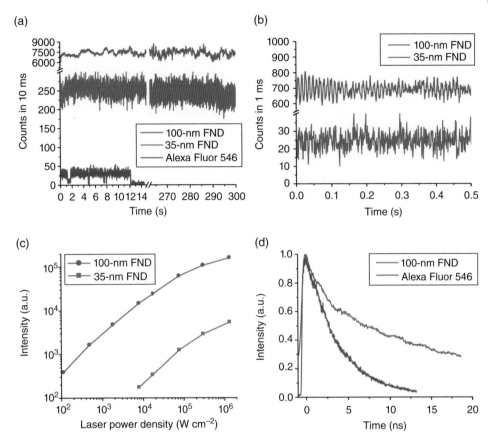

Figure 7.2 Spectroscopic characterization of 35- and 100-nm FNDs. (a) Typical time traces of the fluorescence from a single 100-nm FND, a single 35-nm FND, and a single Alexa Fluor 546 dye molecule attached to a single double-stranded DNA molecule. (b) Time traces of the fluorescence from a single 100-nm FND and a single 35-nm FND acquired with a time resolution of 1 ms, showing no photoblinking behavior. (c) Plot of the fluorescence intensity as a function of the laser power density over the range of 1×10^2–1×10^6 W cm^{-2} for 35- and 100-nm FNDs. (d) Fluorescence lifetime measurements of 100-nm FNDs and Alexa Fluor 546 dye molecules. Fitting the time traces with two exponential decays for the FND reveals a fast component of 1.7 ns (4%) and a slow component of 17 ns (96%). The latter is approximately 4 times longer than that (~4 ns) of Alexa Fluor 546. *Source:* Reprinted with permission from Ref. [7].

covalently linked to DNA molecules blinked randomly and photobleached rapidly within 12 seconds. The exceptional photostability of FNDs was also reflected in the power dependence measurement of fluorescence intensity (Figure 7.2c). The observed intensity scaled nearly linearly with the laser power density from 1×10^2 to 1×10^5 W cm^{-2} and gradually reached its saturation at 1×10^6 W cm^{-2}. Both the FND particles retained their fluorescence characteristics even under such high-power laser excitation for several minutes [7]. This remarkable characteristic renders FNDs useful as fiducial markers to correct microscope drifts. A correction of the drifts with both accuracy and precision of better than 5 nm is achievable by tracking continuously the positions of single FNDs (diameter ~100 nm) on a glass substrate for hours [8].

(a)

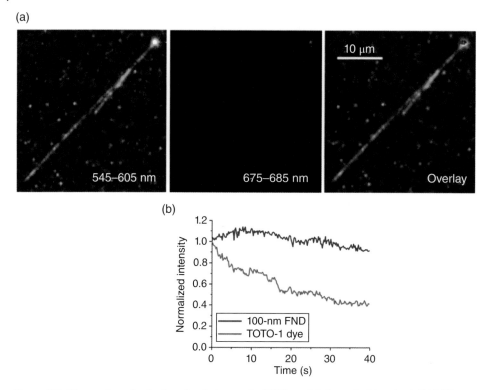

(b)

Figure 7.3 Observation of a single poly-L-lysine-coated FND particle bound with a single T4 DNA molecule on an amine-terminated glass substrate. (a) Dual-view fluorescence images of a single DNA/FND complex. An overlay (right) of the images from the shorter (545–605 nm) and longer (675–685 nm) wavelength channels shows that the T4 DNA molecule is wrapped around the 100-nm FND particle and stretched to a V-shape configuration. (b) Fluorescence decays of the 100-nm FND particle and the TOTO-1 dyes intercalated in the T4 DNA molecule. *Source:* Adapted with permission from Ref. [7].

The exceptional photostability of the NV⁻ center also allows easy identification of individual FNDs in a noisy background environment. An example of this is given in Figure 7.3, where a single poly-L-lysine-coated FND particle was wrapped around by a single T4 DNA molecule fluorescently labeled with TOTO-1 dye and attached to an amine-terminated glass surface [7]. With a channel combing method, the T4 DNA molecule was stretched to have a V-shape configuration and then fluorescently imaged. Two fluorescence detection channels (545–605 nm and 675–685 nm) were monitored simultaneously to reveal the interaction between the positively charged FND and the negatively charged DNA (Figure 7.3a). Despite that the image was overwhelmed by the fluorescence signals of the TOTO-1-labeled DNA molecule, the presence of FND in the complex could be readily confirmed by detecting its red fluorescence after prolonged illumination of the complex for more than 40 seconds, where a substantial decrease of the fluorescence intensity occurred for the TOTO-1 dye molecules excited under the same conditions as that of the FND (Figure 7.3b).

Another way to improve the sensitivity of detecting single FNDs in a noisy environment is to utilize the time-gating techniques. Figure 7.2d displays a time trace obtained by co-adding the data of 30 individual 100-nm FND particles spin-coated on a glass coverslip. The long-lived fluorescence lifetime is in qualitative agreement with the ensemble measurement (Figure 6.11a). The slow component of the fluorescence decay has a lifetime of 17 ns, which is significantly shorter than that (~26 ns) of FNDs in aqueous solution due to the larger refractive index of the glass substrate than water. Such an energy relaxation behavior is markedly different from that of molecular fluorophores (such as ~4 ns for Alexa Flour 546 in the same figure), making it easy to isolate the FND emission from other background signals (such as cellular endogenous fluorescence) using various time-gating techniques including *fluorescence lifetime imaging microscopy* (FLIM) [9]. Exploiting this distinct photophysical property, Faklaris et al. [10] were able to observe FNDs in the cytoplasm of cells at the single-particle and single-defect level. More discussion about the techniques can be found in Section 8.2.4.

7.1.2   Spectroscopic Properties

Table 3.2 lists some spectroscopic properties of vacancy-related color centers in bulk diamonds. The values of these parameters may change as the sizes of the crystals are reduced to the nanoscale. We saw this effect in Section 6.2.3, where the NV^- centers in FNDs have a considerably longer fluorescence lifetime than those in bulk diamonds (~20 ns vs. ~11 ns). Similarly, the quantum efficiency of the NV^- emission is another parameter that may be affected by the particle size. The efficiency is expected to be lower in FNDs than in bulk diamonds due to the presence of structural defects and/or chemical impurities on the surface, which may quench the fluorescence. This size effect is more prominent for smaller particles, again, because of the large surface-area-to-volume ratios inherited by the nanoparticles. It was reported by Mohtashami and Koenderink [11] that the quantum efficiencies of NV^- in the individual 25-nm FND particles were widely distributed between 0 and 20%. The distribution was even wider (over 10–90%) in larger FNDs (such as 100 nm), suggesting a high degree of structural heterogeneity in the samples.

An intrinsic property of the NV^- center is the absorption cross-section of the $^3A_2 \rightarrow {}^3E$ transition (Figure 3.8). The cross-section (σ, in unit of cm^2) represents the probability of a light absorption process and is related to the molar extinction coefficient (ε, in unit of $M^{-1} cm^{-1}$) by [12].

$$\sigma = \frac{2303}{N_A}\varepsilon = 3.82\times10^{-21}\varepsilon, \tag{7.1}$$

where N_A is the Avogadro constant. Using a direct absorption method, Wee et al. [13] obtained $\sigma = 3.1 \times 10^{-17}$ cm^2 or $\varepsilon = 8 \times 10^3$ $M^{-1}cm^{-1}$ at the wavelength of 532 nm for the NV^- centers in bulk diamond. The method, however, is inapplicable for FNDs since the sample thickness is ill-defined. Taking an entirely different approach, Chapman and Plakhotnik [14] showed that it is possible to quantitatively characterize the absorption cross-section of FNDs using a wide-field epifluorescence microscope under pulsed excitation. They used laser pulses with widths shorter than the timescale of any relaxation process in the system and separated the pulses by a span of time longer than the lifetime of any non-ground state. From a measurement for the saturation curves of the excitation at different laser energy densities for the individual 35-nm particles, a value of $\sigma = 9.5 \times 10^{-17}$ cm^2 or $\varepsilon = 2.5 \times 10^4 M^{-1}cm^{-1}$ was obtained for the 532 nm excitation. Plakhotnik and Aman [15] have recently provided a thorough discussion on the origin of the discrepancy between the measured cross-sections for the NV⁻ centers in bulk diamonds and FNDs. Note that while the reported absorption cross-section of FNDs is about one-half that of fluorescein at 490 nm in aqueous solution at neutral pH [16], it is about 3 orders of magnitude higher than that of the Ti^{3+} ions in a sapphire crystal at 500 nm [17].

7.1.3 Color Center Numbers

The ability to detect single FNDs allows a precise measurement for the number of NV⁻ centers present in the individual particles [18, 19]. This is achievable by performing photon correlation spectroscopy, which measures the correlation functions of the emitters' fluorescence intensities with a *Hanbury–Brown–Twiss interferometer* [20, 21]. Figure 7.4a shows a schematic diagram of the interferometer, in which the emitted light is split with a 50/50 beamsplitter onto two single photon counting detectors that are connected to a photocount correlator and then a time interval analyzer [22]. The photon correlation function is measured electronically as [23]

$$g^{(2)}(\tau) = \frac{\langle I_{d1}(t) I_{d2}(t+\tau) \rangle}{\langle I_{d1}(t) \rangle \langle I_{d2}(t) \rangle}, \tag{7.2}$$

where τ is the inter-photon time, $I_{d1}(t)$ and $I_{d2}(t+\tau)$ are the recorded photocounts in detectors 1 and 2 at times t and $t+\tau$, respectively, and the bracket denotes the average over the detection time interval. For a constant light source with $\langle I(t) \rangle = \langle I(t+\tau) \rangle$, this measurement has been shown to be equivalent to the measurement of the normalized autocorrelation function of intensity or the normalized second-order correlation function of intensity as [24]

$$g^{(2)}(\tau) = \frac{\langle I(t) I(t+\tau) \rangle}{\langle I(t) \rangle^2}. \tag{7.3}$$

In a simple two-level quantum system, where only one photon is emitted at a time, the probability of finding two consecutive photons is 0 at the zero lag time, i.e. $g^{(2)}(0) = 0$, as it takes a finite amount of time for the single emitter to be excited again and emit the second photon. The probability, however, increases as the time lag between the two emitted photons increases, i.e. $g^{(2)}(\tau) > 0$ at $\tau \neq 0$. This leads to a dip around the zero delay ($\tau = 0$) in the time trace of $g^{(2)}(\tau)$, an effect known as *antibunching* [25]. The technique has been applied to observe the photon antibunching in the fluorescence of single color centers in diamonds for single-photon source applications [20, 21]. In a system with N independent and identical emitters, the autocorrelation function at the zero lag time is [23]

$$g^{(2)}(0) = 1 - \frac{1}{N}. \tag{7.4}$$

The depth of the dip thus provides a measure for the number of NV⁻ centers.

In addition to the interferometer, the instrument shown in Figure 7.4a also consists of a confocal fluorescence microscope integrated with an *atomic force microscope* (AFM). The AFM records topography images of the sample (Figure 7.4b), providing a

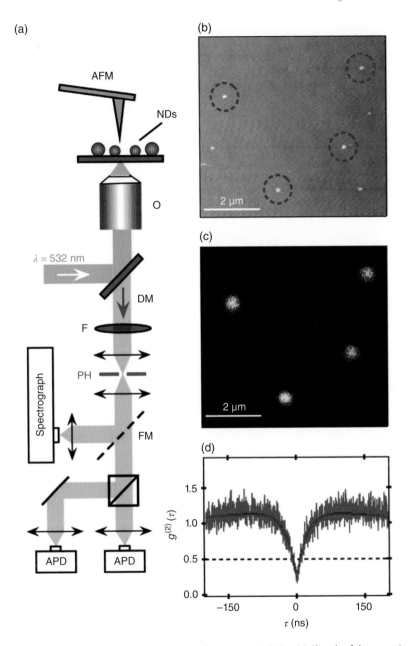

Figure 7.4 Measuring the numbers of color centers in FNDs. (a) Sketch of the experimental setup combining an atomic force microscope (AFM) with an inverted confocal microscope: NDs, nanodiamonds; O, oil immersion microscope objective; DM, dichroic beamsplitter; F, long-pass filter; PH, pinhole; FM, flip mirror directing the collected photoluminescence either to an imaging spectrograph equipped with a back-illuminated cooled charge-coupled-device array, or to a Hanbury–Brown–Twiss interferometer consisting of two silicon avalanche photodiodes (APDs) placed on the output ports of a 50/50 beamsplitter. (b, c) Topography image (b) and photoluminescence raster scan (c) of the same sample. Red dotted circles indicate NDs hosting NV$^-$ defects. (d) Typical second-order autocorrelation function $g^{(2)}(\tau)$ recorded for a single NV$^-$ defect. *Source:* Reprinted with permission from Ref. [22]. Reproduced with permission of American Physical Society.

direct in situ measurement of the FND size, whereas the fluorescence microscope identifies the particles (Figure 7.4c) and determines the number of NV⁻ by measuring the histograms of the photon correlations as a function of time delay. For each FND, both the photon antibunching and the particle size are measured, allowing researchers to correlate the size with the number of NV⁻ centers. Figure 7.4d shows an example of the anticorrelation effect near the zero delay. The result of $g^{(2)}(0) < 0.5$ reveals that there is only a single-photon emitter in the sample. The method provides accurate measures of the color center numbers only at $N \leq 10$ and, therefore, is best suited for the characterization of FNDs of 20 nm or smaller in size [26, 27].

7.2 Single Particle Tracking

7.2.1 Tracking in Solution

Now that we have learned the methods of detecting single FNDs deposited on a glass substrate, a stationary platform as described in the previous section, the next question is how to detect them in solution, an environment that is more relevant to practical uses? Perhaps, what we should really be asking is "Can FNDs be tracked individually in solution?" Performing the tracking in solution is challenging because particles are mobile the entire time due to, among others, the Brownian motions. Therefore, it would require a continuous monitoring of FNDs over the three dimensions (3D) to avoid losing the track of fluorescence signals.

SPT is a technique developed to observe the motion of a single molecule (or particle) in a medium where the coordinates (x, y, z) of the target molecule (or particle) in space are recorded for a series of time steps. Based on the measured 3D trajectory $(x(t), y(t), z(t))$ over time, the *mean square displacements* (MSDs) are calculated as [28]

$$\text{MSD}(n\delta t) = \frac{1}{N-1-n} \sum_{j=1}^{N-1-n} \left\{ \left[x(j\delta t + n\delta t) - x(j\delta t) \right]^2 + \left[y(j\delta t + n\delta t) - y(j\delta t) \right]^2 + \left[z(j\delta t + n\delta t) - z(j\delta t) \right]^2 \right\},$$

(7.5)

where N is the total number of frames in the video-recording sequence, and n and j are positive integers with n determining the time increment as $\Delta t = n\delta t$. In the simplified cases where molecules (or particles) undergo random diffusion and/or systematic transport at uniform velocity in the medium, the MSD in Eq. (7.5) becomes [29]

$$\text{MSD}(\Delta t) = 6D\Delta t + v^2 (\Delta t)^2,$$

(7.6)

where D is the diffusion coefficient and v is the drift rate.

Experimentally, a viable way to perform real-time 3D-SPT of a fluorescent bead in solution is to use a microscopy objective mounted vertically on a z-motion piezoelectric translational stage and an *electron-multiplying charge-coupled device* (EMCCD) camera to acquire the images (Figure 7.5a). The fluorescence image data are then analyzed in real time with computer software. Chang et al. [30] developed such software, which adjusted the image plane of the microscope objective lens by changing the position of

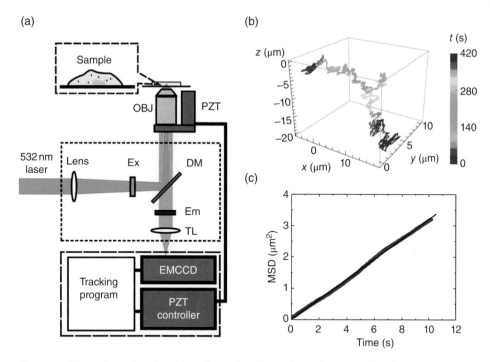

Figure 7.5 Three-dimensional tracking of a single FND in solution by a wide-field fluorescence microscope. (a) Schematic diagram of the experimental setup: OBJ, objective; Ex, excitation filter; DM, dichroic mirror; Em, emission filter; TL, tube lens; EMCCD, electron-multiplying charge-coupled device; PZT, piezoelectric translational stage. (b) Trajectory of a single FND moving freely in 80% glycerol–water solution. Time traces are shown in pseudo-color. (c) Mean square displacement (MSD) of the tracked FND. Black line is the best fit of the experimental data to Eq. (7.6) in text.

the piezoelectric translational stage to maintain the maximum intensity and contrast of the image. Continuous monitoring of the targeted nanoprobe was enabled by the servo-control feedback system that could track the depth of field of the microscope objective for hours. Figure 7.5b displays a typical 3D trajectory of a single FND particle suspended in 80% glycerol–water solution with a viscosity of 60.1 cP [31]. A diffusion coefficient of $D = 5.85 \times 10^{-2}\,\mu m^2\,s^{-1}$ was determined from its trajectory with the drift rate close to 0 (Figure 7.5c). The size of this tracked particle was 127 nm in diameter as calculated from the Stokes–Einstein equation [32]

$$D = \frac{k_B T}{6\pi\eta a},\tag{7.7}$$

where k_B is the Boltzmann constant, T is the absolute temperature, η is the viscosity, and a is the particle radius.

Another way to carry out SPT is offered by confocal fluorescence microscopy, equipped with a feedback-based control of a three-axis piezo-scanning stage mounted with a 100× microscope objective [33]. A user-developed software maintains constant fluorescence intensities at the center ($x = y = 0$) as well as four selected points (e.g. $x = \pm 50\,nm$ and $y = \pm 50\,nm$) of the fluorescence spots observed with an EMCCD camera.

Real-time tracking of the movement of an interrogated FND can be made indefinitely as long as it moves less than the maximum tracking speed and stays within the observation volume, which is typically $30 \times 30 \times 10\,\mu m^3$.

7.2.2 Tracking in Cells

A cell is a complex and dynamic biological system. Over the past few decades, a number of microscopic techniques have been developed to study the biological dynamics in cells at the molecular level, including SPT, *fluorescence correlation spectroscopy* (FCS), and fluorescence recovery after photobleaching, etc. [34]. Prevailing over other methods, SPT is capable of direct monitoring of the movement of molecules or vesicles in biological matrices [35, 36]. Important information pertaining to Brownian diffusions, confined motions, and directed transports can be deduced from the measured trajectories. The technique enables high-precision longitudinal tracking of the fate of the molecules or vesicles transported in cells as well as whole organisms. First developed in the late 1980s, SPT in combination with superresolution fluorescence imaging (cf., Section 10.2) has become one of the key technologies for quantitative analysis of intra- and inter-cellular processes on the nanoscale today [37].

Fu et al. [7] were among the first ones to conduct SPT of FNDs in live cells. They fed HeLa cells with FND particles of 35 nm in diameter and tracked their motions individually inside the cells in two dimensions. Later, the group applied the 3D-SPT technique (Figure 7.5a) to monitor a single FND particle inside the HeLa cell in three dimensions (Figure 7.6a) [30]. Analysis of the trajectories for more than 200 seconds found a diffusion coefficient of $D = 3.1 \times 10^{-3}\,\mu m^2\,s^{-1}$ (Figure 7.6b), a slow diffusion rate that compares well with those of other nanoparticles such as quantum dots confined within endosomal or lysosomal compartments. Further applications of the technique allowed the research team to observe the receptor-mediated endocytosis of single folate-conjugated FNDs by HeLa cells in real time [38]. The tracking, carried out continuously over 400 seconds, revealed confined diffusion of the targeted particles on cell membrane and active transport of the same particles in the cells.

The slow movement of FNDs inside living cells opens an opportunity to track their rotational motions over time. This is achievable by recording the spin resonance spectra of the individual FND particles using the *optically detected magnetic resonance* (ODMR) technique. As discussed in Section 3.4, in the presence of an external magnetic field, the ODMR peaks of a NV$^-$ center will split into two Zeeman components ($m_s = +1$ and $m_s = -1$) and the magnitude of the splitting depends on the orientation of the N–V axis with respect to the direction of the magnetic field. From a detailed analysis of the splittings in the ODMR spectra as a function of time, information about the rotational dynamics of the FND particles can be deduced. McGuinness et al. [39] performed a proof-of-principle experiment by applying a uniform magnetic field to 50-nm FNDs internalized by HeLa cells. Much credited to the enduring photostability of the NV$^-$ centers, the researchers were able to monitor continuously the reorientational dynamics of these FNDs entrapped in the endosomal or lysosomal compartments over 16 hours (cf., captions in Figure 7.7 for details).

The few examples cited above, and many more, have proven that FND is an effective fluorescent marker for intracellular tracking. Now, can we expand the same strategy to study intercellular activities? Furthermore, for practical applications, can FND do more

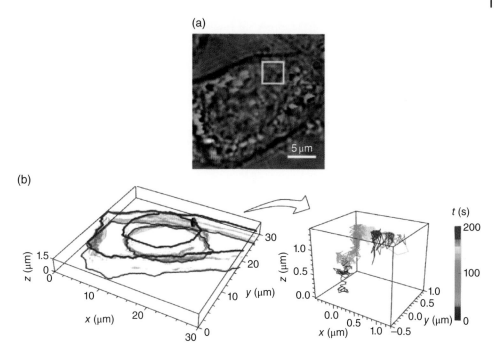

Figure 7.6 Three-dimensional tracking of a single FND in a living cell. (a) Merged bright-field and epifluorescence (red) image of the cell after FND uptake. (b) Location (left) and 3D trajectory (right) of the single FND surrounded by a yellow box in (a) over a time span of 200 seconds. *Source:* Adapted with permission from Ref. [30]. Reproduced with permission of Nature Publishing Group.

than just being a marker? For example, can FND serve as a biocompatible vehicle to transport proteins between cells?

To observe the events associated with intercellular dynamics, Epperla et al. [40] applied the FND-based SPT technique to track endosomal vehicles transported through membrane *tunneling nanotubes* (TNTs) formed between cells such as human embryonic kidney cells (HEK293T). Discovered only recently, TNTs are intercellular communication pathways through which the transport of proteins and other cytoplasmic components can occur. The nanotubes are typically 30 μm in length and 1 μm in diameter, which is large enough for 100-nm FNDs to travel through. In this study, the surface of FNDs was first modified by covalent conjugation with positively charged poly-arginine (PA). Enhanced green fluorescent protein (EGFP) was then noncovalently attached to the PA-grafted FND to form an EGFP–PA–FND complex, which was finally coated with bovine serum albumin (BSA) to avoid aggregation in cell medium, a step essential to ensure the internalization of EGFP–PA–FNDs by cells in isolated form.

Figure 7.8a and b show an example of the TNT formation and the active transport of FND-containing vehicles through the intercellular connection of HEK293T cells, observed by confocal fluorescence microscopy. Two-color imaging allowed simultaneous observations of EGFP and NV⁻ fluorescence, indicating that FNDs could mediate protein transport from the cytoplasm of one cell towards the other through the TNTs. Interestingly, the EGFP-loaded FNDs exhibited a molecular-motor-mediated motion inside the nanotubes with stop-and-go phases. Moreover, some particles moved along

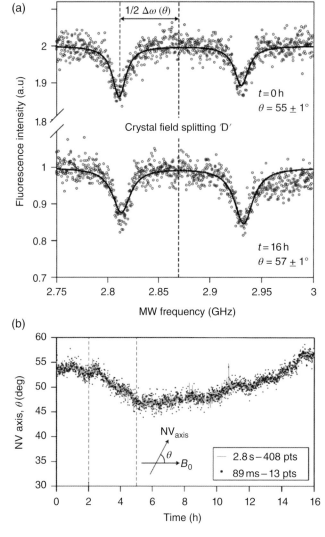

Figure 7.7 Orientation tracking of a single FND in a living cell. (a) Changes in orientation of the NV crystallographic axis (also the NV⁻ spin orientation) relative to an external magnetic field. (b) Measurement of the orientation of the internalized FND as a function of time. *Source:* Adapted with permission from Ref. [39]. Reproduced with permission of Nature Publishing Group.

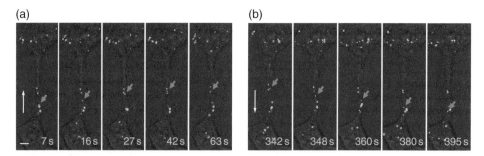

Figure 7.8 Tracking of a single EGFP-conjugated FND particle along the connecting membrane nanotube of two living cells. (a) Forward and (b) backward movement of two different particles. White arrows indicate the directions of the particle movement and orange arrows denote the particles of interest in the frames. Scale bar: 5 μm. *Source:* Reprinted with permission from Ref. [40]. Reproduced with permission of John Wiley & Sons.

the same TNT in both directions, i.e. the to-and-fro motions. Following a detailed MSD analysis (Eq. (7.5)), a diffusion coefficient of $D = 0.08\,\mu\text{m}^2\,\text{s}^{-1}$ and an active transport velocity of $v = 0.54\,\mu\text{m s}^{-1}$ were determined for BSA-coated FNDs. The studies demonstrate that surface-functionalized FNDs are capable of delivering biomolecules from one cell to another, while the entire delivery process can be monitored continuously over an extended period of time.

An innovative application of the FND-based SPT technique is to investigate intraneuronal transport abnormalities induced by brain disease-related genetic risk factors (Figure 7.9a) [41]. The investigation was motivated by the fact that common brain diseases, such as autism and Alzheimer's disease, strike more than 1% of the world's population. The diseases have a highly polygenic architecture and display subtle changes in gene expression. Studies have shown that the abnormalities in intraneuronal transport are linked to the genetic risk factors found in patients [42]. However, the current methods used to measure the neuronal transport process cannot detect these slight structural modifications. In need of a highly sensitive detection and tracking technique, Haziza et al. [41] developed a novel approach based on SPT of FNDs inside branches of dissociated neurons. A combination of bright emission, extraordinary photostability, and the absence of cytotoxicity of the nanoprobe made it possible to perform a high-throughput long-term tracking in living neurons with a 12-nm spatial resolution in a 50-ms time interval (Figure 7.9b and c). The research team applied the FND-tracking assay on two transgenic mouse lines mimicking the slight protein concentration changes (~30%) found in the brains of patients. In both cases, the FND-enabled method proved to have sufficient sensitivity for detecting any subtle changes in the intraneuronal transport.

In addition to SPT, FCS also serves as a useful tool to probe FNDs with high contrast [43–45]. Instead of measuring the autocorrelation function of the fluorescence intensity itself (Eq. (7.3)), the technique analyzes the fluctuation of the fluorescence intensity as [46]:

$$G(\tau) = \frac{\langle \delta I(t)\delta I(t+\tau)\rangle}{\langle I(t)\rangle^2} = \frac{\langle I(t)I(t+\tau)\rangle}{\langle I(t)\rangle^2} - 1, \tag{7.8}$$

where $\delta I(t) = I(t) - \langle I(t)\rangle$ is the deviation from the mean intensity. For a single fluorescent particle suspended in solution and illuminated by a tightly focused continuous-wave laser with a Gaussian beam profile, the autocorrelation function is given by [46]

$$G(\tau) = \frac{1}{1+\dfrac{\tau}{\tau_\text{D}}} \times \frac{1}{\sqrt{1+\dfrac{\tau}{s^2\tau_\text{D}}}}, \tag{7.9}$$

where τ_D is the average diffusion time of the particle in the focal volume and s is the structure parameter determined by calibration against fluorescent molecules (such as rhodamine B) of known sizes. Figure 7.10 shows typical FCS curves of 40-nm FNDs inside a live HeLa cell [44]. The fluorescence intensity data were collected by two-photon excitation of the particles with a femtosecond 1060 nm laser (cf., Section 8.2.3 for further discussion). With this technique, Hui et al. [44] found that the diffusion of the particles in the cytoplasm of a HeLa cell could be enhanced by more than one order of magnitude when the 40-nm FNDs were properly encapsulated in a lipid layer (Section 4.3.1).

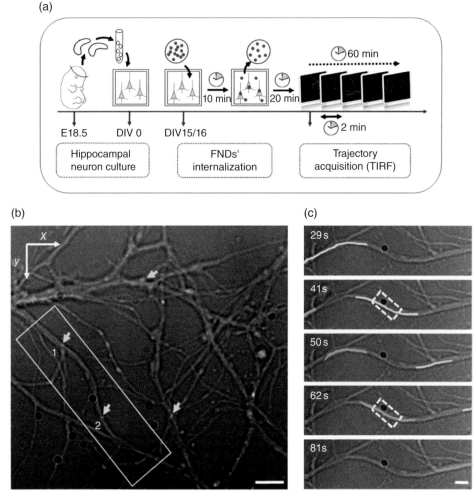

Figure 7.9 Intraneuronal transport monitoring by FND tracking. (a) Experimental pipeline from hippocampal neuron culture dissociated from E18.5 mouse embryo to endosome trajectory acquisition using pseudo-TIRF microscopy. DIV, day *in vitro*; TIRF, total internal reflection fluorescence. (b) Transmission white-light illumination image of the neuronal branches merged with the fluorescence channel showing four FNDs moving within dendrites (yellow arrows). (c) Superimposition onto a white light image of the positions of these two FNDs (1 in yellow; 2 in green), determined by particle tracking, with a persistence of 10 seconds, at different time points. Scale bars: 5 μm in (b) and 1 μm in (c). *Source:* Reprinted with permission from Ref. [41]. Reproduced with permission of Nature Publishing Group.

Finally, one may wonder if it is worthy of combining both FCS and SPT together. Putting technical difficulties aside, will the combined techniques provide us a powerful tool in exploring more dynamics details at the level of a single nanoparticle in a cell? It turns out that the combination of FCS and SPT has enabled researchers to probe both short-term and long-term dynamics of the individual lipid-coated FNDs in aqueous solution and in complex biological medium [44]. An example of this is given by Liu et al. [45]

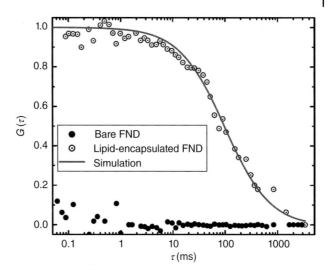

Figure 7.10 Two-photon fluorescence correlation spectroscopy of bare and lipid-encapsulated 40-nm FNDs in live HeLa cells. The solid line is the simulated result using Eq. (7.9) in text. *Source:* Reprinted with permission from Ref. [44].

who conjugated FNDs with transforming growth factor beta (TGF-β), a multifunctional cytokine (Section 5.1). They showed that these bioconjugated nanoparticles could be directed specifically towards membrane TGF-β receptors. With FCS, the researchers first estimated the average diameter of the BSA-coated TGF-FND particles to be 46 nm. By using small molecule inhibitors and monitoring their effects on the 3D motions of these receptors in human lung cancer cells (HCC827) with SPT, the immobilized TGF-β was found to be essential for active signaling, suggesting that the tagging of TGF-β by FND with a size in the range of 40 nm did not significantly alter the outcome of the transmembrane signaling dynamics in living cells.

7.2.3 Tracking in Organisms

Living organisms provide another challenging platform to conduct SPT. An early attempt by Chang et al. [47] employed 100-nm FNDs as nanoprobes to explore the cytoplasmic dynamics in zebrafish yolk cells. Introducing BSA-coated FNDs into the yolk of a zebrafish embryo at the one-cell stage, the researchers found that the microinjected FNDs traveled unidirectional in a stop-and-go traffic within the yolk cell with a velocity of 0.19–0.40 μm s^{-1} along the axial streaming. Interestingly and importantly, these particles could be incorporated into the dividing cells and dispersed in the fish's body when the embryos developed into larvae and subsequently into adult fishes. No apparent abnormalities were detected for the FND-treated fishes throughout their embryogenesis, which supported the notion that the FNDs are useful as photostable, biocompatible, and nontoxic markers in small animal models.

The SPT technique has also been applied to investigate the embryogenesis of the fruit fly, *Drosophila melanogaster*, which is another model organism for genetic studies. Using FNDs coated with BSA to avoid aggregation in cell medium, Simpson et al. [48] microinjected the particles into the embryos to investigate the FND diffusion in both furrow periplasm and subnuclear periplasms by SPT (Figure 7.11). Specifically, they observed both the free diffusion and driven motion of the particles in the blastoderm cells at the posterior end of the *Drosophila* embryos during cellularization. From MSD

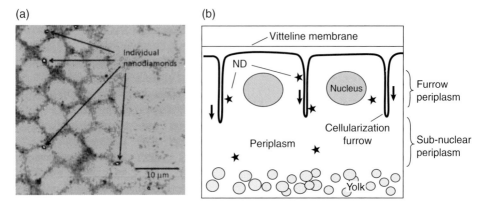

Figure 7.11 Tracking of single FNDs in a *Drosophila* embryo. (a) Confocal fluorescence image of FNDs in the blastoderm cells during stage 5 of development. (b) Schematic drawing of the structure of an early stage 5 embryo, showing the cellularization furrows introgressing between nuclei, which invade the yolk-free periplasm during the later syncytial divisions indicated as arrows. *Source:* Adapted with permission from Ref. [48].

analysis of the trajectories for 130-nm FND particles, a mean diffusion coefficient of $6 \times 10^{-3}\,\mu m^2\,s^{-1}$ and a mean driven velocity of $0.13\,\mu m\,s^{-1}$ in the furrow periplasm were determined. In the subnuclear periplasm region, the corresponding diffusion coefficient and driven velocity were $6.3 \times 10^{-2}\,\mu m^2\,s^{-1}$ and $0.27\,\mu m\,s^{-1}$, respectively. The differences between these values corresponded well to the observation that the cytoskeletal networks were less congested in the subnuclear periplasm than in the furrow periplasm, where the diffusion of FNDs could be significantly hindered.

Pushing the frontier forward, Harada, Shirakawa, and coworkers made the first demonstration for SPT of FNDs in whole organisms based on the ODMR technique [49]. Specifically, the research team injected FNDs into two model organisms: *Caenorhabditis elegans* (*C. elegans*) and mice. They then incorporated 2.87-GHz microwave radiation that matches the spin resonance of the NV⁻ centers into the injected FNDs to induce a fluorescence intensity change by approximately 10% (Figure 3.9). Following the acquisition for a pair of wide-field fluorescence images with the microwave radiation on and off, pixel-by-pixel subtraction between these two images allowed the removal of background signals. The researchers utilized this background-free technique to conduct long-term tracking of single 200-nm FNDs in *C. elegans* (Section 5.4), achieving high image contrast. Additionally, the rotational motions of the particles inside the worms were also observed [50]. Their results showed that the technique was capable of eliminating nearly entire extraneous fluorescence, making it possible for real-time tracking of single FNDs in whole organisms. This selective imaging method is expected to be applicable to a wide range of biological systems both *in vitro* and *in vivo*.

An alternative approach to achieve background-free detection of FNDs in whole organisms for SPT is by time gating as discussed in Section 7.1.1. Kuo et al. [51] first demonstrated the feasibility by conducting *in vivo* imaging of single FNDs in *C. elegans*. They then tracked in real time the intercellular transport of yolk lipoproteins conjugated with FNDs in the same organism [52]. The work started with noncovalent coating of 100-nm FNDs with green fluorescent protein (GFP)-tagged yolk lipoprotein

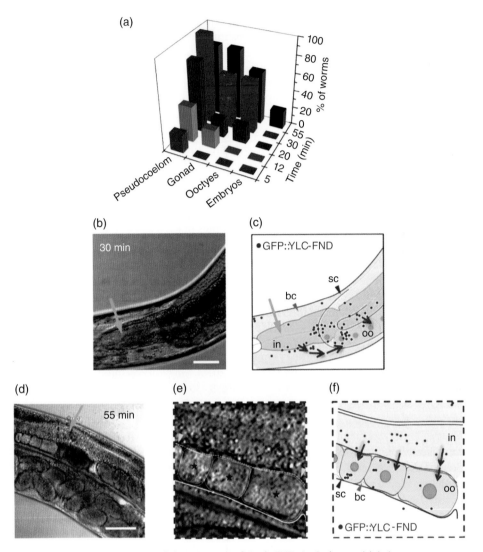

Figure 7.12 Tracking the intercellular transport of single FNDs in *C. elegans*. (a) A time-course localization analysis of GFP::YLC-FNDs in specific tissues and organs of the worms over 55 minutes post-injection. (b) A merged bright-field and time-gated fluorescence image of a representative worm at 30 minutes after injection of GFP::YLC-FNDs into the intestinal cell. The blue arrow indicates the injection site. (c) A cartoon illustrating the excretion of GFP::YLC-FNDs from the intestine (in) to the body cavity (bc) and after passing through the sheath cells (sc), the excreted GFP::YLC-FNDs enter oocytes (oo). (d) A merged bright-field and time-gated fluorescence image of a representative worm at 55 minutes after injection of GFP::YLC-FNDs into the intestinal cell. (e, f) An enlarged image (e) and a cartoon (f) of the area in the red box in (d), showing the presence of GFP::YLC-FNDs in both intestine (with a boundary marked by the yellow dotted line) and oocytes (with boundaries marked by thin white lines and nuclei labeled with blue stars). Scale bars: 50 μm. *Source:* Reprinted with permission from Ref. [52]. Reproduced with permission of Elsevier.

complexes (YLCs) by physical adsorption to form complexes (denoted as GFP::YLC-FNDs). After microinjection of the complexes into the anterior intestinal cells near the pharynx, time-lapse images were acquired by gating the fluorescence signals at the time longer than 10 ns, which greatly enhanced the image contrast of FNDs through background removal. A total of more than 25 worms were examined individually to identify the location of GFP::YLC-FNDs in the specific cells or organs over a 55-min time period after injection (Figure 7.12). It was discovered that immediately following administration, the GFP::YLC-FND particles first emerged in the posterior intestine, followed by the appearance in the body cavity and the loop region of the gonad at the post-injection time of 5 and 12 min, respectively. Twenty minutes later, the NV⁻ fluorescence became visible in the oocytes. Finally, the embryos bearing FNDs could be found at 55 min after injection, indicating that the FNDs did not affect the fertilization ability of the oocytes or the subsequent development of the zygotes into embryos.

In this chapter, we have surveyed several examples in each category of tracking single FNDs, whether it is in cells or organisms. The results all point to one conclusion that FNDs hold an excellent promise as biocompatible and specific endogenous protein tags to study intracellular and intercellular processes in living organisms. There are more to come in the following chapters.

References

1 Moerner, W.E., Shechtman, Y., and Wang, Q. (2015). Single-molecule spectroscopy and imaging over the decades. *Faraday Discuss* 184: 9–36.

2 Orrit, M. and Bernard, J. (1990). Single pentacene molecules detected by fluorescence excitation in a p-terphenyl crystal. *Phys Rev Lett* 65: 2716–2719.

3 Gruber, A., Drabenstedt, A., Tietz, C. et al. (1997). Scanning confocal optical microscopy and magnetic resonance on single defect centers. *Science* 276: 2012–2014.

4 Schirhagl, R., Chang, K., Loretz, M., and Degen, C.L. (2014). Nitrogen-vacancy centers in diamond: nanoscale sensors for physics and biology. *Annu Rev Phys Chem* 65: 83–105.

5 Pawley, J.B. (ed.) *Handbook of Biological Confocal Microscopy*, 3e. Berlin: Springer.

6 Hui, Y.Y., Hsiao, W.W.W., Haziza, S. et al. (2017). Single particle tracking of fluorescent nanodiamonds in cells and organisms. *Curr Opin Solid State Mater Sci* 21: 35–42.

7 Fu, C.C., Lee, H.Y., Chen, K. et al. (2007). Characterization and application of single fluorescent nanodiamonds as cellular biomarkers. *Proc Natl Acad Sci USA* 104: 727–732.

8 Colomb, W., Czerski, J., Sau, J.D., and Sarkar, S.K. (2017). Estimation of microscope drift using fluorescent nanodiamonds as fiducial markers. *J Microsc* 266: 298–306.

9 Chang, C.W., Sud, D., and Mycek, M.A. (2007). Fluorescence lifetime imaging microscopy. *Method Cell Biol* 81: 495–524.

10 Faklaris, O., Garrot, D., Joshi, V. et al. (2008). Detection of single photoluminescent diamond nanoparticles in cells and study of the internalization pathway. *Small* 4: 223–2239.

11 Mohtashami, A. and Koenderink, A.F. (2013). Suitability of nanodiamond nitrogen-vacancy centers for spontaneous emission control experiments. *New J Phys* 15: 043017.

12 Lakowicz, J.R. (2006). *Principles of Fluorescence Spectroscopy*, 3e. Springer.

13 Wee, T.L., Tzeng, Y.K., Han, C.C. et al. (2007). Two-photon excited fluorescence of nitrogen-vacancy centers in proton-irradiated type Ib diamond. *J Phys Chem A* 111: 9379–9386.

14 Chapman, R. and Plakhotnik, T. (2011). Quantitative luminescence microscopy on nitrogen-vacancy centers in diamond: saturation effects under pulsed excitation. *Chem Phys Lett* 507: 190–194.

15 Plakhotnik, T. and Aman, H. (2018). NV-centers in nanodiamonds: how good they are. *Diam Relat Mater* 82: 87–95.

16 Sjoback, R., Nygren, J., and Kubista, M. (1995). Absorption and fluorescence properties of fluorescein. *Spectrochimica Acta Part A* 51: L7–L21.

17 Moulton, P.F. (1986). Spectroscopic and laser characteristics of Ti:Al$_2$O$_3$. *J Opt Soc Am B* 3: 125–133.

18 Hui, Y.Y., Chang, Y.R., Lim, T.S. et al. (2009). Quantifying the number of color centers in single fluorescent nanodiamonds by photon correlation spectroscopy and Monte Carlo simulation. *Appl Phys Lett* 94: 013104.

19 Hui, Y.Y., Chang, Y.R., Mohan, N. et al. (2011). Polarization modulation spectroscopy of single fluorescent nanodiamonds with multiple nitrogen-vacancy centers. *J Phys Chem A* 115: 1878–1884.

20 Brouri, R., Beveratos, A., Poizat, J.P., and Grangier, P. (2000). Photon antibunching in the fluorescence of individual color centers in diamond. *Opt Lett* 25: 1294–1296.

21 Kurtsiefer, C., Mayer, S., Zarda, P., and Weinfurter, H. (2000). Stable solid-state source of single photons. *Phys Rev Lett* 85: 290–293.

22 Rondin, L., Dantelle, G., Slablab, A. et al. (2010). Surface-induced charge state conversion of nitrogen-vacancy defects in nanodiamonds. *Phys Rev B* 82: 115449.

23 Kitson, S.C., Jonsson, P., Rarity, J.G., and Tapster, P.R. (1998). Intensity fluctuation spectroscopy of small numbers of dye molecules in a microcavity. *Phys Rev A* 58: 620–627.

24 Arecchi, F.T., Corti, M., Degiorgio, V., and Donati, S. (1971). Measurements of light intensity correlations in the subnanosecond region by photomultipliers. *Opt Commun* 3: 284–288.

25 Kimble, H.J., Dagenais, M., and Mandel, L. (1977). Photon antibunching in resonance fluorescence. *Phys Rev Lett* 39: 691–695.

26 Mohan, N., Tzeng, Y.K., Yang, L. et al. (2010). Sub-20-nm fluorescent nanodiamonds as photostable biolabels and fluorescence resonance energy transfer donors. *Adv Mater* 22: 842–847.

27 Tisler, J., Balasubramanian, G., Naydenov, B. et al. (2010). Fluorescence and spin properties of defects in single digit nanodiamonds. *ACS Nano* 3: 1959–1965.

28 Kusumi, A., Sako, Y., and Yamamoto, M. (1993). Confined lateral diffusion of membrane receptors as studied by single particle tracking (nanovid microscopy). Effects of calcium-induced differentiation in cultured epithelial cells. *Biophys J* 65: 2021–2040.

29 Qian, H., Sheetz, P., and Elson, E.L. (1991). Single-particle tracking: analysis of diffusion and flow in 2-dimensional systems. *Biophys J* 60: 910–921.

30 Chang, Y.R., Lee, H.Y., Chen, K. et al. (2008). Mass production and dynamic imaging of fluorescent nanodiamonds. *Nat Nanotechnol* 3: 284–288.

31 Dorsey, N.E. (1940). *Properties of Ordinary Water-Substance*, 184. New York: Reinhold.

32 Atkins, P. and de Paula, J. (2006). *Atkins' Physical Chemistry*. Oxford: Oxford University Press.

33 Tzeng, Y.K., Tsai, P.C., Liu, H.Y. et al. (2015). Time-resolved luminescence nanothermometry with nitrogen-vacancy centers in nanodiamonds. *Nano Lett* 15: 3945–3952.

34 Lippincott-Schwartz, J., Snapp, E., and Kenworthy, A. (2001). Studying protein dynamics in living cells. *Nat Rev Mol Cell Biol* 2: 444–456.

35 Saxton, M.J. and Jacobson, K. (1997). Single-particle tracking: applications to membrane dynamics. *Annu Rev Biophys Biomol Struct* 26: 373–399.

36 Levi, V. and Gratton, E. (2007). Exploring dynamics in living cells by tracking single particles. *Cell Biochem Biophys* 48: 1–15.

37 Cognet, L., Leduc, C., and Lounis, B. (2014). Advances in live-cell single-particle tracking and dynamic super-resolution imaging. *Curr Opin Chem Biol* 20: 78–85.

38 Zhang, B., Li, Y., Fang, C.Y. et al. (2009). Receptor-mediated cellular uptake of folate-conjugated fluorescent nanodiamonds: a combined ensemble and single-particle study. *Small* 5: 2716–2721.

39 McGuinness, L.P., Yan, Y., Stacey, A. et al. (2011). Quantum measurement and orientation tracking of fluorescent nanodiamonds inside living cells. *Nat Nanotechnol* 6: 358–363.

40 Epperla, C.P., Mohan, N., Chang, C.W. et al. (2015). Nanodiamond-mediated intercellular transport of proteins through membrane tunneling nanotubes. *Small* 11: 6097–6105.

41 Haziza, S., Mohan, N., Loe-Mie, Y. et al. (2017). Fluorescent nanodiamond tracking reveals intraneuronal transport abnormalities induced by brain-disease-related genetic risk factors. *Nat Nanotechnol* 12: 322–328.

42 Millecamps, S. and Julien, J.P. (2013). Axonal transport deficits and neurodegenerative diseases. *Nat Rev Neurosci* 14: 161–176.

43 Neugart, F., Zappe, A., Jelezko, F. et al. (2007). Dynamics of diamond nanoparticles in solution and cells. *Nano Lett* 7: 3588–3591.

44 Hui, Y.Y., Zhang, B.L., Chang, Y.C. et al. (2010). Two-photon fluorescence correlation spectroscopy of lipid-encapsulated fluorescent nanodiamonds in living cells. *Opt Express* 18: 5896–5905.

45 Liu, W., Yu, F., Yang, J. et al. (2016). 3D single-molecule imaging of transmembrane signaling by targeting nanodiamonds. *Adv Funct Mater* 26: 365–375.

46 Schwille, P., Haupts, U., Maiti, S., and Webb, W.W. (1999). Molecular dynamics in living cells observed by fluorescence correlation spectroscopy with one- and two-photon excitation. *Biophys J* 77: 2251–2265.

47 Chang, C.C., Zhang, B., Li, C.Y. et al. (2012). Exploring cytoplasmic dynamics in zebrafish yolk cells by single particle tracking of fluorescent nanodiamonds. *Proc SPIE* 8272: 827205.

48 Simpson, D.A., Thompson, A.J., Kowarsky, M. et al. (2014). *In vivo* imaging and tracking of individual nanodiamonds in drosophila melanogaster embryos. *Biomed Opt Express* 5: 1250–1261.

49 Igarashi, R., Yoshinari, Y., Yokota, H. et al. (2012). Real-time background-free selective imaging of fluorescent nanodiamonds *in vivo*. *Nano Lett* 12: 5726–5732.

50 Yoshinari, Y., Mori, S., Igarashi, R. et al. (2015). Optically detected magnetic resonance of nanodiamonds *in vivo*; implementation of selective imaging and fast sampling. *J Nanosci Nanotechnol* 15: 1014–1021.

51 Kuo, Y., Hsu, T.Y., Wu, Y.C. et al. (2013). Fluorescence lifetime imaging microscopy of nanodiamonds *in vivo*. *Proc SPIE* 8635: 863503.

52 Kuo, Y., Hsu, T.Y., Wu, Y.C., and Chang, H.C. (2013). Fluorescent nanodiamond as a probe for the intercellular transport of proteins *in vivo*. *Biomaterials* 34: 8352–8360.

8

Cell Labeling and Fluorescence Imaging

Nanocarbons are a family consisting of fullerenes, carbon nanotubes, graphenes, nanodiamonds, and other variations (Section 1.2). *Fluorescent nanodiamonds* (FNDs) joined the family in 2005 [1]. While receiving less attention than other members of the family, FNDs are gaining popularity as a novel nanoparticle platform for biomedical applications in recent years [2]. An ideal biomedical nanoparticle platform should possess three functionalities: targeting, imaging, and therapeutic [3]. Of course, the nanoparticles must be endowed with high biocompatibility and low cytotoxicity. As evident from all the discussions so far in this book, surface-functionalized FND clearly meets all these requirements [4]. Applying the platform to biomedical studies is expected to enhance our understanding of the pathophysiological basis of disease, open up more sophisticated diagnostic opportunities, and lead to improved therapies and treatments.

In the previous chapters, we have discussed some core features of FNDs, including their superior magneto-optical properties (Chapter 3), the ability to be functionalized with diversified bioactive groups (Chapter 4), the inherent biocompatibility (Chapter 5), as well as the high sensitivity detection at the single particle level (Chapter 7). This chapter focuses on arguably the most promising application of the FND-based platform in biomedicine: *cell labeling and imaging*. The modalities to achieve the fluorescence imaging of FND-labeled cells *in vitro* at various scales are presented. How the methods developed in this chapter can be applied for *in vivo* cell tracking will be discussed in the next chapter.

8.1 Cell Labeling

Before looking into any technical details, we should probably start with some basic questions about the subject: What is cell labeling? Why labeling? What can be achieved by labeling cells? How to label the cells? How does it fit in the current biomedical development? The following discussions intend to address these issues.

Cell labeling, in a layman's term, is conducted for the visualization of cells and cellular structures as well as the tracking and modulation of proteins, nucleic acids, and other components in cells. Fluorescence microscopy is one of the two key instruments used by biologists to characterize the outcomes of the labeling. The other is flow

Fluorescent Nanodiamonds, First Edition. Huan-Cheng Chang, Wesley Wei-Wen Hsiao and Meng-Chih Su.
© 2019 John Wiley & Sons Ltd. Published 2019 by John Wiley & Sons Ltd.

cytometry or *fluorescence-activated cell sorting* (FACS), which is a laser-based biophysical method for cell counting, cell sorting, and biomarker detection [5]. Unlike fluorescence microscopy that detects cells on a glass slide, FACS probes laser-induced fluorescence from suspended cells in a stream of solution passing through a capillary tube. It allows simultaneous multiparametric analysis of cells in high speed, about several thousand cells in minutes. Signals collected by FACS include forward-scattered light (FSC), side-scattered light (SSC), and side-fluorescence light (SFL). The FSC intensity is related to the cell size, the SSC intensity provides information about the cell content (such as nucleus and granules), and the SFL intensity is correlated with the amount of biomarkers present in the cells, which represents a quantitative measurement for the cell labeling efficiency. The technique provides complimentary information to those of fluorescence microscopy.

8.1.1 Nonspecific Labeling

A cell can be labeled with FNDs either specifically or nonspecifically. Surface modification (Chapter 4) plays a central role in determining these two types of labeling. FNDs with a size in the range of 100 nm are often chosen for the cell labeling application because these particles can be produced in a bulk quantity and their fluorescence intensities are sufficiently high for quantitative FACS analysis. Moreover, they are readily usable for nonspecific labeling right after purification by air oxidation and acid wash. As discussed in Section 4.1, FNDs after these two purification steps are surface-functionalized with a variety of oxygen-containing groups, including the carboxyl moiety. The carboxylated particles, which are negatively charged, are well dispersed in water at neutral pH (and, therefore, *hydrophilic*) with high stability. However, they are prone to aggregation in phosphate-buffered saline (PBS) and cell mediums [6] such as the Dulbecco's modified Eagle medium (DMEM), which contains amino acids, salts, glucose, and vitamins as the nutrients. Despite the aggregation problem, these surface-oxidized FND particles are highly biocompatible and useful for cell tracking as illustrated below and in the next chapter.

To conduct the labeling, FNDs after extensive acid washes are first incubated with cells in culture medium without fetal bovine serum (FBS) to facilitate their uptake by endocytosis [7, 8]. Cells are then detached from the dish surface using trypsin, followed by homogenization in PBS and fluorescence intensity measurement by FACS to quantify the amounts of FNDs internalized. Figure 8.1 shows a typical result of the flow cytometric analysis of FND-labeled HeLa cells, where the amount of FNDs endocytosed by the cells increased nearly linearly with the concentration ($1-100\,\mu g\,ml^{-1}$) of the particles in the medium [7]. The half-life (i.e. the time required to reach half of the maximum accumulation value) was approximately two hours, as revealed by a kinetics study of the uptake (inset of Figure 8.1). In contrast, if the cells were incubated with FNDs in cell medium containing 10% FBS under the same experimental conditions, the cellular uptake efficiency would decrease by about 10-fold, partially, due to severe agglomeration of the FND particles in the medium.

A distinct feature of this nanoparticle-based labeling method is that the surface-oxidized FNDs are spontaneously and avidly taken up by a wide range of cells in the absence of FBS and thus is categorized as *nonspecific labeling*. Virtually all adherent cells, including cancer and stem cell lines [7–28] as well as primary cells [28–32]

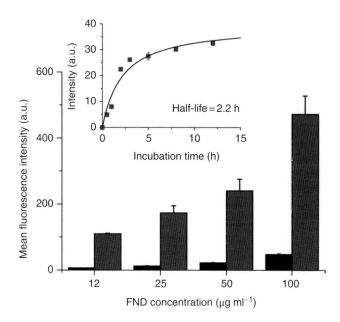

Figure 8.1 Flow cytometric analysis of 100-nm FNDs internalized by HeLa cells as a function of particle concentration after three hours of incubation with (black) or without (blue) 10% FBS in cell medium. Inset: Kinetics of the uptake of FNDs (without FBS in medium) at the particle concentration of 25 μg ml^{-1}. *Source:* Reprinted with permission from Ref. [7]. Reproduced with permission of John Wiley & Sons.

(Table 8.1), can be labeled with this method for flow cytometric analysis. To assess the health hazard of the labeling, Vaijayanthimala et al. [8] performed biocompatibility testing with mouse 3T3-L1 pre-adipocytes and 489-2 osteoprogenitors and found that the internalized 100-nm FND particles did not affect cellular proliferation when monitored at both clonal and population levels. More importantly, no detrimental effect on the *in vitro* adipogenic differentiation of 3T3-L1 cells and osteogenic differentiation of 489-2 cells was observed. In cases where FBS is absolutely required in medium during cell culture, the amount of uptake can be recovered by coating the FND surface with poly-L-lysine (PLL), which binds readily with the negatively charged cell membrane by electrostatic attractions [8]. Compared with other cationic polyelectrolytes such as polyethylenimine (PEI), PLL is more biocompatible and less toxic. Other variations of the nonspecific labeling method include coating FNDs with serum albumin to avoid their aggregation in cell medium (Section 4.2.1) so as to achieve more uniform labeling of the cells [19–21].

In FACS analysis of cells labeled with red FNDs (containing NV$^-$ centers), the fluorescence signals are typically collected in the far-red channel (wavelength > 600 nm), where the cell autofluorescence is weak. A signal-to-background ratio of 100 : 1 can be achieved when cells are labeled at a particle concentration of 100 μg ml^{-1} (Figure 8.2a). In addition to fluorescence emission, ironically, the light scattering from FNDs is also useful for cell tracking and sorting applications. Diamond has the highest refractive index of all transparent minerals (Table 2.1), with strong light scattering from ultraviolet to the infrared regions [33]. In an experiment to image diamond nanoparticles individually in mammalian epithelial cells, Smith et al. [34] estimated that the intensity of the

Table 8.1 Cells successfully labeled with FNDs by endocytosis [7–32].

Source	Cell line	Primary cell
Human	HeLa cervical cancer cells	Mesenchymal stem cells
	HEK293 embryonic kidney cells	Monocytes
	A547 alveolar basal epithelial cells	
	HepG2 liver cancer cells	
	U-87 MG glioblastoma	
	NT2/D1 embryonal carcinoma stem cells	
Mouse	3T3-L1 pre-adipocytes	Lung stem/progenitor cells
	489-2 osteoprogenitors	Hippocampal neurons
	MCF-7 breast cancer cells	
	MDA-MB-231 breast cancer cells	
	ASB145-1R breast cancer cells	
	P19 embryonal carcinoma stem cells	
	RAW264.7 cells	

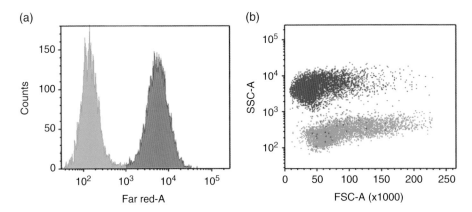

Figure 8.2 Flow cytometric analysis of the 1 : 1 mixture of FND-labeled (magenta) and unlabeled (green) 489-2.1 cells in the (a) far-red fluorescence and (b) side scattering channels. *Source:* Adapted with permission from Ref. [18]. Reproduced with permission of John Wiley & Sons.

elastically scattered light from a single 55-nm particle was about 300-fold greater than that of a cell organelle of similar size. The same finding was manifested in Figure 8.2b, where the internalization of the 100-nm FND particles markedly changed the light scattering characteristics of the cells, resulting in a dramatic increase in SSC and a concomitant decrease in FSC. The changes suggested that there was an increase in cellular granularity due to feeding of the 489-2 cells with FNDs. Such a significant enhancement in the SSC signals makes light scattering a convenient tool for the identification, isolation, and recovery of the FND-labeled stem cells from the recipient's tissue; all can be done on a standard cell-sorting machine.

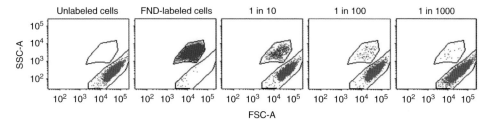

Figure 8.3 Flow cytometric analysis of FND-labeled and unlabeled 489-2.1 cells utilizing the light scattering property of FNDs at various cell number ratios of 1 : 10, 1 : 100, and 1 : 1000. Green data points represent the major live cell population and the blue data points represent the gated FND-positive cells. *Source:* Reprinted with permission from Ref. [18]. Reproduced with permission of John Wiley & Sons.

To further develop FNDs into a nanoparticle-based platform for practical biological applications, Vaijayanthimala et al. [8] prepared and tested mixtures of FND-labeled cells and unlabeled cells at the number ratio of 1 : 10, 1 : 100, and 1 : 1000. When using the far-red fluorescence channel for cell sorting, the researchers detected a non-negligible background (~1%) derived from the 594-nm excitation of the unlabeled live cells in the high fluorescence intensity region, which made sorting of the FND-labeled cells in the 1 : 1000 mixture exceedingly difficult. The SSC channel, in contrast, was relatively clean and free of background (<0.005%) from cell autofluorescence. By properly gating the signals in both FSC and SSC channels (Figure 8.3), the researchers were able to obtain a sorted population containing 83.2% viable FND-labeled cells from a starting 1 : 1000 mixed cell sample after one round of cell sorting. All of this was made possible by the unique photophysical and chemical properties of FNDs.

8.1.2   Specific Labeling

The surface of FNDs provides a versatile platform to conjugate various types of molecules through chemical modification [35]. What is more, because the color centers are deeply embedded in the diamond lattice, all the strategies described in Chapter 4 can be safely applied to FNDs without significantly altering the color centers' optical absorption and emission properties. Therefore, considerable research efforts have been made to conjugate bioactive molecules onto the FND surface for specific cell labeling and targeting [7, 16, 17, 21, 24, 25, 36, 37]. These molecules include folic acid [7] and transferrin [16, 17]. However, such conjugation often leads to aggregation of the bioconjugated FNDs in physiological medium such as PBS (Section 8.1.1) and the labeling that follows may turn out to be nonspecific. A way to circumvent this problem is to block the empty sites on the FND surface with bovine serum albumin (BSA), which is an excellent stabilizing agent in preventing nanoparticle aggregation (Section 4.2.1). A minor caution worthy of noting: Although the BSA coating can be simply made on the FND surface by physical adsorption, the protein molecule is relatively large (with a molecular weight of 66 kD) and thus may interfere with some specific labeling processes if the targeting ligands are significantly smaller in size.

Recognizing the unique characteristic of serum albumin, Chang et al. [21] coated the surface of FNDs with BSA chemically modified with carbohydrates (known as

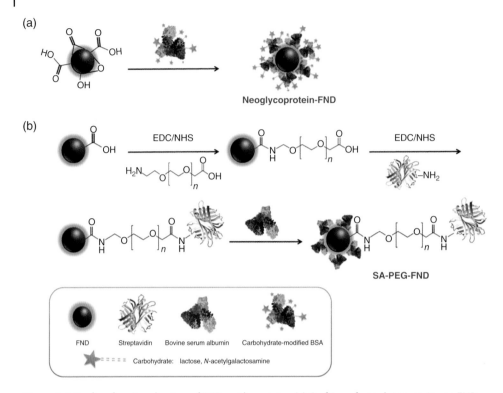

Figure 8.4 Surface functionalization of FNDs with proteins. (a) Grafting of neoglycoproteins on FND. The proteins are attached to the surface of acid-washed FND by physical adsorption. (b) Grafting of streptavidin on FND. The grafting starts with activation of the surface carboxyl groups on FND with EDC and NHS, forming amine-reactive terminus. The FND is then conjugated with carboxyl PEG amines via carboxyl-to-amine cross-linking. Further activation leads to covalent coupling between of the carboxyl groups of PEG-FND and the primary amine groups (–NH$_2$) of streptavidin through amide bond formation. Finally, the SA-conjugated PEG-FND is noncovalently covered with BSA to prevent particle aggregation in PBS. *Source:* Reprinted with permission from Ref. [21]. Reproduced with permission of John Wiley & Sons.

neoglycoproteins), such as galactose (Gal), *N*-acetylgalactosamine (GalNAc), and lactose (Lac), for hepatic targeting (Figure 8.4a). The coating, conducted by noncovalent conjugation, effectively prevented particle aggregation in PBS and cell medium, providing FNDs a high specific targeting ability for asialoglycoprotein receptors (ASGPRs) on HepG2 cells (a human hepatoma cell line). Flow cytometric analysis revealed that the amount of the particle uptake increased steadily with the increasing concentration of the bioconjugated FNDs (Figure 8.5). Only particles coated with the neoglycoproteins or glycoproteins could be effectively taken up by the cells (and, hence, *specific labeling*). At the concentration of 50 μg ml^{-1}, the mean fluorescence intensity of the HepG2 cells with the Lac-BSA-FND labeling was approximately 10-fold greater than that of the same cells labeled by BSA-FNDs alone. To further examine the labeling specificity, a competition assay was conducted using free ligands to inhibit the binding of Lac-BSA-FNDs with ASGPRs. As shown in Figure 8.6, the uptake of Lac-BSA-FND by HepG2 cells was significantly suppressed in the presence of lactose at various incubation times. At the particle concentration of 50 μg ml^{-1} and the incubation time of 2.5 hours, Lac-BSA-FND

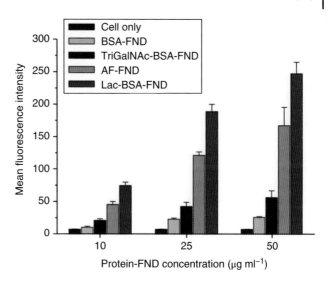

Figure 8.5 Flow cytometric analysis of the uptakes of protein-conjugated FNDs by HepG2 cells. Mean fluorescence intensities as a function of the concentration (0, 10, 25, and 50 μg ml⁻¹) of the labeling agent, as annotated in the figure, reflect different levels of the uptake of FNDs conjugated with BSA, neoglycoprotein, or glycoprotein. *Source:* Reprinted with permission from Ref. [21]. Reproduced with permission of John Wiley & Sons.

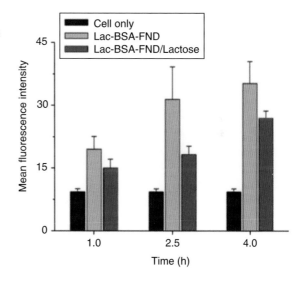

Figure 8.6 Competition assays for the uptake of Lac-BSA-FND by HepG2 cells measured by flow cytometry. The cellular uptake of Lac-BSA-FND (50 μg ml⁻¹) is significantly suppressed due to the presence of lactose (0.3 M), showing the competition for binding with ASGPRs by free lactose. *Source:* Reprinted with permission from Ref. [21]. Reproduced with permission of John Wiley & Sons.

alone displayed a mean fluorescence intensity of roughly 30, which was reduced to by half in the presence of lactose, indicating a high degree of hepatic targeting efficacy.

Antibody-based targeting is the second method of specific labeling as illustrated in Figure 8.4b with heterobifunctional poly(ethylene glycol) (PEG) as the linker for surface functionalization (Section 4.2.2). In this method, FNDs are first PEGylated and then covalently conjugated with streptavidin (SA), to which biotin-labeled antibodies of interest are linked [20]. It involves the reaction between the carboxyl groups on FNDs and the amino groups of NH_2-PEG-COOH, followed by covalent conjugation of the bifunctional PEG with SA through amide bond formation. The purpose of conjugating SA is that the protein can bind with any biotin-labeled antibodies of interest through the *avidin–biotin interaction*, the strongest known noncovalent bonding in biology [38]. Finally, the SA-conjugated FNDs are noncovalently coated with BSA to prevent aggregation in biological buffer. Chang et al. [21] demonstrated the high targeting specificity of the bioconjugated FNDs with biotinylated antibodies and then with the surface antigens on HepG2 cells and the breast cancer cell lines, MCF-7, MDA-MB-231, and ASB145-1R. In these experiments, it was crucial to ensure good dispersibility of the SA-conjugated FND in the high-salt media, since particle agglomeration and/or precipitation were always a concern. With the BSA coating, we now have the necessary safeguard to carry out the specific labeling.

Taking a different approach, Cigler and coworkers [24] encapsulated FNDs in 10–20 nm-thick translucent silica shells (Section 4.3.2). They then applied the silica-encapsulated FNDs, coated with a biocompatible *N*-(2-hydroxypropyl)methacrylamide copolymer shell, for cancer-cell targeting and fluorescence imaging. Through click chemistry, the copolymer could bear both Alexa Fluor 488 and cyclic peptides, Arg-Gly-Asp (RGD). These RGD-conjugated particles were able to selectively target integrin $\alpha_v\beta_3$ receptors on human glioblastoma cells with a high internalization efficacy as confirmed by flow cytometric analysis.

As part of an ongoing effort to improve the efficacy of the specific labeling, Hsieh et al. [37] employed biotinylated lipid-coated FNDs (bL-FND) to specifically target cell surface antigens. With a lipid layer composed of biotinylated PEGylated 1,2-distearoyl-sn-glycero-3-phosphoethanolamine, they synthesized the bL-FND particles by using the solvent evaporation method described in Section 4.3.1. Their results showed that the encapsulation of FNDs in the biotinylated lipids enabled the particles not only stable dispersion in high ionic strength buffers but also high target specificity. A viable application of the technique was demonstrated with biotin-mediated immunostaining of CD44 antigens on fixed human cells. Thanks to the bright fluorescence and enduring photostability of 100-nm bL-FNDs, the positions of the CD44 antigens could be located by *correlative light-electron microscopy* (CLEM) with accuracy better than 50 nm, limited mainly by the size of the FND particles (cf., Section 10.4 for further discussion).

8.2 Fluorescence Imaging

8.2.1 Epifluorescence and Confocal Fluorescence

In this chapter, we have so far learned how to make the cells *visible* by labeling with FNDs. Next, we will have to find ways to *see* them by fluorescence imaging. We will soon find out that FNDs are a promising candidate for probing both temporal and spatial events in live cells. Particularly for the NV$^-$ centers, their fluorescence emission band (peaking at ~690 nm) is well separated from that (peaking at ~500 nm) of the cell autofluorescence derived from endogenous fluorophores such as flavins and NAD(P)H [9, 39], making it easy to observe red FNDs in cells by epifluorescence imaging with a mercury vapor lamp as the light source. A representative example is given in Figure 8.7 for surface-oxidized red FNDs internalized by HeLa cells. Optical section images along the *z* direction show that most of the particles are distributed in the cytoplasm but not within the nuclei. To identify the exact location of the individual FND particles (~100 nm in diameter), confocal fluorescence microscopy can serve the purpose well. The microscopy provides better optical resolution and contrast due to the addition of a spatial pinhole

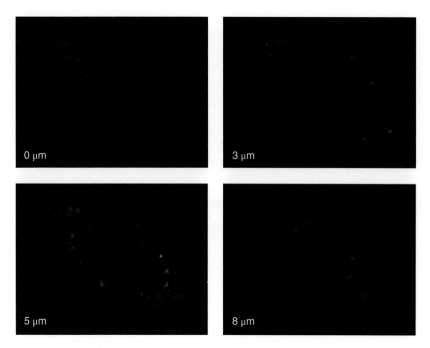

Figure 8.7 Vertical cross-section scans (0–8 µm) of the wide-field epifluorescence image of a HeLa cell after FND uptake by endocytosis. As evidenced by the images taken at 3–5 µm, most of the FND particles are distributed in the cytoplasm and do not enter the nucleus of the cell.

placed at the confocal plane of the lens to eliminate out-of-focus light [40]. The technique enables three-dimensional structural reconstruction from the images acquired at different depths of focus within an object. By performing colocalization analysis using endosomal/lysosomal markers through confocal fluorescence imaging, Faklaris et al. [14, 15] confirmed that the internalized FND particles were predominantly trapped in the endocytic vesicles (endosomes in Figure 8.8a and lysosomes in Figure 8.8b), leaving only some smaller particles free in the cytosol. The majority of the FND particles were located at the perinuclear region.

Equipped with excellent photostability and biocompatibility, FNDs can be used to identify directly and specifically the labeled cells over an extended period of time by fluorescence imaging. A proof-of-principle experiment conducted by Chang et al. [21] used SA-PEG-FNDs to probe specifically CD44 antigens on the surface of the ASB145-1R cells. In this experiment, cells were first labeled with biotinylated anti-CD44 and then stained with either SA-PEG-FND or DyLight488-SA at 4 °C, after which the cells were washed with PBS and incubated at 37 °C to facilitate endocytosis. Figure 8.9 compares confocal fluorescence images of the cells with dye- and FND-labeled CD44 on their surfaces before (panels a–h) and after (panels i–k) endocytosis. As noted, the dyed samples could not last long enough for a valid experiment to complete due to photobleaching (Figure 8.9d). In contrast, the FND fluorescence intensities remained at nearly the same level throughout the entire time of the experiment (Figure 8.9h),

(a)

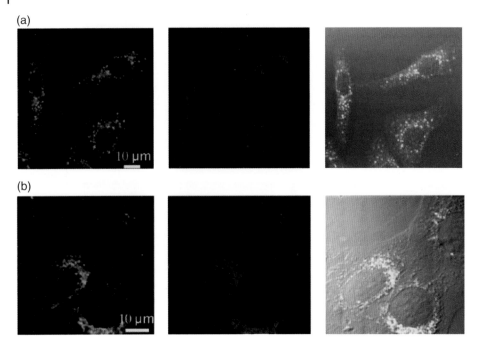

(b)

Figure 8.8 (a) Colocalization study of FNDs with early endosomes labeled with EEA1-FITC fluorescent conjugates. (b) Colocalization of FNDs with lysosomes labeled with green LysoTracker. FNDs colocalized with endosomes or lysosomes appear in yellow in the merged fluorescence scans. Images were acquired by confocal fluorescence raster scans of HeLa cells incubated with FNDs (10 µg ml^{-1}) in normal (control) conditions, then fixed, followed by labeling with endosomal or lysosomal markers. From left to middle: raster scan in the green channel (500–530 nm) showing the endocytic compartments and in the red channel (600–750 nm) showing the FNDs. Images on the right represent the merged green and red scans. *Source:* Reprinted with permission from Ref. [15]. Reproduced with permission of American Chemical Society.

indicating that the nanoprobes were not only non-photobleaching but also resistant to degradation by enzymes and acids in endosomes and/or lysosomes.

8.2.2 Total Internal Reflection Fluorescence

Total internal reflection fluorescence microscopy (TIRFM) is a sensitive method to detect FNDs attached to cell membrane. Using an *evanescent wave* generated by total internal reflection, TIRFM can selectively excite fluorophores in a region (typically in a few hundred nanometer thick) immediately adjacent to the glass-water interface [41]. The technique has found a wide range of bioimaging applications, particularly in areas where it is necessary to view biomolecules attached to a planar surface or to study the position and dynamics of biomolecules and organelles in live cells near the contact regions with the glass substrate. Because the evanescent wave intensity decays exponentially as the distance grows from the water-solid interface, nearly background-free detection can be achieved with this technique.

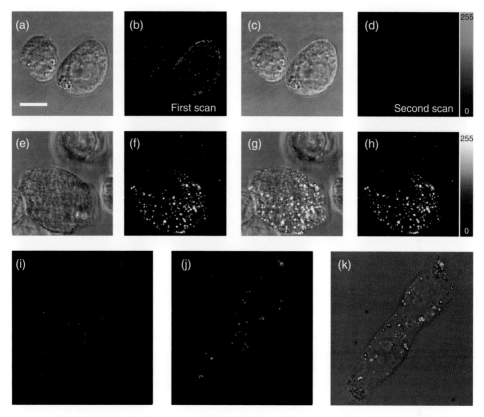

Figure 8.9 Bright-field and confocal fluorescence images of live ASB145-1R cells labeled with DyLight488-SA or SA-PEG-FND before and after endocytosis. The cells were first stained with biotinylated anti-CD44 antibody and then incubated with DyLight488-SA (a–d) or SA-PEG-FND (e–h) at 4 °C. Panels (a) and (e) are bright-field images, (b), (d), (f), and (h) are Z-stacked confocal fluorescence images, and (c) and (g) are merged bright-field and Z-stacked confocal fluorescence images of the cells before endocytosis. The contrast in photostability between these two types of fluorophores is evident in (d) and (h), where the organic dye photobleached during the second scan of the specimen. FND-labeled CD44 on the cell surface was then internalized through endocytosis at 37 °C (i–k). The orange spots in the merged fluorescence image (k) evidence that the FND (red, i) are colocalized with endosome/lysosomes, which were stained by the endo-lysosomal marker, Rab7 (green, j). Scale bar: 10 μm. *Source:* Reprinted with permission from Ref. [21]. Reproduced with permission of John Wiley & Sons.

Sotoma et al. [25] applied TIRFM to track the diffusion trajectory of FND-labeled proteins on membrane surface. The targeted protein was IL18Rα, a membranous receptor protein for interleukin-18, which is one of the most representative pro-inflammatory cytokines expressed in a variety of cancers [42]. The researchers first modified 30-nm FNDs with hyperbranched polyglycerol (HPG) to suppress nonspecific interaction with biomolecules and self-aggregation. The HPG-coated nanoparticles were then conjugated with ampicillin (Amp), which can covalently bind with mutated β-lactamase-tag (BL-tag) via the β-lactam ring [43]. To demonstrate the specific labeling capability of the nanoparticle bioconjugates, they constructed a plasmid DNA encoding BL-tag-fused IL18Rα and established a human embryonic kidney cell line (HEK293) that could express this protein on the plasma membrane. Labeling of IL-18Rα with FND-HPG-Amp allowed them to track the protein on cell membrane by TIRFM for more than eight seconds without any sign of photobleaching. The high sensitivity of TIRFM has also enabled background-free tracking of single 35-nm FNDs in axons of hippocampal neurons to study brain-related disease, as illustrated in Figure 7.9 [31].

8.2.3 Two-Photon Excitation Fluorescence

In the field of biomedicine, *two-photon excitation* (TPE) microscopy represents an advanced imaging modality that has attracted considerable attention for *in vivo* imaging applications [44, 45]. The technique has several distinct advantages over conventional *one-photon excitation* (OPE) microscopy. First, the absence of out-of-focus absorption significantly improves the image contrast of the target specimens [44]. Second, the employment of red and near-infrared light in TPE produces lower autofluorescence and scattering signals than those in OPE. Third, TPE microscopy provides optical sectioning at a depth of up to several hundred micrometers, which is about one order of magnitude deeper than that of OPE [45]. However, the trade-off is that the method requires an ultrafast laser for the excitation as explained below.

To conduct TPE fluorescence spectroscopy of the NV^- centers in bulk diamond, Wee et al. [46] used a picosecond mode-locked laser operating at 1064 nm as the excitation source. Based on the corresponding one-photon absorption cross section of 3.1×10^{-17} cm^2 at 532 nm (Section 7.1.2), they determined the two-photon absorption cross section of a single NV^- center at 1064 nm to be 0.45×10^{-50} cm^4 s photon^{-1}, which was about 1/30th that of a dye molecule like rhodamine B. Further studies over a wider wavelength range found the two-photon absorption cross sections to fall in the range of $(0.1–0.5) \times 10^{-50}$ cm^4 s photon^{-1} at 800–1040 nm [47]. These values are three orders of magnitude smaller than that of typical CdS quantum dots. Although the two-photon absorption cross section of NV^- is small, the deficiency can be compensated by increasing the number of NV^- centers in the particles [48]. For example, a 100-nm FND may contain up to 1000 NV^- centers, making it highly competitive as a TPE marker.

As a biomedical imaging technique, TPE microscopy outperforms OPE in that it generates better contrast images because of the dramatic reduction of autofluorescence and light scattering from the specimens. The same outstanding feature is also present for TPE of FNDs in cells. Figure 8.10a displays confocal scanning images of a fixed HeLa cell labeled with 140-nm FNDs and probed with both OPE and TPE microscopies. The three images in each row are axial slices of the cell acquired at three different positions in each excitation, showing the cellular uptake of the particles (labeled with red and yellow boxes). A comparison between the results of these two excitations clearly indicates that the TPE images have a better contrast. Using residual light scattering and cell autofluorescence signals, Chang et al. [13] reconstructed a three-dimensional image of the cell, as shown in Figure 8.10b. No evidence was found for the entry of the particles into the nucleus. The research team further analyzed the intensity profiles of the FNDs residing in the cytoplasm (Figure 8.10c) and concluded that both the lateral and axial cross sections of the TPE images were only half those of the corresponding OPE images, proving the advantages of using TPE in fluorescence imaging.

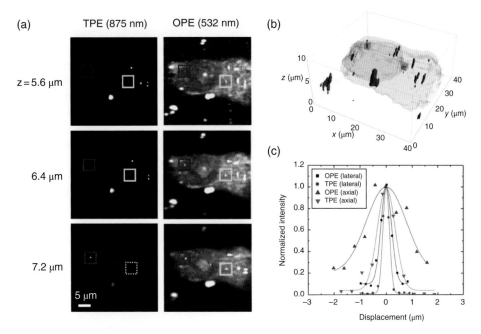

Figure 8.10 One-photon and two-photon excited fluorescence images of 140-nm FNDs in a fixed HeLa cell. (a) Typical OPE and TPE confocal fluorescence images of the same cell. (b) Three-dimensional TPE image of the cell and the internalized FNDs. (c) Lateral and axial cross-sections of the FND labeled with a red box in (a). Note that the resolution of TPE for this particular particle was approximately 300 nm in the lateral direction and approximately 800 nm in the axial direction, both of which are close to their theoretical diffraction limits. *Source:* Reprinted with permission from Ref. [13]. Reproduced with permission of Nature Publishing Group.

In another study, Hui et al. [49] employed a femtosecond Nd: glass laser operating at 1060 nm and 72 MHz for TPE of lipid-encapsulated FNDs in live HeLa cells. By using this technique, they were able to detect single FND particles of approximately 40 nm in size with good image contrast in the cytoplasm and track their motions by fluorescence correlation spectroscopy (Figure 7.10). Again, a better image contrast of the FNDs in cells was achieved by TPE as it outperformed OPE in eliminating background fluorescence. The excited volume of TPE was small due to the quadratic dependence of the fluorescence intensity on the intensity of the incident light. As a result, the photon-mediated damage of the live cells was largely reduced. Finally, an added benefit of using the 1060-nm excitation was that it gave characteristic signatures of only NV^- centers in the fluorescence spectra, discriminating all NV^0 emissions that would have needed TPE at 875 nm instead.

8.2.4 Time-Gated Fluorescence

Suggested by its own name, *fluorescence lifetime imaging microscopy* (FLIM) is a time-resolved imaging technique. The method utilizes the differences in average fluorescence lifetimes of molecules or particles to generate high-contrast images [50]. As already discussed in Section 3.4, the average fluorescence lifetime of the NV^- defects in bulk diamond is 11.6 ns (Table 3.2), whereas the decay time extends to about 20 ns if the same centers are hosted in FNDs either spin-coated on a glass substrate [51, 52], suspended in water [48], or injected into living organisms [53]. Such a difference in lifetime stems from the large change in refractive index of the surrounding medium when replacing a bulk crystal with a nanometer-sized crystal. Nevertheless, whether in a bulk diamond or in an FND, the lifetimes of the NV⁻ centers are all significantly longer than those (0.3–6.8 ns) of endogenous and exogenous fluorophores commonly used in cell biology [50]. This allows researchers to sequester the FND emission from the huge ensemble of autofluorescence backgrounds in cells and tissues by employing various time-gating techniques to improve the image contrast.

Faklaris et al. [14] first leveraged this method to acquire the confocal fluorescence images of sub-50 nm FNDs, each containing only a single NV⁻ center, in HeLa cells (Figure 8.11a and b). Significant improvement of the image contrast was achieved by using a picosecond pulsed laser (532 nm) as the excitation source and detecting emitted photons at 15–53 ns after the laser pulses (Figure 8.11c). Their result demonstrated that FNDs could be easily detected at the single-particle level in high cell autofluorescence

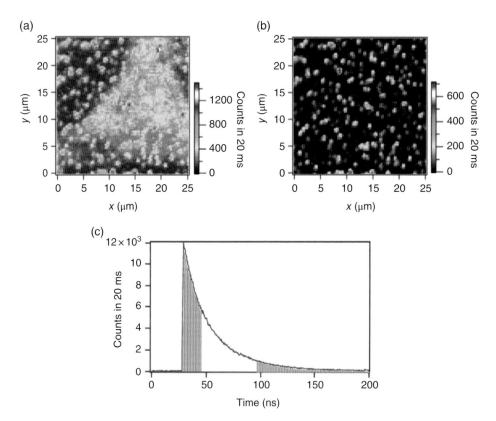

Figure 8.11 Time-resolved confocal fluorescence images of a fixed cell containing FNDs. (a) Raster-scan image obtained by detecting all photons, displaying NV⁻ fluorescence together with cell autofluorescence. (b) Time-gated raster scan constructed from photons detected between 15 and 53 ns after pulsed laser excitation. Scan area: $25 \times 25 \, \mu m^2$. (c) Fluorescence time decay from one of the FNDs shown in (a). *Source:* Adapted with permission from Ref. [14]. Reproduced with permission of John Wiley & Sons.

background by time-gated imaging, unambiguously locating the positions of the nano-particles inside the cells.

In another study, Hui et al. [22] conducted wide-field time-gated fluorescence imaging of FND-labeled cancer cells in whole blood with a sub-nanosecond laser for excitation followed by detection of an *intensified charge-coupled device* (ICCD). Figure 8.12a and b display fluorescence images of HeLa cells labeled with 100-nm FNDs by endocytosis. The images were acquired either with or without time gating after 599-nm laser excitation. The image contrast, defined as the ratio between FND and background fluorescence signals, was poor in Figure 8.12a because the hemoglobin molecules in human blood (with a concentration of ~150 mg ml^{-1}) also emitted red fluorescence upon

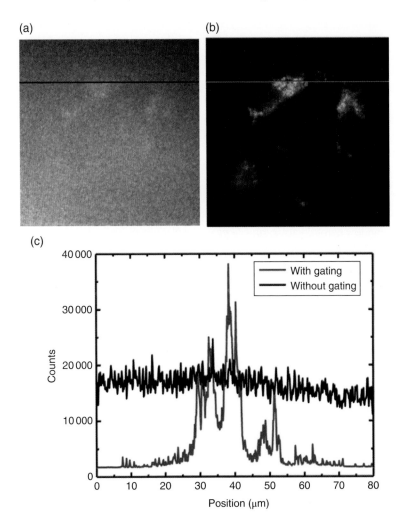

Figure 8.12 (a, b) Wide-field fluorescence images of FND-labeled HeLa cells attached to a coverglass slide and immersed in human blood without (a) or with (b) time gating. The exposure times used for the fluorescence imaging with a 100× oil-immersion objective lens in (a) and (b) are 0.1 and 0.3 seconds, respectively. (c) Intensity profiles along the black and red color lines denoted in (a) and (b), respectively. *Source:* Reprinted with permission from Ref. [22].

excitation with the orange-red light. Fortunately, as the fluorescence lifetime of hemoglobin was shorter than 1 ns, the background noise could be effectively filtered out temporally by setting the ICCD shutter delay time at 10 ns (Figure 8.12b). An improvement of the image contrast by a factor of approximately 20 was achieved when the time gating function was turned on (Figure 8.12c). Further discussion of the technique for *in vivo* imaging can be found in the next chapter.

A natural extension of the time-gated fluorescence imaging technology is the time-gated flow cytometry. Hui et al. [22] proved the concept by detecting individual FND-labeled HeLa cells in blood, which flowed through a microchannel of 50 μm in width and 50 μm depth. The dimensions of this channel were intentionally chosen to mimic those of capillaries in mice and rats. With sequential readout of the signals from an ICCD camera, the optimal frame rate of the recording was 23 Hz (or 43.5 ms frame^{-1}). The researchers demonstrated that the time-gating technique could effectively suppress background fluorescence signals (and thus enhance the visibility of FNDs), allowing for direct observation of the flow of FND-labeled HeLa cells in human blood at a concentration of lower than 8000 cells ml^{-1} in the capillary-sized microchannels. The same principle can be applied to achieving background-free detection of FND-labeled cells in a standard flow cytometer equipped with a femtosecond supercontinuum white light laser system [54]. It enables FACS analysis of cells with ultra-high sensitivity and selectivity.

References

1 Yu, S.J., Kang, M.W., Chang, H.C. et al. (2005). Bright fluorescent nanodiamonds: no photobleaching and low cytotoxicity. *J Am Chem Soc* 127: 17604–17605.

2 Hong, G., Diao, S., Antaris, A.L., and Dai, H. (2015). Carbon nanomaterials for biological imaging and nanomedicinal therapy. *Chem Rev* 115: 10816–10906.

3 Zhang, L., Gu, F.X., Chan, J.M. et al. (2008). Nanoparticles in medicine: therapeutic applications and developments. *Clin Pharmacol Ther* 83: 761–769.

4 Vaijayanthimala, V. and Chang, H.C. (2009). Functionalized fluorescent nanodiamonds for biomedical applications. *Nanomedicine* 4: 47–55.

5 Herzenberg, L.A., Parks, D., Sahaf, B. et al. (2002). The history and future of the fluorescence activated cell sorter and flow cytometry: a view from Stanford. *Clin Chem* 48: 1819–1827.

6 Hemelaar, S.R., Nagl, A., Bigot, F. et al. (2017). The interaction of fluorescent nanodiamond probes with cellular media. *Microchim Acta* 184: 1001–1009.

7 Zhang, B.L., Li, Y.Q., Fang, C.Y. et al. (2009). Receptor-mediated cellular uptake of folate-conjugated fluorescent nanodiamonds: a combined ensemble and single-particle study. *Small* 5: 2716–2721.

8 Vaijayanthimala, V., Tzeng, Y.K., Chang, H.C., and Li, C.L. (2009). The biocompatibility of fluorescent nanodiamonds and their mechanism of cellular uptake. *Nanotechnology* 20: 425103.

9 Fu, C.C., Lee, H.Y., Chen, K. et al. (2007). Characterization and application of single fluorescent nanodiamonds as cellular biomarkers. *Proc Natl Acad Sci USA* 104: 727–732.

10 Neugart, F., Zappe, A., Jelezko, F. et al. (2007). Dynamics of diamond nanoparticles in solution and cells. *Nano Lett* 7: 3588–3591.

11 Chao, J.I., Perevedentseva, E., Chung, P.H. et al. (2007). Nanometer-sized diamond particle as a probe for biolabeling. *Biophys J* 93: 2199–2208.

12 Liu, K.K., Cheng, C.L., Chang, C.C., and Chao, J.I. (2007). Biocompatible and detectable carboxylated nanodiamond on human cell. *Nanotechnology* 18: 325102.

13 Chang, Y.R., Lee, H.Y., Chen, K. et al. (2008). Mass production and dynamic imaging of fluorescent nanodiamonds. *Nat Nanotechnol* 3: 284–288.

14 Faklaris, O., Garrot, D., Joshi, V. et al. (2008). Detection of single photoluminescent diamond nanoparticles in cells and study of the internalization pathway. *Small* 4: 2236–2239.

15 Faklaris, O., Joshi, V., Irinopoulou, T. et al. (2009). Photoluminescent diamond nanoparticles for cell labeling: study of the uptake mechanism in mammalian cells. *ACS Nano* 3: 3955–3962.

16 Weng, M.F., Chiang, S.Y., Wang, N.S., and Niu, H. (2009). Fluorescent nanodiamonds for specifically targeted bioimaging: application to the interaction of transferrin with transferrin receptor. *Diam Relat Mater* 18: 587–591.

17 Li, Y. and Zhou, X. (2010). Transferrin-coupled fluorescence nanodiamonds as targeting intracellular transporters: an investigation of the uptake mechanism. *Diam Relat Mater* 19: 1163–1167.

18 Fang, C.Y., Vaijayanthimala, V., Cheng, C.A. et al. (2011). The exocytosis of fluorescent nanodiamond and its use as a long-term cell tracker. *Small* 7: 3363–3370.

19 Tzeng, Y.K., Faklaris, O., Chang, B.M. et al. (2011). Superresolution imaging of albumin-conjugated fluorescent nanodiamonds in cells by stimulated emission depletion. *Angew Chem Int Ed* 50: 2262–2265.

20 Lee, J.W., Lee, S., Jang, S. et al. (2013). Preparation of non-aggregated fluorescent nanodiamonds (FNDs) by non-covalent coating with a block copolymer and proteins for enhancement of intracellular uptake. *Mol Biosyst* 9: 1004–1011.

21 Chang, B.M., Lin, H.H., Su, L.J. et al. (2013). Highly fluorescent nanodiamonds protein-functionalized for cell labeling and targeting. *Adv Funct Mater* 23: 5737–5745.

22 Hui, Y.Y., Su, L.J., Chen, O.Y. et al. (2014). Wide-field imaging and flow cytometric analysis of cancer cells in blood by fluorescent nanodiamond labeling and time gating. *Sci Rep* 4: 5574.

23 Hsu, T.C., Liu, K.K., Chang, H.C. et al. (2014). Labeling of neuronal differentiation and neuron cells with biocompatible fluorescent nanodiamonds. *Sci Rep* 4: 5004.

24 Slegerova, J., Hajek, M., Rehor, I. et al. (2015). Designing the nanobiointerface of fluorescent nanodiamonds: highly selective targeting of glioma cancer cells. *Nanoscale* 7: 415–420.

25 Sotoma, S., Iimura, J., Igarashi, R. et al. (2016). Selective labeling of proteins on living cell membranes using fluorescent nanodiamond probes. *Nanomaterials* 6: 56.

26 Liu, K.K., Qiu, W.R., Raj, E.N. et al. (2017). Ubiquitin-coated nanodiamonds bind to autophagy receptors for entry into the selective autophagy pathway. *Autophagy* 13: 187–200.

27 Prabhakar, N., Khan, M.H., Peurla, M. et al. (2017). Intracellular trafficking of fluorescent nanodiamonds and regulation of their cellular toxicity. *ACS Omega* 2: 2689–2693.

28 Suarez-Kelly, L.P., Campbell, A.R., Rampersaud, I.V. et al. (2017). Fluorescent nanodiamonds engage innate immune effector cells: a potential vehicle for targeted anti-tumor immunotherapy. *Nanomedicine* 13: 909–920.

29 Wu, T.J., Tzeng, Y.K., Chang, W.W. et al. (2013). Tracking the engraftment and regenerative capabilities of transplanted lung stem cells using fluorescent nanodiamonds. *Nat Nanotechnol* 8: 682–689.

30 Huang, Y.A., Kao, C.W., Liu, K.K. et al. (2014). The effect of fluorescent nanodiamonds on neuronal survival and morphogenesis. *Sci Rep* 4: 6919.

31 Haziza, S., Mohan, N., Loe-Mie, Y. et al. (2017). Fluorescent nanodiamond tracking reveals intraneuronal transport abnormalities induced by brain disease-related genetic risk factors. *Nat Nanotechnol* 12: 322–328.

32 Su, L.J., Wu, M.H., Hui, Y.Y. et al. (2017). Fluorescent nanodiamonds enable quantitative tracking of human mesenchymal stem cells in miniature pigs. *Sci Rep* 7: 45607.

33 Edwards, D.F. and Philipp, H.R. (1985). Cubic carbon diamond. In: *Handbook of Optical Constants of Solids* (ed. E.D. Polik), 665–673. Academic Press.

34 Smith, B.R., Niebert, M., Plakhotnik, T., and Zvyagin, A.V. (2007). Transfection and imaging of diamond nanocrystals as scattering optical labels. *J Lumin* 127: 260–263.

35 Krueger, A. and Lang, D. (2012). Functionality is key: recent progress in the surface modification of nanodiamond. *Adv Funct Mater* 22: 890–906.

36 Pham, D.M., Epperla, C.P., Hsieh, C.L. et al. (2017). Glycosaminoglycans-specific cell targeting and imaging using fluorescent nanodiamonds coated with viral envelope proteins. *Anal Chem* 89: 6527–6534.

37 Hsieh, F.J., Chen, Y.W., Huang, Y.K. et al. (2018). Correlative light-electron microscopy of lipid-encapsulated fluorescent nanodiamonds for nanometric localization of cell surface antigens. *Anal Chem* 90: 1566–1571.

38 Diamandis, E.P. and Christopoulos, T.K. (1991). The biotin-(strept)avidin system: principles and applications in biotechnology. *Clin Chem* 37: 625–636.

39 Billinton, N. and Knight, A.W. (2001). Seeing the wood through the trees: a review of techniques for distinguishing green fluorescent protein from endogenous autofluorescence. *Anal Biochem* 291: 175–197.

40 Inoué, S. (2006). Foundations of confocal scanned imaging in light microscopy. In: *Handbook of Biological Confocal Microscopy*, 3e (ed. J.B. Pawley), 1–19. Springer.

41 Axelrod, D. (2008). Total internal reflection fluorescence microscopy. *Method Cell Biol* 89: 169–221.

42 Gracie, J.A., Robertson, S.E., and McInnes, I.B. (2003). Interleukin-18. *J Leukoc Biol* 73: 213–224.

43 Watanabe, S., Mizukami, S., Hori, Y., and Kikuchi, K. (2010). Multicolor protein labeling in living cells using mutant β-lactamase-tag technology. *Bioconjug Chem* 21: 2320–2326.

44 Xu, C., Zipfel, W., Shear, J.B. et al. (1996). Multiphoton fluorescence excitation: new spectral windows for biological nonlinear microscopy. *Proc Natl Acad Sci USA* 93: 10763–10768.

45 Helmchen, F. and Denk, W. (2005). Deep tissue two-photon microscopy. *Nat Methods* 2: 932–940.

46 Wee, T.L., Tzeng, Y.K., Han, C.C. et al. (2007). Two-photon excited fluorescence of nitrogen-vacancy centers in proton-irradiated type Ib diamond. *J Phys Chem A* 111: 9379–9386.

47 Ivanov, I.P., Li, X., Dolan, P.R., and Gu, M. (2013). Nonlinear absorption properties of the charge states of nitrogen-vacancy centers in nanodiamonds. *Opt Lett* 38: 1358–1360.

48 Su, L.J., Fang, C.Y., Chang, Y.T. et al. (2013). Creation of high density ensembles of nitrogen-vacancy centers in nitrogen-rich type Ib nanodiamonds. *Nanotechnology* 24: 315702.

49 Hui, Y.Y., Zhang, B.L., Chang, Y.C. et al. (2010). Two-photon fluorescence correlation spectroscopy of lipid-encapsulated fluorescent nanodiamonds in living cells. *Opt Express* 18: 5896–5905.

50 Chang, C.W., Sud, D., and Mycek, M.A. (2007). Fluorescence lifetime imaging microscopy. *Method Cell Biol* 81: 495–524.

51 Beveratos, A., Brouri, R., Gacoin, T. et al. (2001). Nonclassical radiation from diamond nanocrystals. *Phys Rev A* 64: 061802.

52 Storteboom, J., Dolan, P., Castelletto, S. et al. (2015). Lifetime investigation of single nitrogen vacancy centres in nanodiamonds. *Opt Express* 23: 11327–11333.

53 Kuo, Y., Hsu, T.Y., Wu, Y.C. et al. (2013). Fluorescence lifetime imaging microscopy of nanodiamonds *in vivo. Proc SPIE* 8635: 863503.

54 Jin, D., Piper, J.A., Leif, R.C. et al. (2009). Time-gated flow cytometry: an ultra-high selectivity method to recover ultra-rare-event μ-targets in high-background biosamples. *J Biomed Opt* 14: 024023.

9

Cell Tracking and Deep Tissue Imaging

Cell labeling, both specific and nonspecific, discussed in the previous chapter belongs in part to the field of *cell therapy*, defined as "administration of live whole cells or maturation of a specific cell population in a patient for the treatment of a disease" [1]. Cell therapy is an emerging field in the medical research as a promising treatment option for human injuries and diseases [2]. While the first use of cells for therapeutic purposes can be traced to the nineteenth century [3], much is still lacking in our knowledge of the therapy, including the biodistribution, pharmacokinetics, and pharmacodynamics of transplanted human cells *in vivo* [4]. Such knowledge is essential for detailed understanding of the cells' development, fate, and contribution to regenerating tissues prior to clinical trials.

The advent of nanotechnology in the 1980s (Section 1.1) makes it possible to produce innovative fluorescent markers to solve this long-standing cell tracking issue. High expectation has been raised for using functional nanomaterials to overcome limitations inherited in conventional therapeutic and diagnostic approaches [5]. However, in this type of application, it is critical to ensure that the cell labeling technique is robust and safe to use. Inorganic fluorescent nanoparticles such as quantum dots have been considered as a potential candidate owing to their superior optical properties [6]. But there is always a concern about the inherent toxicity of this cadmium-based compound [7], which may impede its real-world applications in nanomedicine. As a new star in the field, *fluorescent nanodiamonds* (FNDs) with their superlative optical properties and excellent biocompatibility provide a favorable and appealing alternative [8–11]. In this chapter, we discuss the research and development of FNDs as cell trackers and their use in deep tissue imaging. The first step in cell tracking with this new technology is the uptake of FNDs by cells.

9.1 Cellular Uptake

9.1.1 Uptake Mechanism

We have discussed in Section 8.1 how a cell can be labeled with FNDs by endocytosis. Cellular uptake by endocytosis is an energy-expending process and the major pathways include clathrin-mediated endocytosis, caveolar-type endocytosis, phagocytosis, and macropinocytosis [12]. During endocytosis, the engulfed nanoparticles are pinched off from the cell membrane and then enclosed within the membrane-bound endosomes,

Fluorescent Nanodiamonds, First Edition. Huan-Cheng Chang, Wesley Wei-Wen Hsiao and Meng-Chih Su.
© 2019 John Wiley & Sons Ltd. Published 2019 by John Wiley & Sons Ltd.

Table 9.1 Biochemical and physiological actions of some representative endocytosis inhibitors.

Name of inhibitors	Biochemical and physiological actions
Sodium azide and 2-deoxyglucose	Depletion of intracellular ATP
Phenylarsine oxide	A clathrin inhibitor
Filipin	A claveolae-specific inhibitor
Cytochalasin D	An inhibitor of actin-dependent endocytosis through exerting its effect by depolymerizing the cellular actin microfilament network
Nocodazole	An inhibitor of microtubule-dependent endocytosis through effective blocking of tubulin self-assembly

Source: From Ref. [20]. Reproduced with permission of Springer.

through which they are carried into the cell interior along a network of tubulovesicular membranous structure [13]. For cultured mammalian cells, almost one-half of the endocytic uptakes can go through non-clathrin processes. Although it has been reported that some nanoparticles may slip through cell membrane and/or directly penetrate into cells, the mechanism is still difficult to confirm.

To optimize the cell labeling conditions for FNDs by endocytosis, it is necessary to know their exact uptake mechanism. Chang and coworkers [14], as well as other research groups [15–19], have methodically examined the uptake mechanism and pathways for FNDs through a battery of metabolic and cytoskeletal inhibitors, with their functions listed in Table 9.1 [20]. In this series of experiments, cells were first treated with inhibitors for 30 minutes, washed, and then incubated with FNDs for 3 hours. After another thorough wash to remove free FNDs, cells were harvested and the amount of the internalized FNDs was determined by flow cytometry (Section 8.1). Figure 9.1a displays a typical result for HeLa cells nonspecifically labeled with bare FNDs. It was found that preincubation of the cells with sodium azide and 2-deoxyglucose and phenylarsine oxide blocked the FND uptake down to a level comparable to that of control cells without the FND treatment. In contrast, preincubation with filipin resulted in a level of FND uptake similar to those without the inhibitor treatment. The observations strongly suggested that the FND uptake was through an energy-dependent clathrin-mediated endocytosis and the claveolae-mediated pathway was not involved in the uptake process. Cytochalasin D [21] and nocodazole [22] were next tested for their effectiveness in blocking the cellular uptake of FNDs. As seen in Figure 9.1a, pretreatment of the cells with these two chemicals decreased the FND uptake close to the control cell level, an indication that both microfilaments and microtubules also played important roles in the endocytic process.

With HeLa cells as the model cell line, the above results indicate that the cellular uptake of FND is (i) energy dependent, (ii) claveolae independent, (iii) clathrin-mediated, and (iv) involving an intact cellular microfilament and microtubule architecture. To examine whether the uptake mechanism of bare FNDs was cell-type specific, a well-established murine pre-adipocyte cell line (3T3-L1) was chosen for a comparative study. Analogous to the cases of HeLa cells, preincubation with a combination of

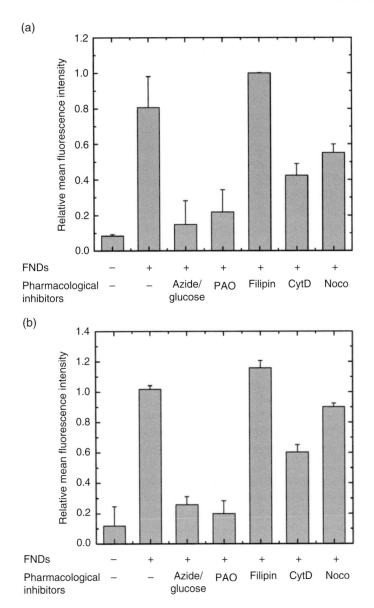

Figure 9.1 Flow cytometric analysis of the mechanism of FND uptakes by (a) human HeLa cells and (b) murine 3T3-L1 pre-adipocytes. Cells were preincubated with various inhibitors for specified time, followed by incubation with FNDs for three hours before cell harvest for the analysis. PAO, phenylarsine oxide; CytD, cytochalasin D; Noco, nocodazole. The first two columns in each figure represent control experiments. *Source*: Reprinted with permission from Ref. [14].

metabolic inhibitors, sodium azide, 2-deoxyglucose, and phenylarsine oxide, reduced the FND uptake level to that of control cells (Figure 9.1b). In addition, treatment with filipin did not appreciably influence the internalization of FNDs, whereas the pretreatment with cytochalasin D and nocodazole led to a modest decrease of the FND uptake by 3T3-L1 cells. It led to the conclusion that the overall FND uptake mechanism and pathways were similar to each other between these two different types of cell lines. However, more work involving other kinds of cells is needed before any final conclusion can be made on the specificity in the mechanism of FND uptake.

9.1.2 Entrapment

Once internalized by cells, FNDs are primarily trapped in endosomes and lysosomes. A lysosome is a membrane-bound organelle containing hydrolytic enzymes that can degrade virtually all kinds of biomolecules. The pH value of the lysosome lumen is approximately 4.8, optimal for the enzymes involved in the hydrolysis [23]. Although the pH is low and the enzymatic activity is high within these cellular compartments, the internalized FNDs should stay intact therein because of their extreme chemical inertness. A number of colocalization studies of FNDs and endosomal/lysosomal markers by confocal fluorescence imaging have confirmed the occurrence of the entrapment (Section 8.2.1), while a report claimed that some prickly FNDs could undergo quick endosomal escape due to the shape effect [19, 24]. Tested over a wide range of cancer and stem cells, including human cervical cancer cells [25], human liver carcinoma cells [19, 24], mouse pre-adipocytes [14], mouse lung stem/progenitor cells [26], mouse primary neurons [27], and human mesenchymal stem cells (MSCs) [28], neither entrapped nor escaped FNDs have shown any measurable cytotoxic and detrimental effects. Indeed, FND is a user-friendly agent for cell labeling applications.

In a recent study, Liu et al. [29] investigated in detail the molecular endocytic mechanisms and cellular trafficking pathways of 100-nm FNDs in cells. The research focused on the *autophagy* pathway, a key mechanism for decomposing cellular components or foreign pathogens [30]. Although studies have shown that FNDs can be taken up by cells and subsequently trapped in endosomes and lysosomes, their intracellular processing through selective autophagy [31] has not been known. To address this question, the research team first fed adenocarcinomic human alveolar basal epithelial cells (A549 cells) with bare FNDs and then isolated protein–FND complexes from the cell lysates by modified sucrose density gradient centrifugation. Based on the Western blotting analysis at different time points, the FND particles were found first coated with ubiquitin (Ub), which is a 76-amino acid protein, and then bound to the autophagy receptor p62, which interacts with ATG8 (autophagy-related protein 8)/LC3 (light-chain 3) to enter the selective autophagy pathway. In addition, the selective autophagy receptors OPTN and NDP52 were also involved in the decomposition of the ubiquitinated particles. The FNDs were finally deposited in lysosomes during cell division. Figure 9.2 displays a model of how Ub-coated NDs can be conjugated with autophagy receptors, a process essential for the nanoparticles to enter the selective autophagy pathway. It is expected that a detailed understanding of the molecular mechanism of the cellular uptake will greatly facilitate the use of the carbon-based nanomaterials as cell tracking devices and drug delivery vehicles for biomedical applications (Chapter 13).

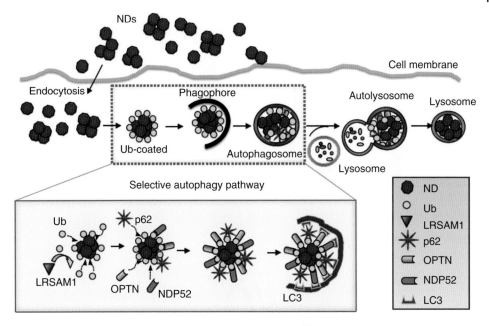

Figure 9.2 Schematic illustration of a proposed model on how Ub-coated FNDs bind with autophagy receptors when entering the selective autophagy pathway, leading to entrapment in the lysosome. *Source:* Reprinted with permission from Ref. [29]. Reproduced with permission of Taylor & Francis.

9.1.3   Quantification

In Figure 9.1, data collected from flow cytometric measurements show that a substantial amount of FND particles can be taken up by HeLa cells in the absence of serum. However, it is known that flow cytometry provides only a relative measure for the degree of cell labeling. What is the absolute amount of FNDs entrapped in cells? Su et al. [28] applied the *magnetically modulated fluorescence* (MMF) technique discussed in Section 6.2.4 to address this issue. In this proof-of-principle experiment, FNDs were first coated with human serum albumin (HSA) by physical adsorption (Section 4.2.1) and then fed to A549 cells and HeLa cells at the particle concentrations of $10–200\,\mu g\,ml^{-1}$. The cellular uptake was later quantified by laser-induced fluorescence after removal of untaken HSA-FND particles in the medium. To conduct the analysis, a standard calibration curve was first prepared by plotting the measured fluorescence intensity against the FND concentration gradient. The amounts of internalized FNDs were then determined from the fluorescence intensity measurements for the FND-labeled cells (1×10^{6} cells ml^{-1}) after sonication in water for one hour to break up their plasma membrane in a cuvette. For A549 cells incubated with $100\,\mu g\,ml^{-1}$ HSA-FND at $37\,°C$ for 4 hours, an average weight of 6.0 pg (or $\sim 3.2 \times 10^{3}$ particles) was measured for the internalized FNDs. The weight decreased monotonically as the HSA-FND concentration was reduced, until eventually reaching to zero (Figure 9.3a). The result was in good correlation with flow cytometric analysis of the same cells incubated under the same conditions (Figure 9.3b). Compared with A549, HeLa cells were able to take up twice more particles, reflecting a distinguishable difference in the endocytic behavior between these two types of human cell lines.

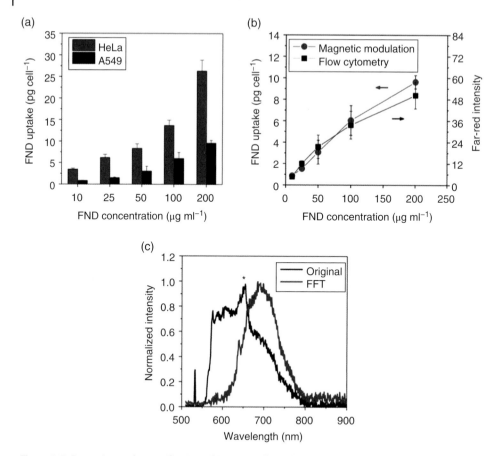

Figure 9.3 Detection and quantification of FNDs in cells and tissues. (a) Dose-dependent uptakes of HSA-coated FNDs by A549 cells and HeLa cells in culture, analyzed by MMF. (b) Comparison of the uptakes of HSA-coated FNDs by A549 cells, analyzed by both MMF and flow cytometry. (c) Normalized fluorescence spectrum of FNDs (5 µg ml^{-1}) in an acid digest of pig liver tissue (black) and its FFT spectrum (red) after demodulation at $f = 2$ Hz. The asterisk denotes the Raman peaks of water. FFT, fast Fourier transform. *Source:* Adapted with permission from Ref. [28].

The high detection sensitivity provided by the magnetic modulation method also enables researchers to quantify the amount of FNDs in organs and tissues. In the same experiment as discussed earlier, Su et al. [28] spiked FNDs (25 µg) in a sample solution (5 ml) prepared by digestion of pig liver tissue (1 g) in a concentrated HNO$_3$/H$_2$O$_2$ mixture at 140 °C for several hours. Fluorescence intensities were then measured directly for the FNDs in the tissue digests without extraction or other separation procedures to avoid loss of the particles during otherwise required centrifugation or filtration treatments [32, 33]. Due to the nanomaterial's chemical robustness, FNDs could still preserve their unique magneto-optical properties, allowing the magnetic modulation to recover their fluorescence signals from the acid digests at a concentration as low as 1 µg ml^{-1}. The recovery rate was more than 90%, despite the fact that the spectra were overwhelmed by the Raman peaks of water, residual chemicals, and cell lysates (Figure 9.3c). It should be noted that such a high level of quantitative analysis would not be possible with molecular fluorophores like organic dyes and fluorescent proteins because of their lack of chemical stability in strong acids.

9.2 Cell Tracking

9.2.1 Tracking *In Vitro*

In using nanoparticles as a cell tracking device, it is preferable to label the cells with a high-density ensemble of the agents. The reason is that when cells divide, the amount of nanoparticles in each labeled cell will decrease by half, a destined fate that could potentially impede any long-term cell tracking. Consider a cell with a doubling time of 24 hours in culture [34]. It is possible to conduct the tracking up to 10 days if the cell is initially labeled with 1024 (or 2^{10}) FND particles that can be detected individually in cellular compartments by confocal fluorescence microscopy (Section 7.1). Figure 9.4a displays snapshots of a real-time tracking of some internalized FNDs through the cell cycle of a FND-labeled HeLa cell by bright-field/epifluorescence microscopy [24]. These particles were predominantly entrapped in lysosomes for disposal but, surprisingly, no significant exocytosis of the lysosomal FNDs was found during cell proliferation. The result could be rationalized by the FND's chemical inertness and high biocompatibility, and that particle excretion was an energy-expending process. A combination of these unique features made it possible for the research team to track continuously the fate of the FND-labeled HeLa cells over eight generations by flow cytometry (Figure 9.4b), confirming the long-term labeling and tracking capabilities of the FND-based platform for cell division, proliferation, and differentiation studies.

The FND-enabled technology, derived mainly from the particle's extraordinary properties, soon finds applications in the studies of slow-proliferating and quiescent cancer stem cells (CSCs), which have long been considered to be a source of tumor initiation [35]. In the past, there was a lack of effective tools among the traditional materials for CSC identification and isolation even *in vitro*. Now, being both chemically and photophysically stable, FNDs become an excellent choice to serve the purposes for long-term tracking. Prior to the cell finding/tracking experiments, Lin et al. [36] first carried out genotoxicity tests with comet and micronucleus assays for human fibroblasts and breast cancer cells to confirm that FNDs neither caused DNA damage nor impaired cell growth (Section 5.1). They then employed AS-B145-1R breast cancer cells as the model cell line for CSCs and compared in parallel the performance of FNDs and the commonly used cell trackers, carboxyfluorescein diacetate succinimidyl ester (CFSE) and 5-ethynyl-2′-deoxyuridine (EdU). Their results indicated that the nanotechnology-based method was capable of quantitative analysis of the FND-labeled cells by flow cytometry and also outperformed CFSE and EdU in the comparative long-term tracking capability (Figure 9.5). It is anticipated that further integration of the FND-enabled cell tracking platform with the functional assays of protein markers will greatly enhance our knowledge of the properties of CSCs both *in vitro* and *in vivo*.

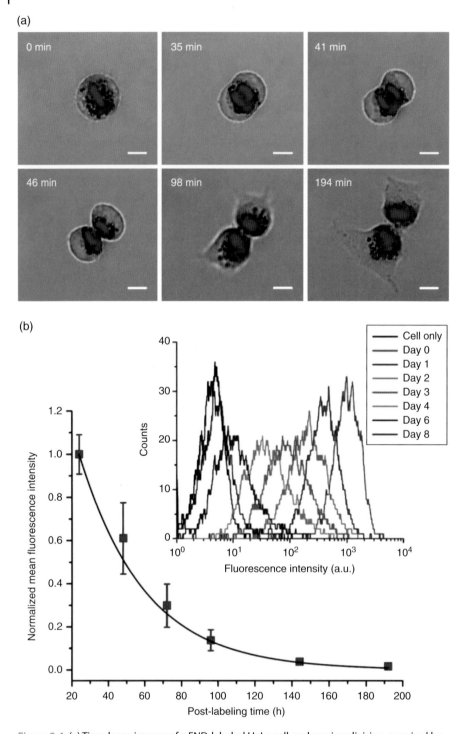

Figure 9.4 (a) Time-lapse images of a FND-labeled HeLa cell undergoing division, acquired by differential interference contrast and epifluorescence microscopy. (b) Long-term tracking of FND-labeled HeLa cells over eight days by flow cytometry. The fluorescence intensity of each cell decreases exponentially with time due to cell proliferation. *Source:* Adapted with permission from Ref. [25]. Reproduced with permission of John Wiley & Sons.

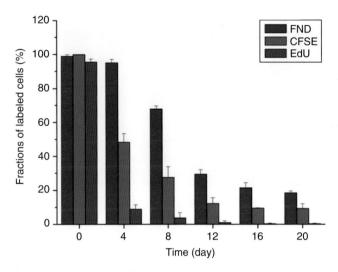

Figure 9.5 Comparison of the long-term tracking capability of EdU, CFSE, and FND. The assays were conducted for the mammospheres generated from AS-B145-1R cells labeled separately with EdU, CFSE, and FND and then dissociated for flow cytometric analysis with a 4-day period for 20 days. *Source:* Reprinted with permission from Ref. [36].

Table 9.2 Characteristics of an ideal imaging technology for stem cell tracking during clinical trial.

Number	Characteristics
1.	Biocompatible, safe, and nontoxic
2.	No genetic modification or perturbation to the stem cell
3.	Single-cell detection at any anatomic location
4.	Quantification of cell number
5.	Minimal or no dilution with cell division
6.	Minimal or no transfer of contrast agent to nonstem cells
7.	Noninvasive imaging in the living subject over months to years
8.	No requirement for injectable contrast agent

Source: From Ref. [37].

9.2.2 Tracking *In Vivo*

Stem cells are a group of undifferentiated biological cells with the ability to self-renew and differentiate into various specialized cells [23]. These cells are delicate and fragile, and their properties including growth rate and differentiation capacity are prone to the interference by fluorescent labeling and gene transfection. Therefore, an ideal method for tracking stem cells must be biocompatible and nontoxic, require no genetic modification, have single-cell detection sensitivity, and permit quantification of cell numbers at any anatomic location [37]. FNDs are among the few nanomaterials that can fulfill just about all these requirements listed in Table 9.2.

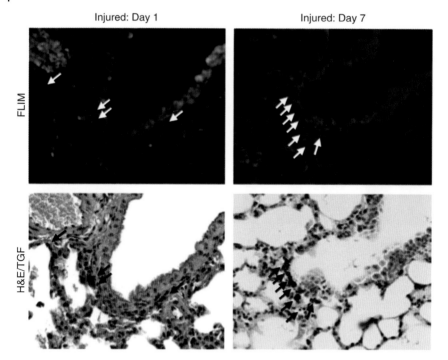

Figure 9.6 Tracking the engraftment and regenerative capability of transplanted lung stem cells using FNDs in a lung-injured mouse model. Lung tissue sections were examined on day 1 and 7 after intravenous injection of the FND-labeled lung stem cells. Arrows indicate the identified cells. *Source:* Adapted with permission from Ref. [26]. Reproduced with permission of Nature Publishing Group.

Wu et al. [26] pioneered the application of the FND labeling technique to track the engraftment and regenerative capability of lung stem cells (LSCs) transplanted into mice. *In vitro* experiments first verified that the FND labeling did not eliminate the properties of the LSCs extracted from neonatal mouse lung tissues to self-renew and differentiate into type I and type II pneumocytes. The FND-labeled LSCs were then injected into lung-injured mice via intravenous administration. Finally, mice were sacrificed and organs were collected at different time points to search for the injected LSCs by fluorescence microscopy. However, as commonly encountered in tissue section imaging, the high-level autofluorescence noises in the background prevent clear identification of the target cells in the specimens. To improve the image contrast, *fluorescence lifetime imaging microscopy* (FLIM) serves as an ideal approach because the NV⁻ centers in FNDs have a significantly longer emission lifetime than that of autofluorescence (Section 8.2.4). Figure 9.6 shows FLIM images of the lung tissue collected on day 1 and 7 [26]. The injected cells could be easily discerned, even when the cells were stained with hematoxylin and eosin (H&E) to aid histological examination. It was found that the FND-labeled LSCs preferentially resided at the terminal bronchioles of the lungs on day 7 after intravenous administration, as confirmed by *time-gated fluorescence* (TGF) imaging of the tissue sections of naphthalene-injured mice with single-cell resolution. Moreover, the damaged lung cells could rapidly recover after transplantation of the FND-labeled LSCs into the mice. The study offered new insights into the components that limited the acceptance of the transplanted stem cells as well as the mechanism of their regeneration within a host. The FND labeling method is simple and general, holding great potential for monitoring the homing of different kinds of stem cells (such as MSCs) in larger animal models (such as miniature pigs) for preclinical experimentation with important implications for future stem cell therapy and cancer treatment.

Very recently, a new platform has been developed using 100-nm FNDs for quantitative tracking of human placenta-derived MSCs in miniature pigs. Su et al. [28] labeled the cells with HSA-coated FNDs and determined with the MMF technique that each MSC could take up more than 60000 FND particles without noticeable excretion. At this remarkable loading capacity, no detrimental effects of the particles on the human stem cells were found, granting FND safe to use. The use of pigs as the animal model is highly forward-thinking because the animal shares many physiological similarities with humans and offer several distinct advantages (including size and availability) over other non-human primates for preclinical experimentation. With the magnetic modulation to achieve background-free detection, together with the TGF imaging as mentioned earlier, the research team was able to determine the numbers and the precise positions of the transplanted FND-labeled cells in organs and tissues of the miniature pigs after intravenous administration. The same method is applicable to single-cell imaging and quantitative tracking of human stem/progenitor cells in rodents, rabbits, monkeys, and other non-human primate animal models as well [38].

9.3 Deep Tissue Imaging

9.3.1 Wide-Field Fluorescence Imaging

Apart from playing the role as fluorescent labels for cells, FNDs can also be used as beacons to indicate the positions of tumors in patients. At present, the most commonly used imaging modalities in hospitals are magnetic resonance imaging, X-ray computed tomography, and ultrasound. These modalities have resolution high enough to provide anatomically relevant details but are low in sensitivity and difficult to be operated along with surgical treatment. A number of model organisms, including mice [39–44], rats [39], and pigs [28], have been deployed to evaluate the potential use of FNDs for such real-world biomedical applications. However, as discussed below, deep tissue imaging remains a challenge for FNDs in these animal models, clearly deserving further investigations.

Among the five different types of vacancy-related defects (Table 3.2), the NV⁻ center is most appealing for deep tissue imaging because it emits bright fluorescence in the far-red range. Nearly 70% of the fluorescence lies in the near-infrared (NIR) window from 650 to 1350 nm (Figure 9.7) [39], where light has the maximum penetration depth in biological tissue [45]. Another type of defect, GR1, is also a potential candidate because its emission band peaks at around 800 nm (Figure 3.5). While the fluorescence quantum efficiency of GR1 is low (only ~1%) [46], the deficiency can be compensated by increasing the vacancy density in the particles if necessary.

To assess the feasibility of deep tissue imaging with the NV⁻ centers, we have recently employed *fluorescent microdiamonds* (FMDs) of 400 μm in diameter and chicken breast tissue of different thickness for testing. Specifically, we covered two FMD particles (separated by ~200 μm) with 0-, 1.5-, 3-, or 5-mm thick chicken breast tissue and then imaged the samples with both bright-field and fluorescence microscopy. As shown in Figure 9.8a–d, the particles could be clearly seen under bright-field

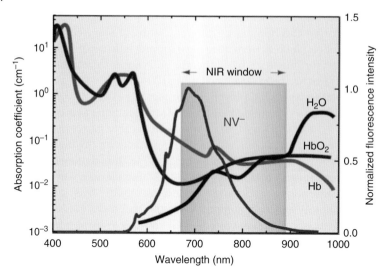

Figure 9.7 Comparison of the fluorescence spectrum (red curve) of FNDs with the near-infrared window of biological tissues. The black and gray curves are the absorption spectra of H$_2$O, oxygen-bound hemoglobin (HbO$_2$), and hemoglobin (Hb), respectively. *Source:* Reprinted with permission from Ref. [39].

imaging only when the chicken breast tissue was thinner than 1.5 mm. There was no problem of detecting the fluorescence emission of the NV$^-$ centers even when the chicken breast thickness increased to 5 mm (Figure 9.8e–h), showing the high sensitivity of this optical imaging method. However, although a spatial resolution of approximately 200 μm could be maintained at the tissue thickness of 3 mm, only a blurred light spot was found in the image taken for the 5-mm tissue due to the severe light scattering effect. The poor resolution reflected the limitation of the direct optical methods for deep tissue imaging. The limitation, however, can be lifted by exploiting the magnetic properties of NV$^-$ as discussed in the next section.

In an attempt to visualize FNDs in mice and rats, Vaijayanthimala et al. [39] used particles containing high-density ensembles (~10 ppm) of NV$^-$ centers to investigate their imaging capability and long-term stability in these model animals. The research team introduced 100-nm FNDs into rats through both intraperitoneal injection (Figure 9.9a) and subcutaneous injection (Figure 9.9b), followed by tracking the particles with a standard *in vivo* fluorescence imaging system. In order to facilitate the detection of FNDs with minimal interference from the tissue's background autofluorescence, the specimen was excited at 605 nm and its fluorescence was collected at 780 nm. Additionally, to improve the contrast, the primary images were corrected by subtracting the scaled background images taken by 430 nm excitation. It was found that the intensity of the FNDs in rats after subcutaneous injection stayed nearly the same for more than 37 days. Moreover, measurements of water consumption, fodder consumption, body weight, and organ index all showed no significant differences between the control and FND-treated groups of the rats subjected to intraperitoneal injection over five months (with a dose of 5 mg kg^{-1} body weight per week). The studies suggest that FNDs may serve as an ideal long-term marker for small animal models like mice and rats.

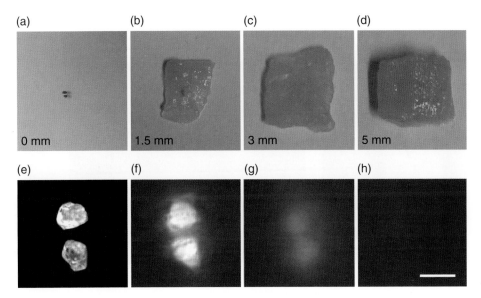

Figure 9.8 (a–d) Bright-field images of two FMD particles covered with chicken breast tissues with thickness of 0, 1.5, 3, and 5 mm, respectively. (e–h) Corresponding fluorescence images of FMDs in (a–d) illuminated by a continuous-wave laser operating at 637 nm.

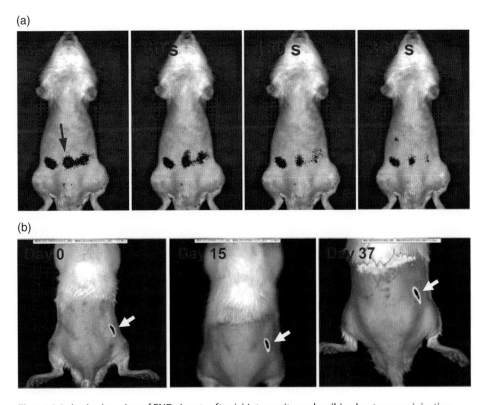

Figure 9.9 *In vivo* imaging of FNDs in rats after (a) intraperitoneal or (b) subcutaneous injection. The times indicated in both panels are the time points post-injection. White and blue arrows indicate the sites of injection. *Source:* Reprinted with permission from Ref. [39].

(a)   (b)

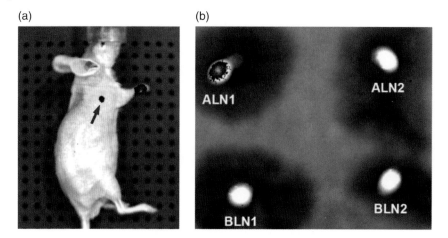

Figure 9.10 *In vivo* and *ex vivo* lymph node imaging of a nude mouse after intradermal injection. (a) Image showing the accumulation of BSA-FND particles in the right axillary lymph node (indicated by the blue arrow) on day 8. Note that most of the injected BSA-FND particles remain at the injection site. (b) *Ex vivo* fluorescence image of four extracted lymph nodes, where ALN1 and ALN2 are the lymph nodes located at the right and left axilla, respectively, and BLN1 and BLN2 are the lymph nodes located at the right and left brachial region, respectively. *Source:* Reprinted with permission from Ref. [39].

A more clinically relevant application of FNDs is the sentinel lymph node imaging. Sentinel lymph node is the hypothetical first lymph node (or group of nodes) draining a cancer, and thus its imaging is one of the most important and routine procedures in any cancer treatment [47]. In this study, FNDs were first administered into mice by intradermal injection and then monitored using a preclinical *in vivo* imaging system. Figure 9.10a and b show, respectively, the results of *in vivo* and *ex vivo* lymph node imaging of a nude mouse after intradermal injection with FNDs. Accumulation of the FND particles in the right axillary lymph node could be readily identified on day 8 after injection. The optical imaging, in combination with transmission electron microscopy, confirmed that the intradermally administered FNDs could be drained from the injection sites by macrophages and, in turn, selectively accumulated in the axillary lymph nodes of the treated mice.

Research along this line has also applied FNDs as a fluorescent tag to assess the disposition of aluminum oxyhydroxide (alum) *in vivo* [43]. Alum, with the chemical formula $AlO(OH)$, is a crystalline compound widely used as an immunologic adjuvant of vaccines. Although millions of people have received alum-adjuvanted vaccines to date [48], the long-term fate, residence time, accumulation, and impact of the particles in the human body are still not well understood. Alum particles are neither fluorescent nor magnetic, thereby requiring tagging with highly stable fluorophores in order to be tracked over a long period of time. Eidi et al. [43] conjugated hyperbranched polyglycerol with FNDs (~80 nm in diameter) to allow noncovalent coupling with the alum particles (~3 µm in diameter). The conjugates showed specificities comparable to those of the whole reference vaccine (anti-hepatitis B vaccine) in terms of particle sizes and zeta potentials (Section 4.1). Following the injection of the alum-FND conjugates in the tibialis muscle of mice for 21 days, the researchers sacrificed the animals and were able to find these conjugates in the sections of muscle, draining lymph nodes, spleen, liver, and brain tissue by wide-field fluorescence imaging. The detection of FNDs was easy to implement, allowing detailed assessment for the biodistribution of the adjuvants in tissues and organs down to the subcellular level.

9.3.2 Optically Detected Magnetic Resonance Imaging

Aiming at developing *optically detected magnetic resonance* (ODMR) into an *in vivo* imaging technique, Hegyi and Yablonovitch [49, 50] have set up a functional prototype instrument for high-resolution imaging of FNDs in tissue. The technique exploits the unique magneto-optic properties of FNDs, as elaborated in Section 3.4. It detects the fluorescence signals at the magnetic field-free point, where the crystal field splitting of NV$^-$ is in resonance with the external 2.87-GHz microwave radiation (Figure 3.9). In the presence of an amplitude-modified microwave field, a change of the fluorescence intensity (measured via a lock-in amplifier) proportional to the FND concentration results only at the field-free region. Sweeping the field-free point across the FND-containing tissue (or a living organism), one can obtain a quantitative "map" of the FND concentration by recording the change in fluorescence intensity as a function of position. The feasibility of this method was demonstrated by the researchers with multiple FND phantoms within chicken breast (Figure 9.11a and b), where the fluorescence of the NV$^-$ centers was selectively detected for samples located at the magnetic-field-free region,

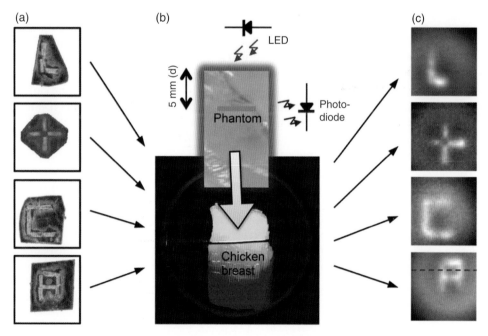

Figure 9.11 NV-based deep tissue imaging. (a) FND phantoms made of double-sticky tape and FNDs. (b) Chicken breast illuminated by a LED. The gray stripe on the piece of chicken breast represents the edge of the phantom, which is placed inside the chicken breast, 5 mm back from the front surface, facing the LED operating at 621 nm. Fluorescence is collected off to the side. (c) FND phantoms imaged outside of the chicken breast. *Source:* Reprinted with permission from Ref. [49]. Reproduced with permission of American Chemical Society.

achieving a sensitivity equivalent to 740 pg of FNDs in a 100-s measurement time and a spatial resolution of 800 µm over a 1-cm^2 field of view (Figure 9.11c). More significantly, a spatial resolution in the range of 100 µm was achievable even when the FND samples were situated about 12 mm below the chicken breast after deconvolution of images by point-spread functions. Such a resolution represents a remarkable advancement in all-optical deep tissue imaging.

The ODMR technique has similarly been applied to improve the image contrast of FNDs injected into the mouse model. In this protocol as detailed in Section 7.2.3, wide-field fluorescence images were first acquired either with or without 2.87-GHz microwave irradiation. Subtraction between the signals in these two images at every pixel led to selective detection of FNDs. With this technique, Igarashi et al. [40] successfully removed the background fluorescence signals from the images and achieved a high image contrast for long-term tracking of FND aggregates in the flank of a nude mouse at a depth of 400–500 µm. The protocol is applicable to an array of living systems, from single cells to whole organisms.

9.3.3 Time-Gated Fluorescence Imaging

The concept of single particle tracking introduced in Section 7.2 can be readily adapted to tracking single cells *in vivo*. Hui et al. [41] reported a new approach to achieve background-free fluorescence imaging of single FND-labeled cells in a mouse model by using an *intensified charge-coupled device* (ICCD) as the detector. ICCD is a highly sensitive camera capable of high-speed gating operations for wide-field imaging. The gating helps enhance the image contrast by suppressing the background fluorescence signals with shorter lifetimes than that of FNDs. Employing a nanosecond ICCD and a Raman shifter, the team demonstrated the first application of the device for wide-field fluorescence imaging of FND-labeled cells in mice. They used mouse lung cancer cells as the model cell line and introduced the cells labeled with 100-nm FNDs into mice via tail vein injection. TGF imaging was then conducted to establish a quiet background for the cells in the main blood vessel of the mouse's ear. Shown in Figure 9.12a is a typical bright-field and time-gated fluorescence image of the ear tissue. About 10 minutes after intravenous injection of the cells into the mouse, bright objects moving at an average speed of 0.4 µm s^{-1} could be clearly identified in the vessel (Figure 9.12b and c). The speed was lower than the blood flow velocity by almost three orders of magnitude, suggesting that the motion was most likely associated with the rolling, instead of coaxial flowing, of the FND-labeled cells in the blood vessel. This time-gating technique is promising for wide-field imaging and tracking of transplanted stem cells in small animal models for real-time observations of tissue repair and regeneration.

9.3.4 Magnetically Modulated Fluorescence Imaging

The latest advance in the field is led by the development of a wide-field background-free imaging technique based on magnetic modulation of the FND fluorescence as discussed in Section 6.2.4. Sarkar et al. [42] developed two methods to achieve this goal. The first method relied on pairwise subtraction of the images with and without a magnetic field, while the second was a phase-sensitive lock-in detection of the intensity-modulated FND

(a)　　　　　　　　　　　　(b)　　　　　　　　　　　　(c)

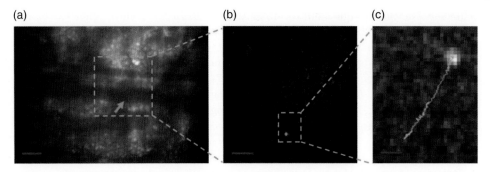

Figure 9.12 Tracking of single FND-labeled cells in a mouse ear. (a) Bright-field image of a mouse ear tissue. The green arrow indicates the position of an FND-labeled mouse lung cancer cell in the blood vessel of approximately 50 μm in diameter. (b) Enlarged view of the fluorescence image of the square green region in (a). The bright spot corresponds to the FND-labeled cell. (c) Enlarged view of the fluorescence image of the rectangular green region in (b), showing the trajectory of the FND-labeled lung cancer cell moving in the vessel. Scale bars: 100 nm in (a), 50 nm in (b), and 10 μm in (c). *Source:* Reprinted with permission from Ref. [41]. Reproduced with permission of Nature Publishing Group.

emission. They reported an improvement in the signal-to-noise ratio of up to 100-fold by conducting sentinel lymph mode mapping in a mouse model. These simple and relatively straightforward approaches to achieve background-free imaging of FNDs can be readily incorporated into existing microscopes or animal imaging systems. The technique has great potential to significantly improve and extend the FND-based imaging capabilities in diverse platforms, suitable for clinical applications. Compared with optically modulated fluorescence (Section 6.2.4), this approach is more attractive for whole animal imaging because of the efficient penetration of magnetic fields through tissues. The development of this and more background-free imaging modalities paves the way for the use of NV-containing diamond particles as biocompatible fiducial markers for image-guided therapy based on optical means [51].

References

1 American Society of Gene and Cell Therapy (2000). Gene therapy and cell therapy defined. http://www.asgct.org/general-public/educational-resources/gene-therapy--and-cell-therapy-defined (accessed 16 April 2018).

2 Trounson, A. and McDonald, C. (2015). Stem cell therapies in clinical trials: progress and challenges. *Cell Stem Cell* 17: 11–22.

3 Lefrère, J.J. and Berche, P. (2010). Doctor Brown-Sequard's therapy. *Ann Endocrinol (Paris)* 71: 69–75.

4 Sensebé, L. and Fleury-Cappellesso, S. (2013). Biodistribution of mesenchymal stem/stromal cells in a preclinical setting. *Stem Cells Int* 2013: 678063.

5 Zhang, L., Gu, F.X., Chan, J.M. et al. (2008). Nanoparticles in medicine: therapeutic applications and developments. *Clin Pharmacol Ther* 83: 761–769.

6 Medintz, I.L., Uyeda, H.T., Goldman, E.R., and Mattoussi, H. (2005). Quantum dot bioconjugates for imaging, labelling and sensing. *Nat Mater* 4: 435–446.

7 Tsoi, K.M., Dai, Q., Alman, B.A., and Chan, W.C. (2013). Are quantum dots toxic? Exploring the discrepancy between cell culture and animal studies. *Acc Chem Res* 46: 662–671.

8 Vaijayanthimala, V. and Chang, H.C. (2009). Functionalized fluorescent nanodiamonds for biomedical applications. *Nanomedicine* 4: 47–55.

9 Xing, Y. and Dai, L. (2009). Nanodiamonds for nanomedicine. *Nanomedicine* 4: 207–218.

10 Perevedentseva, E., Lin, Y.C., Jani, M., and Cheng, C.L. (2013). Biomedical applications of nanodiamonds in imaging and therapy. *Nanomedicine* 8: 2041–2060.

11 Kaur, R. and Badea, I. (2013). Nanodiamonds as novel nanomaterials for biomedical applications: drug delivery and imaging systems. *Int J Nanomedicine* 8: 203–220.

12 Doherty, G.J. and McMahon, H.T. (2009). Mechanisms of endocytosis. *Annu Rev Biochem* 78: 857–902.

13 Maxfield, F.R. and McGraw, T.E. (2004). Endocytic recycling. *Nat Rev Mol Cell Biol* 5: 121–132.

14 Vaijayanthimala, V., Tzeng, Y.K., Chang, H.C., and Li, C.L. (2009). The biocompatibility of fluorescent nanodiamonds and their mechanism of cellular uptake. *Nanotechnology* 20: 425103.

15 Faklaris, O., Garrot, D., Joshi, V. et al. (2008). Detection of single photoluminescent diamond nanoparticles in cells and study of the internalization pathway. *Small* 4: 2236–2239.

16 Faklaris, O., Joshi, V., Irinopoulou, T. et al. (2009). Photoluminescent diamond nanoparticles for cell labeling: study of the uptake mechanism in mammalian cells. *ACS Nano* 3: 3955–3962.

17 Zhang, B.L., Li, Y.Q., Fang, C.Y. et al. (2009). Receptor-mediated cellular uptake of folate-conjugated fluorescent nanodiamonds: a combined ensemble and single-particle study. *Small* 5: 2716–2721.

18 Perevedentseva, E., Hong, S.F., Huang, K.J. et al. (2013). Nanodiamond internalization in cells and the cell uptake mechanism. *J Nanopart Res* 15: 1384.

19 Chu, Z., Miu, K., Lung, P. et al. (2015). Rapid endosomal escape of prickly nanodiamonds: implications for gene delivery. *Sci Rep* 5: 11661.

20 Ivanov, A.I. (2008). Pharmacological inhibition of endocytic pathways: is it specific enough to be useful? *Methods Mol Biol* 440: 15–33.

21 Brenner, S.L. and Korn, E.D. (1979). Substoichiometric concentrations of cytochalasin D inhibit actin polymerization. Additional evidence for an F-actin treadmill. *J Biol Chem* 254: 9982–9985.

22 Samson, F., Donoso, J.A., Heller-Bettinger, I. et al. (1979). Nocodazole action on tubulin assembly, axonal ultrastructure and fast axoplasmic transport. *J Pharmacol Exp Ther* 208: 411–417.

23 Alberts, B., Johnson, A., Lewis, J. et al. (2014). *Molecular Biology of the Cell*, 6e. Garland Science.

24 Chu, Z., Zhang, S., Zhang, B. et al. (2014). Unambiguous observation of shape effects on cellular fate of nanoparticles. *Sci Rep* 4: 4495.

25 Fang, C.Y., Vaijayanthimala, V., Cheng, C.A. et al. (2011). The exocytosis of fluorescent nanodiamond and its use as a long-term cell tracker. *Small* 7: 3363–3370.

26 Wu, T.J., Tzeng, Y.K., Chang, W.W. et al. (2013). Tracking the engraftment and regenerative capabilities of transplanted lung stem cells using fluorescent nanodiamonds. *Nat Nanotechnol* 8: 682–689.

27 Huang, Y.A., Kao, C.W., Liu, K.K. et al. (2014). The effect of fluorescent nanodiamonds on neuronal survival and morphogenesis. *Sci Rep* 4: 6919.

28 Su, L.J., Wu, M.H., Hui, Y.Y. et al. (2017). Fluorescent nanodiamonds enable quantitative tracking of human mesenchymal stem cells in miniature pigs. *Sci Rep* 7: 45607.

29 Liu, K.K., Qiu, W.R., Raj, E.N. et al. (2017). Ubiquitin-coated nanodiamonds bind to autophagy receptors for entry into the selective autophagy pathway. *Autophagy* 13: 187–200.

30 Mizushima, N. and Komatsu, M. (2011). Autophagy: renovation of cells and tissues. *Cell* 147: 728–741.

31 Svenning, S. and Johansen, T. (2013). Selective autophagy. *Essays Biochem* 55: 79–92.

32 Yuan, Y., Chen, Y.W., Liu, J.H. et al. (2009). Biodistribution and fate of nanodiamonds *in vivo*. *Diam Relat Mater* 18: 95–100.

33 Doudrick, K., Corson, N., Oberdörster, G. et al. (2013). Extraction and quantification of carbon nanotubes in biological matrices with application to rat lung tissue. *ACS Nano* 7: 8849–8856.

34 Freshney, R.I. (2006). Basic principles of cell culture. In: *Culture of Cells for Tissue Engineering* (ed. G. Vunjak-Novakovic and R.I. Freshney), 1–22. New York: Wiley.

35 Dean, M., Fojo, T., and Bates, S. (2005). Tumour stem cells and drug resistance. *Nat Rev Cancer* 5: 275–284.

36 Lin, H.H., Lee, H.W., Lin, R.J. et al. (2015). Tracking and finding slow-proliferating/quiescent cancer stem cells with fluorescent nanodiamonds. *Small* 11: 4394–4402.

37 Frangioni, J.V. and Hajjar, R.J. (2004). *In vivo* tracking of stem cells for clinical trials in cardiovascular disease. *Circulation* 110: 3378–3383.

38 Moore, L., Yang, J., Lan, T.T. et al. (2016). Biocompatibility assessment of detonation nanodiamond in non-human primates and rats using histological, hematologic, and urine analysis. *ACS Nano* 10: 7385–7400.

39 Vaijayanthimala, V., Cheng, P.Y., Yeh, S.H. et al. (2012). The long-term stability and biocompatibility of fluorescent nanodiamond as an *in vivo* contrast agent. *Biomaterials* 33: 7794–7802.

40 Igarashi, R., Yoshinari, Y., Yokota, H. et al. (2012). Real-time background-free selective imaging of fluorescent nanodiamonds *in vivo*. *Nano Lett* 12: 5726–5732.

41 Hui, Y.Y., Su, L.J., Chen, O.Y. et al. (2014). Wide-field imaging and flow cytometric analysis of cancer cells in blood by fluorescent nanodiamond labeling and time gating. *Sci Rep* 4: 5574.

42 Sarkar, S.K., Bumb, A., Wu, X. et al. (2014). Wide-field *in vivo* background free imaging by selective magnetic modulation of nanodiamond fluorescence. *Biomed Opt Express* 5: 1190–1202.

43 Eidi, H., David, M., Crépeaux, G. et al. (2015). Fluorescent nanodiamonds as a relevant tag for the assessment of alum adjuvant particle biodisposition. *BMC Med* 13: 144.

44 Suarez-Kelly, L.P., Campbell, A.R., Rampersaud, I.V. et al. (2017). Fluorescent nanodiamonds engage innate immune effector cells: a potential vehicle for targeted anti-tumor immunotherapy. *Nanomedicine* 13: 909–920.

45 Weissleder, R. and Ntziachristos, V. (2003). Shedding light onto live molecular targets. *Nat Med* 9: 123–128.

46 Davies, G., Thomaz, M.F., Nazare, M.H. et al. (1987). The radiative decay time of luminescence from the vacancy in diamond. *J Phys C Solid State Phys* 20: L13–L17.

47 Jakub, J.W., Pendas, S., and Reintgen, D.S. (2003). Current status of sentinel lymph node mapping and biopsy: facts and controversies. *Oncologist* 8: 59–68.

48 Moxon, E.R. and Siegrist, C.A. (2011). The next decade of vaccines: societal and scientific challenges. *Lancet* 378: 348–359.

49 Hegyi, A. and Yablonovitch, E. (2013). Molecular imaging by optically detected electron spin resonance of nitrogen-vacancies in nanodiamonds. *Nano Lett* 13: 1173–1178.

50 Hegyi, A. and Yablonovitch, E. (2014). Nanodiamond molecular imaging with enhanced contrast and expanded field of view. *J Biomed Opt* 19: 011015.

51 Chen, G.T., Sharp, G.C., and Mori, S. (2009). A review of image-guided radiotherapy. *Radiol Phys Technol* 2: 1–12.

10

Nanoscopic Imaging

We have learned a great deal so far about what *fluorescent nanodiamonds* (FNDs) can do, for example, from detecting a single biomolecule (Chapter 7) to imaging a whole organism (Chapter 9). Now the question is how. How to make all these wonders of science? Specifically, in a tiny world of nanoparticles, what tools are necessary to study FNDs? What do we have to know in order to make the best use of these tools?

To "see" FNDs in action, researchers have been relying on fluorescence microscopy, which has made profound advance throughout past decades in enhancing our understanding of molecular and cellular biology. As powerful as it is, however, fluorescence microscopy is limited by the diffraction of light to elucidate detailed structure of the organelles in cells. Considering a microscopic object to consist of diffraction gratings, Ernst Abbe concluded in 1873 that the resolution limit of a microscope is one half the wavelength of the light used for illumination [1],

$$d_{min} = \frac{\lambda}{2\sin\alpha},$$
(10.1)

where d_{min} is the minimum resolvable distance, λ is the wavelength of the light, and α is the half aperture angle of the microscope's objective [2]. For green light of 532 nm in wavelength as an example, the Abbe limit is $d_{min} = \lambda/2 = 266$ nm (or 0.266 μm), which is small compared to most biological cells (~10 μm), but large compared to viruses (~100 nm), proteins (~5 nm), and small biomolecules (<1 nm). Applying Eq. (10.1) to optical imaging of nanoparticles (typically smaller than 100 nm in diameter) suggests that there exist some ambiguities and uncertainties in the identification of single FND particles either spin-coated on glass substrates or endocytosed in cells (Chapter 7). To achieve subdiffraction imaging of the individual FNDs or even their color centers embedded in the diamond matrix with visible light, a variety of superresolution fluorescence methods have been developed and deployed. We discuss in this chapter the research and development of these subdiffraction methods, along with the associated electron microscopic techniques including *cathodoluminescence* (CL) microscopy and *correlative light-electron microscopy* (CLEM).

Fluorescent Nanodiamonds, First Edition. Huan-Cheng Chang, Wesley Wei-Wen Hsiao and Meng-Chih Su.
© 2019 John Wiley & Sons Ltd. Published 2019 by John Wiley & Sons Ltd.

10.1 Diffraction Barrier

To achieve superresolution imaging, we will have to break the *diffraction barrier* as explained here. Diffraction is a fundamental property of light, involving a change in the direction of light waves when passing through an opening or around an obstacle in the light path [3]. A well-known example of light diffraction is the single-slit diffraction, which produces a pattern of varying intensity at the far-field due to the interference of the diffractive waves traveling along different paths. Treating the circular aperture of a lens as a two-dimensional version of the single slit, theoretical modeling with the *Fraunhofer diffraction equation* at the far-field shows that the diffracted light will create a series of concentric rings, known as the *Airy disk* (Figure 10.1), at the focal plane of the converging lens. The diffraction pattern is symmetric about the normal axis and its intensity profile is given by [3, 4]

$$I(r) \propto \left[\frac{J_1(\pi Dr/\lambda f)}{\pi Dr/\lambda f} \right]^2, \tag{10.2}$$

where r is the radius of the Airy disk at the focal plane, D is the aperture diameter, λ is the wavelength of the light, f is the focal length of the lens, and $J_1(x)$ is the *Bessel function* of the first kind. The first zero of the Bessel function occurs at $\pi Dr_1/\lambda f = 3.832$ and therefore

$$r_1 = \frac{1.22\lambda}{D/f} \approx \frac{1.22\lambda}{2NA}, \tag{10.3}$$

where NA is the numerical aperture of the lens. The approximation holds when NA is small, and the equation is known as the *Rayleigh criterion*, which specifies the minimum separation between two point sources that may be resolved into two distinct objects.

According to Eq. (10.1), the diffraction barrier (or the resolution limitation) restricts the ability of a microscope to distinguish between two objects separated by a lateral distance of approximately *half* the wavelength of the light used for imaging. Therefore, to increase the resolution, light with shorter wavelengths (such as ultraviolet and X-ray)

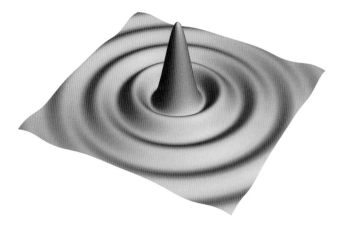

Figure 10.1 An Airy disk resulting from the diffraction of light on a circular aperture.

should be used. However, the application often suffers from the lack of image contrast and the risk of radiation damage, making them unsuitable for live cell imaging. To meet the demands for ultra-sharp resolution, scientists around the world have invested in tremendous amounts of time and effort over the past few decades, and their diligent hard work has led to remarkable progresses, finally breaking the diffraction barrier and bringing optical microscopy to proper use in nanometric dimensions [5–7]. The endeavors were fully recognized in 2014 when the Nobel Prize in Chemistry was awarded to Eric Betzig, Stefan W. Hell, and William E. Moerner "for the development of super-resolved fluorescence microscopy" [8].

10.2 Superresolution Fluorescence Imaging

10.2.1 Stimulated Emission Depletion Microscopy

A landmark in developing subdiffraction imaging is the *stimulated emission depletion* (STED) microscopy. The technique involves the use of two laser beams of different wavelengths and shapes [9]. During a STED measurement, the first laser beam (a Gaussian beam) brings the fluorophore of interest to its excited state. The second laser beam, which has a doughnut shape at the focus of the objective lens, depletes the emission from all excited molecules except those in the middle of the doughnut, resulting in a nanometer-sized volume (Figure 10.2) [10]. Consequently, the final "fluorescent volume" becomes apparently smaller than the diffraction limit of the first laser beam. Naturally, the smaller the fluorescent volume is, the higher the image resolution becomes. Depending on the intensity of the second laser beam (or the STED beam), the

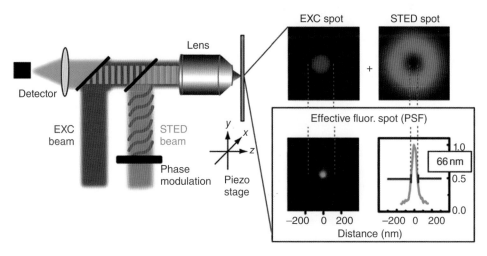

Figure 10.2 Principle of STED microscopy. A blue excitation (EXC) beam is focused to a diffraction-limited excitation spot, shown in the adjacent panel in blue, while the orange STED beam de-excites molecules. The STED beam is phase-modulated to form the focal doughnut shown in the top right panel. Superimposition of the two focal spots confines the area in which emission is only possible in the doughnut center, yielding the effective fluorescent spot of subdiffraction size shown in green in the lower panel. *Source:* Adapted with permission from Ref. [10].

fluorescence emission can be confined to a spot with a diameter much smaller than the wavelength of light.

Hell and coworkers [11] have shown that in the presence of the STED beam, the Abbe's equation should be modified to

$$d \approx \frac{\lambda}{2n \sin \alpha \sqrt{1 + I/I_{sat}}}, \qquad (10.4)$$

where n is the refractive index of the medium, I is the laser intensity, and I_{sat} is the saturation intensity at which the probability of the fluorescence emission is reduced by half. STED microscopy is straightforward, requiring no elaborate post-data processing. In principle, it is the most suitable method for high-resolution imaging of live cells at real time and in three dimensions. However, due to the fact that the diffraction limit is inversely proportional to the square root of $(1 + I/I_{sat})$, a STED beam with a high laser power is needed to effectively improve the spatial resolution. For example, using a STED laser power of $I = 8I_{sat}$ will result in only a factor of 3 in resolution improvement. The demand for such a high laser power severely limits the applicability of the method in bioimaging because most molecular fluorophores (such as fluorescein and fluorescent proteins) photobleach rapidly under intense laser illumination. In contrast, the nitrogen-vacancy (NV⁻) centers in FNDs are perfectly photostable without photobleaching and blinking, representing one of the few exceptions. Therefore, FNDs are well suited for use as a contrast agent for superresolution imaging by STED microscopy [12–20].

STED was first applied to detect single NV⁻ centers in bulk diamond with a remarkable resolution down to 5.8 nm [12]. Using the same technique (532 nm laser pulses for excitation and 775 nm laser pulses for depletion), Han et al. [13] effectively procured high-resolution images of 35 nm FNDs spin-coated on a glass plate, reporting a resolution of 40 nm likely limited by the size of the particles. No photobleaching was found even under STED illumination with a laser power as high as 160 mW. In follow-up experiments, the research team made further efforts to conduct controlled addressing and readout of the NV⁻ spin states to resolve the individual NV⁻ centers in FNDs. It was demonstrated by Arroyo-Camejo et al. [14] that the STED microscopy could detect single NV⁻ centers in FNDs with a resolution of 10 nm. Even multiple adjacent NVs located in a single FND could be imaged individually down to a relative distance of 15 nm (Figure 10.3a). These centers were found located predominantly near the center of the particle as confirmed by superimposing images acquired by STED and scanning electron microscopy (Figure 10.3b).

Inspired by the success of superresolution imaging of NV⁻ centers, Tzeng et al. [15] applied STED to probe single FNDs in fixed HeLa cells. To prevent particle agglomeration in cell medium, acid-treated FNDs (~35 nm in diameter) were first noncovalently coated with bovine serum albumin (BSA) and then delivered to the cell cytoplasm by endocytosis (Section 8.1). Using 532-nm light for excitation and a doughnut-shaped 740-nm laser beam for depletion, they attained a spatial resolution of approximately 40 nm (cf., captions in Figure 10.4 for details). The resolution allowed for the identification of single isolated FND particles in cells and distinguished them from FND aggregates confined in the endosomes. The study provided new insight into the FND uptake mechanism of cells, as discussed in Section 9.1, with single-particle sensitivity and nanometric resolution. Meanwhile, Prabhakar et al. [16] applied a similar STED technique to probe the endosome merging into late endosomes and estimated the drug load taken up by individual cells with silica-coated FNDs as the imaging agents as well as the

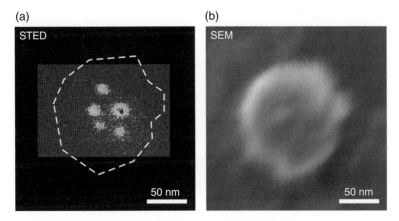

(a)

STED

50 nm

(b)

SEM

50 nm

Figure 10.3 (a) STED and (b) SEM images of a single FND particle. *Source:* Reprinted with permission from Ref. [14]. Reproduced with permission of American Chemical Society.

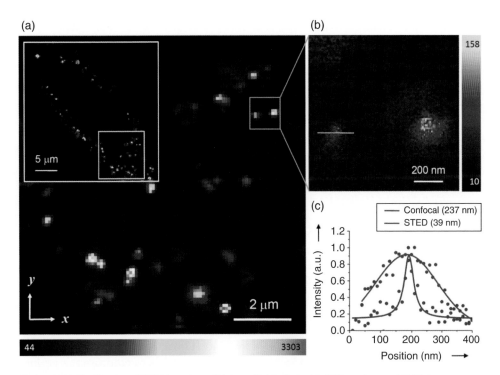

Figure 10.4 Confocal and STED imaging of HeLa cells labeled with BSA-conjugated FNDs by endocytosis. (a) Confocal image acquired by raster scanning of an FND-labeled cell. The fluorescence image of the entire cell is shown in the white box and demonstrates fairly uniform cell labeling by BSA-conjugated FNDs. (b) STED image of single BSA-FND particles enclosed within the green box in (a). (c) Confocal and STED fluorescence intensity profiles of the particle are indicated in (b) with a blue line. Solid curves are best fits to one-dimensional Gaussian (confocal) or Lorentzian (STED) functions. The corresponding full widths at half-maximum are given in parentheses. *Source:* Reprinted with permission from Ref. [15]. Reproduced with permission of John Wiley & Sons.

drug carriers. An improvement of the spatial resolution by a factor of 5 was readily achieved by using a high-power STED laser to deplete the NV^- emission. Such improvements in optical imaging are crucial for revealing the detailed structures of biological macromolecular complexes and assemblies, including cellular organelles and subcellular compartments.

In addition to red FNDs, green FNDs containing the N–V–N centers (or H3 in Section 3.3.3) are also amenable to subdiffraction imaging with STED microscopy. Laporte and Psaltis [19] first measured the absorption cross-section of the H3 centers in green FNDs to be $\sigma = 1.18 \times 10^{-17}$ cm^2 at 470 nm using a saturation method similar to that described in Section 7.1.2 for the NV^- centers. Next, STED images were obtained by employing a 488-nm pulsed laser for excitation and a 590-nm continuous-wave laser for depletion. In order to attain high spatial resolution (better than 70 nm), a depletion laser intensity on the order of 100 MW cm^{-2} was used. The researchers exemplified the application of green FNDs as a non-photobleaching STED marker by imaging 70-nm FNDs taken up into HeLa cells. The availability of more than one photostable color centers of FNDs has opened a door to perform multicolor imaging and spatial correlation studies with STED.

While green and red FNDs show great potentials, superresolution imaging of them in live cells by STED remains a challenge because the use of a high-power STED beam often causes photodamage to the target cells. A variation of STED is ground-state depletion, which requires relatively modest laser intensity (1 MW cm^{-2} or less) [21]. Hell and Kroug [22] first proposed in 1995 that if the doughnut-shaped depletion beam could excite the fluorophores to a long-lived dark state, i.e. a state where little or no fluorescence occurs, images with high resolution could also be obtained. Applying the same principle and taking advantage of the photochromic properties of NV centers (Figure 3.7), Chen et al. [23, 24] developed a technique known as *charge-state depletion microscopy* to achieve superresolution. Specifically, two pulsed laser beams (637 and 532 nm) were used to initialize and switch the charge states, and the third pulsed laser (589 nm) detected the NV^- centers. The experiment started with the photoionization of nearly all NV^- centers to form NV^0 with the 637-nm Gaussian laser beam, only to be followed by a 532-nm doughnut-shaped laser beam for the photoconversion of NV^0 to NV^-, which was finally detected with the 589-nm light. The researchers were able to resolve the individual NV centers in bulk diamond with a resolution as fine as 4.1 nm, where the centers appeared as dark spots in the fluorescence image. In practical bioimaging, it is preferable to have the fluorophores appearing as bright spots, instead of dark spots. This was achievable by changing the roles of the 637- and 532-nm laser beams at the expense of resolution. Despite that the resolution was significantly reduced to 28.6 nm, the imaging required the use of lasers with a power density on the order of 1 MW cm^{-2}, which represented a factor of 100 lower than that used in STED to achieve comparable resolution. The technique makes it possible to perform superresolution imaging of FNDs in live cells.

Finally, FNDs have been applied as photostable nanoprobes to compare the performance between STED microscopy and *structured illumination microscopy* (SIM), which is another useful technique to achieve superresolution imaging [25]. Side-by-side comparison of the resolution of these two methods with 35-nm FNDs at the single particle level showed that STED provided more structural details, whereas SIM offered a larger field of view with a higher imaging speed [18]. However, SIM tended to produce deconvolution smoothing and orientational artifacts due to the post-data processing. In a separate study, FNDs were also employed to investigate the polarization effects in lattice-STED microscopy, where superresolution imaging with 75-nm resolution and a field of view of 7.5 µm × 7.5 µm was achieved [20]. Notably, the exceptional photostability of the particles made it possible to image 35-nm FNDs at different optical lattice configurations and compare the experimental measurements with theoretical simulations.

10.2.2 Saturated Excitation Fluorescence Microscopy

In conventional fluorescence microscopy, the instrument is typically operated in the linear mode, namely, the measured fluorescence intensity is linearly proportional to the excitation energy. However, saturation may occur if the excitation intensity elevates while the number of molecules in the excitation volume stays the same. This nonlinear effect is more prominent in the center than at the edge of the laser focus as the intensity distribution is Airy-like as illustrated in Figure 10.1. One can thus improve the spatial resolution if the nonlinear components in the fluorescence intensity distribution are extracted for imaging. The concept can be realized by temporally modulating the excitation laser intensity at a frequency (ω) and demodulating the measured fluorescence intensity at the corresponding harmonic frequencies (i.e. 2ω, 3ω, …), where the signals are produced due to the nonlinear effect. Theoretical simulations show that high harmonic fluorescence signals contribute most to the resolution improvement and a higher resolution is attained at a higher demodulation frequency (Figure 10.5) [26]. Approximately a twofold improvement is achievable in both lateral and axial directions, if the eighth harmonic signals are detected for imaging.

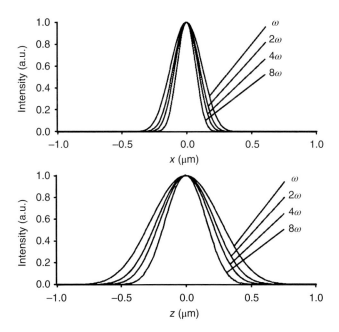

Figure 10.5 Fluorescence intensity profiles on the lateral (*x*) and axial (*z*) axes, calculated for different demodulation frequencies in saturated excitation fluorescence microscopy. The excitation intensities used to obtain the demodulated intensity profiles at ω, 2ω, 4ω, and 8ω are 1.5, 15, 400, and 700 kW cm^{-2}, respectively. *Source:* Reprinted with permission from Ref. [26]. Reproduced with permission of American Physical Society.

An intrinsic difficulty associated with this so-called *saturated excitation fluorescence microscopy* is that it requires high illumination intensities to saturate fluorophores in the excited states. Similar to STED, the high laser intensities often result in rapid photobleaching. Fujita and coworkers [27] utilized FNDs as photostable fluorescent nanoprobes to circumvent this problem. They confirmed that FNDs exhibited a nonlinear fluorescence response under the saturated excitation conditions generated by intense laser light. A lateral resolution of 140 nm could be achieved at the demodulation frequency 4ω and a laser intensity of $25\,\mathrm{kW\,cm^{-2}}$. The exceptionally high photostability of FNDs allowed the researchers to quantify the spatial resolution improvement inherent in this method and demonstrate the scalability of the spatial resolution of this nonlinear superresolution method. The use of FNDs also enabled the research team to perform fluorescence imaging of multicolor-stained macrophage cells with a spatial resolution beyond the diffraction limit.

10.2.3 Deterministic Emitter Switch Microscopy

The capability of detecting single NV^- spins in FNDs by *optically detected magnetic resonance* (ODMR) offers a different route to superresolution imaging. The fundamental principle behind this method is that the degeneracy of the $m_s = \pm 1$ sublevels at the ground state of NV^- is lifted in the presence of a magnetic field via the Zeeman effect (Figure 10.6a), and the energy difference between these two sublevels depends on the strength of the applied magnetic field as well as the orientation of the magnetic moment of the NV^- center with respect to the field. In bulk diamond, the NV^- center has four possible orientations and thus four distinct Zeeman splittings (Figure 10.6b and c). If only one pair of the spin resonance peaks is observed in the ODMR spectrum, it is an indication for the presence of a single NV^- center in the laser probe volume (cf., Figure 3.9). This unique magneto-optical property makes it possible to achieve superresolution of differently oriented centers by multispectral imaging in the microwave domain. Dolde et al. [28] have demonstrated the principle by exploiting the spin addressability for subdiffraction microscopy of two NV^- centers in bulk diamond. Compared with single-spin *stochastic optical reconstruction microscopy* (STORM) [29], which utilizes a low-power 594 nm laser to induce frequent switching between NV^- and NV^0 centers, the ODMR method is deterministic and more straightforward, requiring less post-data processing.

With random orientation on a substrate, FNDs, in principle, have an unlimited number of NV^- centers that may be distinguishably addressed with this ODMR-based superresolution imaging technique. However, in reality, the number is limited by the width of the spin resonance peaks. On the basis of the peak width (typically 20 MHz) and the field-dependent peak shift ($\sim 28\,\mathrm{MHz\,mT^{-1}}$), Chen et al. [30] estimated that at an applied magnetic field of 20 mT in strength, it is possible to resolve more than 50 uniquely addressable NV^- centers within a diffraction-limited spot. However, instead of conducting experiments in confocal mode, a technique called *wide-field deterministic emitter switch microscopy* was developed to achieve high-speed, subdiffraction imaging with low laser intensities across a wide field of view. Under resonant microwave excitation, only the NV^- centers with correct orientation showed a reduction in fluorescence intensities and could be clearly identified by subtraction of the signals with and without the presence of the resonant microwave excitation (Figure 10.6d). By

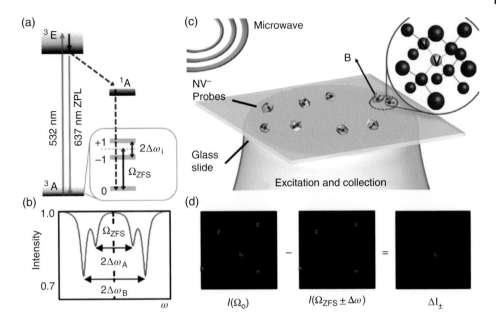

Figure 10.6 (a) Energy level diagram of NV⁻, showing how preferential shelving of the $m_s = \pm$ excited states (^3E) into the dark metastable state (^1A) gives rise to a typical electron spin resonance spectrum. (b) Fluorescence of two NV⁻ centers in the presence of a static magnetic field as a function of applied microwave frequency. The splitting of the two dips ($\Delta\omega_A$ and $\Delta\omega_B$) is given by the projection of the incident magnetic field on the magnetic moment of the NV⁻. The contrast of each dip has an inverted Lorentzian shape. (c) Illustration of NV⁻ centers in a static magnetic field, each having field splitting frequencies corresponding to their uniquely oriented magnetic moments relative to the magnetic field. (d) Schematic diagram of the method for resolving a switchable emitter by taking the difference between two images where a uniquely addressed emitter is dimmed and not dimmed by resonant microwave excitation at $I(\Omega_{ZFS} \pm \Delta\omega)$ and $I(\Omega_o)$, respectively. ZPL, zero-phonon line; ZFS, zero-field splitting. *Source:* Adapted with permission from Ref. [30]. Reproduced with permission of American Physical Society.

using this technique, the research team was able to resolve two FNDs, each containing a single NV⁻ center, physically separated by 55 nm apart and determine their locations down to 12-nm precision over a wide area of $35 \times 35 \, \mu m^2$.

10.2.4 Tip-Enhanced Fluorescence Microscopy

The diffraction barrier, i.e. Eq. (10.3), exists when the diffraction pattern is viewed at the focal plane of a converging lens or when it is viewed at a long distance from the aperture, known as the *far field* [1]. Various near-field techniques have been developed to extend optical microscopy beyond the diffraction limit [31]. The introduction of an aperture probe to enable near-field imaging, known as *near-field scanning optical microscopy*, has proven to be a successful approach [32, 33]; however, the low light throughput from the probe ($\sim 10^{-4}$ for a 100 nm aperture), together with some practical complications in the implementation, prevents widespread use of the technique. To overcome these limitations, researchers in the field have invented an alternative method by replacing the aperture probe with a sharp metal tip to achieve far-field superresolution fluorescence imaging [34].

A metal tip is known to enhance both Raman and fluorescence signals due to plasmon oscillation at the tip [35]. Plasmon oscillation is a collective response of electrons, which can produce a highly localized electric field that in turn can accelerate the radiative and non-radiative decay rates of nearby fluorophores. The field enhancement is so tightly confined to the vicinity of the tip apex that the method is applicable to improve the optical resolution to below the diffraction limit of light. The applicability of the tip-enhance fluorescence microscopy was first demonstrated with dye molecules [34] and quantum dots [36, 37], achieving a spatial resolution in the range of 20 nm.

Hui et al. [38] extended the study to FNDs at the single particle level by applying radially polarized light to excite the plasmon oscillation of the gold tip of an *atomic force microscope* (AFM). With FNDs of approximately 30 nm in diameter (Figure 10.7a) and an experimental setup similar to that displayed in Figure 7.4a, their work showed that the gold tip increased not only the fluorescence intensity of the NV⁻ centers in FNDs but also the optical resolution of the fluorescence image of the particles down to 40 nm (Figure 10.7b and c). An average fluorescence intensity enhancement by a factor of three could be achieved when the tip was placed in close contact with the particle. From a measurement of the fluorescence decay lifetime and the saturation intensity, it was confirmed that the fluorescence enhancement was contributed predominantly by an increase of the radiative decay rate. Using a similar setup, Beams et al. [39] also investigated the near-field optical properties of FNDs and found that the NV⁻ centers in the particles could serve as a sensitive probe of the surrounding electromagnetic mode structure. Equipped with the enhanced sensitivity, the research team attained the local density of states for an optical antenna by fluorescence lifetime imaging of the single NV⁻ centers.

10.3   Cathodoluminescence Imaging

CL is the emission of photons when an electron strikes a luminescent material. It is a tool commonly used to characterize materials with nanometric resolution based on their dispersed fluorescence spectra. Consider an electron beam of 20 keV in energy. The wavelength of the electrons in the beam is $\lambda = 27$ pm, according to the *de Broglie equation* [40],

$$\lambda = \frac{h}{mv} = \frac{h}{\sqrt{2meV}}, \tag{10.5}$$

where h is the Planck constant, m is the mass of an electron, v is the velocity, e is the charge, and V is the acceleration voltage. This calculated wavelength is much smaller than that of visible photons ($\lambda = 400$–800 nm) and thus capable of resolving features that cannot be visualized by optical microscopy. The resolution of an electron microscope is theoretically unlimited, depending only on the energy of the electrons. In a standard *scanning electron microscope* (SEM), which typically operates in the energy range of 1–40 keV, the attainable resolution is on the order of a few tens of nanometers, whereas when using a *transmission electron microscope* (TEM), which typically operates in the energy range of 100–400 keV, nanometer-sized structures can be readily resolved.

A CL microscope typically consists of an SEM, a light collection system, and a fluorescence spectrometer [41]. By scanning the electron beam and analyzing the light

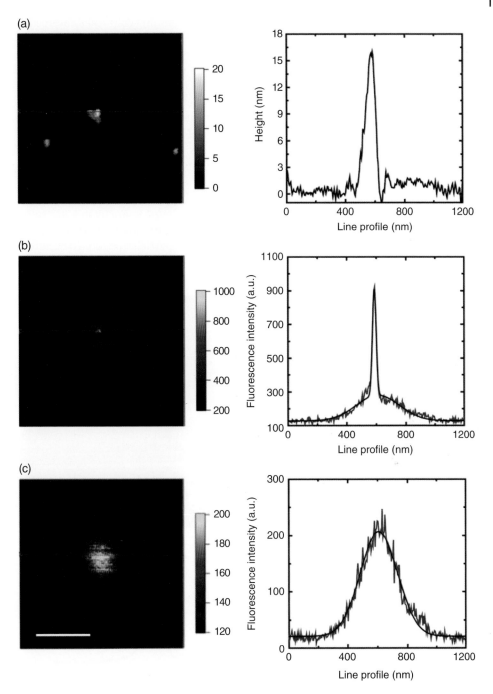

Figure 10.7 (a) AFM image, (b, c) fluorescence images, and their corresponding line profiles of a 30-nm FND particle (b) with and (c) without a gold tip close to it. The resolution of the fluorescence image is significantly improved by the tip enhancement effect from (c) 290 nm to (b) 40 nm. The gray scale in (a) is in unit of nm. Black solid curves are best Gaussian fits to the experimental data. Scale bar: 400 nm. *Source:* Adapted with permission from Ref. [38]. Reproduced with permission of AIP Publishing LLC.

emitted at each point, similar to the mode used in laser scanning confocal fluorescence microscopy, a map of the optical activity of the specimen can be obtained. However, unlike optical photons, the electron beam often causes destruction of the biological samples under examination, leading to low intensity and rapid degradation of the signals of the fluorescent labels. A possible solution to this problem is to use robust nanoparticles containing photostable and spectrally distinct fluorophores as the biolabels. FND is among the few nanoparticles that can meet the stringent requirements. Using FNDs, Glenn et al. [42] demonstrated the value of red FNDs for CL imaging by acquiring well-correlated CL and secondary electron images of the individual FND particles in the size range of 40–80 nm. Much superior to fluorescence imaging, the CL microscopy provided a spatial resolution in the region of 5 nm, surpassing the diffraction limit of light. Similarly, Kociak and coworkers [43–45] studied the spectrally and spatially resolved CL of FNDs using a focused electron beam equipped in a *scanning transmission electron microscope* (STEM). They were able to identify different color centers, including H3 and NV^0, in a mixture of green and red FND particles (Figure 10.8a and b) through careful analysis of the CL spectra (Figure 10.8c and d) [44]. Their work verified

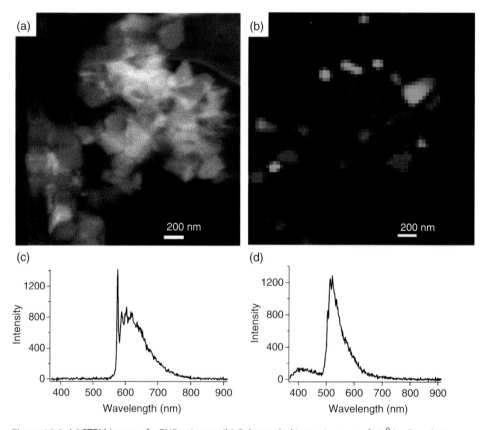

Figure 10.8 (a) STEM image of a FND mixture. (b) Color-coded intensity map of NV^0 (red) and H3 (green) centers in the FNDs. (c, d) Typical CL spectra of FNDs containing NV^0 and H3 centers in (c) and (d), respectively. *Source:* Reprinted with permission from Ref. [44]. Reproduced with permission of John Wiley & Sons.

the feasibility of STEM for ultrahigh-resolution CL imaging of multicolor centers in FNDs. In all these studies, no CL was observed from NV⁻, probably due to the charge state conversion of the centers from NV^- to NV^0 during electron irradiation.

In a separate study, high spatial resolution imaging of FNDs in live cells was performed with a technique called *direct electron beam excitation-assisted fluorescence microscopy*. In this work, Nawa et al. [46] used a focused electron beam to excite both green and red FNDs in HeLa cells cultured on a silicon nitride membrane attached to an SEM. Similar to *atmospheric scanning electron microscopy* [47], the electron beam emitting from the SEM passed through the thin membrane (~50 nm in thickness), irradiated the specimen in solution, and generated CL signals, which were finally collected by an optical microscope for detection. Two-color imaging demonstrated that these two different types of FNDs could be simultaneously observed within the same cells with nanometer resolution (cf., captions in Figure 10.9 for details). The technique is potentially applicable for multicolor immunostaining after proper conjugation of FNDs with antibodies to elucidate various cellular functions (Section 4.2). By using the

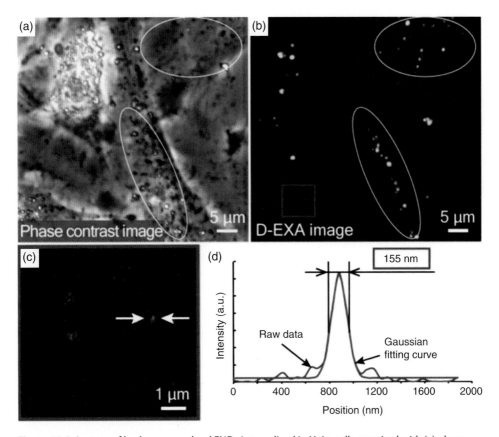

Figure 10.9 Images of both green and red FNDs internalized in HeLa cells, acquired with (a) phase contrast microscopy and (b) direct electron beam excitation-assisted (D-EXA) fluorescence microscopy. The magnified D-EXA image of the area marked with a red square in (c) shows a spatial resolution of 155 nm in (d). *Source:* Reprinted with permission from Ref. [46]. Reproduced with permission of John Wiley & Sons.

bioconjugated FNDs along with these new methods, researchers have been able to achieve molecular localization and, simultaneously, reveal the structural details of cellular organelles with STEM [48].

10.4   Correlative Light-Electron Microscopy

CLEM is a technique recently developed to analyze the same specimen with both *light microscopy* (LM) and *electron microscopy* (EM) [49]. It combines the advantages of the multicolor versatility of LM with the high-resolution power of EM for molecular and cellular biology research. However, different from CL, CLEM images are acquired with two entirely different instruments and therefore colocalization of the same objects at the micrometer or even nanometer scale is crucial. A number of nanomaterials have been developed as fluorescent fiducial markers for CLEM, including dye-labeled nanogolds, dye-doped polystyrene beads, and quantum dots [50]. However, these nanomaterials are not sufficiently stable to allow routine CLEM analysis because, in order to avoid damaging the markers during sample preparation, the specimens must be fixed by high-pressure freezing, which is a highly specialized technique and a time-consuming process. Moreover, the correlative analysis cannot be repeatedly carried out due to fluorescence quenching of these markers by heavy metal staining with reagents, such as OsO_4 and uranyl acetate, required for EM imaging. The use of CLEM for nanometric localization of biological molecules in cells remains a challenge, despite that some successful cases have been reported in the literature [51].

For any molecule or material to serve as a good fluorescent fiducial marker for CLEM, the prerequisites are: highly fluorescent and electron-dense. FND is appealing for this application because the nanomaterial has both high light-emitting capability and a dense carbon core that can be visualized by EM [52–57]. In addition, the detection of the nanomaterial by LM imaging is highly compatible with the post-embedding technique, which involves staining of the specimens with heavy metals for structural preservation in EM. The high compatibility stems from the fact that the fluorescent centers of FNDs are so deeply buried inside the diamond matrix that their fluorescence properties can hardly be altered by the environmental changes (Section 6.2). An added benefit is that background-free detection of FNDs on the grids and in the sections can be achieved by time-gated fluorescence imaging (Section 8.2.4). Therefore, reliable co-registration of LM and EM images is both possible and practical with FNDs as the fluorescent fiducial markers.

An application of CLEM to image and localize the positions of antigens on cell surface was first made by Hsieh et al. [56] using lipid-encapsulated FNDs. The researchers encapsulated surface-oxidized FNDs in lipids, followed by grafting of the lipid layers with biotin molecules (Section 4.3.1). The biotinylated lipid-coated FNDs (bL-FNDs) were then used to target the surface antigen CD44 on HeLa cells by sandwich immunostaining (Section 8.1.2). Figure 10.10a shows a SEM image of HeLa cells labeled with bL-FNDs. Clearly, the SEM image provided considerably more structural information than the corresponding LM images. Despite that some bL-FNDs could be identified under high magnification (Figure 10.10b and c), it was still challenging to unambiguously differentiate bL-FNDs from the complex nanostructures on the cell surface. Taking advantage of the exceptionally high chemical and photophysical stability of

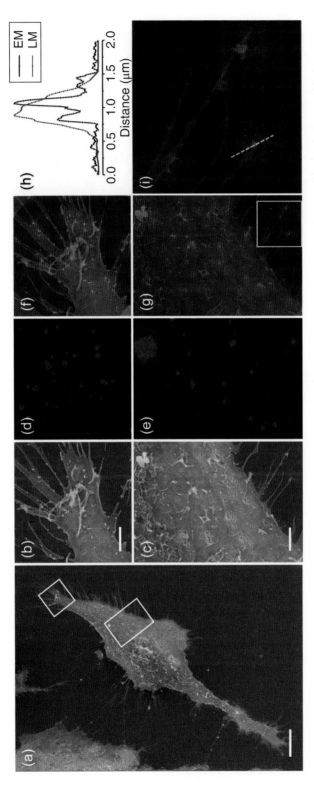

Figure 10.10 (a) SEM images of a HeLa cell labeled with anti-CD44 antibody, neutravidin, and bL-FND. (b–g) Enlarged views of (b, c) SEM images, (d, e) fluorescence images, and (f, g) CLEM images of the white boxes in (a). The tilt angles to achieve complete overlaps of the FNDs are 39° and 1° for (b, d) and (c, e), respectively. (h) Intensity profiles of the white dashed line drawn in panel (i), which is an enlarged view of the white box in (g). Scale bars: 10μm in (a) and 2μm in (b, c). *Source:* Adapted with permission from Ref. [56].

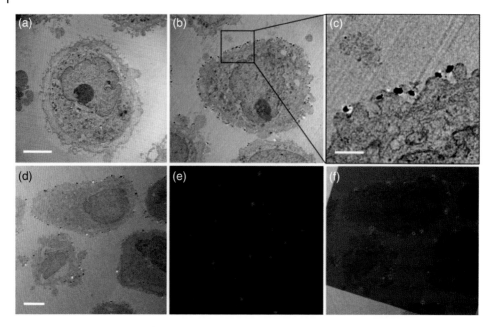

Figure 10.11 (a) TEM image of suspended HeLa cells without labeling. (b–d) TEM images of suspended HeLa cells labeled with biotin-anti-CD44 antibody, neutravidin, and then biotinylated lipid-coated FND. (e, f) Fluorescence and CLEM images of the same sample in (d). The electron energy used to obtain the TEM images is 120 kV. Scale bars: 5 μm in (a, d) and 1 μm in (c).

FNDs as detailed earlier, the research team was able to detect their fluorescence readily even after strong irradiation of the Pt-coated cell samples with 10-kV electrons in SEM (Figure 10.10d and e). By superimposing the LM and EM images with one of them tilted by a certain angle, the locations of the CD44 antigens on the HeLa cell membrane could be determined with an accuracy of 50 nm, which is about 1/5 that of the diffraction limit of light (Figure 10.10f–i).

Apart from SEM, the FND-labeling technique is also compatible with TEM. Figure 10.11a–c displays TEM images of HeLa cells without and with labeling. Many single isolated FNDs were found on the cell surface. To obtain LM images for the same specimen after TEM imaging, the fluorescence signals were collected with a delay time of more than 3 ns after pulsed laser excitation to reduce the background signals arising from the plastic-like resins, in which the samples were embedded. Using this time-gating technique, the researchers were able to identify clearly the individual 100-nm FND particles on the surface of the cells in the thin-sectioned resins. More importantly, the EM image (Figure 10.11d) and the time-gated LM image (Figure 10.11e) were literally superimposed with each other (Figure 10.11f). Up to 30 particles showed excellent co-registration between these two images when the fluorescence image was tilted by 17°. The high brightness and superior photostability of the particles enabled the positions of the CD44 antigens to be located with an accuracy of better than 50 nm, limited mainly by the size of the FND particles. This combined LM and EM analysis, along with the use of smaller FNDs, has opened the door to locate the positions of antigens on cell surface with unprecedented precision and accuracy.

References

1 Lauterbach, M.A. (2012). Finding, defining and breaking the diffraction barrier in microscopy – a historical perspective. *Opt Nanoscopy* 1: 8.

2 Abbe, E. (1873). Beiträge zur Theorie des Mikroskops und der mikroskopischen Wahrnehmung. M. Schultze's. *Archiv für mikroskopische Anatomie* 9: 413–468.

3 Fowles, G.R. (1989). *Introduction to Modern Optics*, 2e. New York: Dover.

4 Dullemond, C.P., van Boekel, R. and Quirrenbach, A. (2010). Chapter 3 Diffraction. Observational Astronomy (MKEP5). http://www.ita.uni-heidelberg.de/~dullemond/lectures/obsastro_2010/Chapter_Diffraction.pdf (accessed 16 April 2018).

5 Betzig, E. (2015). Single molecules, cells, and super-resolution optics (Nobel lecture). *Angew Chem Int Ed* 54: 8034–8053.

6 Hell, S.W. (2015). Nanoscopy with focused light (Nobel lecture). *Angew Chem Int Ed* 54: 8054–8066.

7 Moerner, W.E. (2015). Single-molecule spectroscopy, imaging, and photocontrol: foundations for super-resolution microscopy (Nobel lecture). *Angew Chem Int Ed* 54: 8067–8093.

8 The Nobel Prize in Chemistry 2014. *Nobelprize.org*. Nobel Media AB 2014. https://www.nobelprize.org/nobel_prizes/chemistry/laureates/2014 (accessed 16 April 2018).

9 Hell, S.W. and Wichmann, J. (1994). Breaking the diffraction resolution limit by stimulated emission: stimulated-emission-depletion fluorescence microscopy. *Opt Lett* 19: 780–782.

10 Willig, K.I., Rizzoli, S.O., Westphal, V. et al. (2006). STED microscopy reveals that synaptotagmin remains clustered after synaptic vesicle exocytosis. *Nature* 440: 935–939.

11 Westphal, V. and Hell, S.W. (2005). Nanoscale resolution in the focal plane of an optical microscope. *Phys Rev Lett* 94: 143903.

12 Rittweger, E., Han, K.Y., Irvine, S.E. et al. (2009). STED microscopy reveals crystal colour centres with nanometric resolution. *Nat Photon* 3: 144–147.

13 Han, K.Y., Willig, K.I., Rittweger, E. et al. (2009). Three-dimensional stimulated emission depletion microscopy of nitrogen-vacancy centers in diamond using continuous-wave light. *Nano Lett* 9: 3323–3329.

14 Arroyo-Camejo, S., Adam, M.P., Besbes, M. et al. (2013). Stimulated emission depletion microscopy resolves individual nitrogen vacancy centers in diamond nanocrystals. *ACS Nano* 7: 10912–10919.

15 Tzeng, Y.K., Faklaris, O., Chang, B.M. et al. (2011). Superresolution imaging of albumin-conjugated fluorescent nanodiamonds in cells by stimulated emission depletion. *Angew Chem Int Ed* 50: 2262–2265.

16 Prabhakar, N., Näreoja, T., von Haartman, E. et al. (2013). Core-shell designs of photoluminescent nanodiamonds with porous silica coatings for bioimaging and drug delivery II: application. *Nanoscale* 5: 3713–3722.

17 Lai, N.D., Faklaris, O., Zheng, D. et al. (2013). Quenching nitrogen-vacancy center photoluminescence with infrared pulsed laser. *New J Phys* 15: 033030.

18 Yang, X., Tzeng, Y.K., Zhu, Z. et al. (2014). Sub-diffraction imaging of nitrogen-vacancy centers in diamond by stimulated emission depletion and structured illumination. *RSC Adv* 4: 11305–11310.

19 Laporte, G. and Psaltis, D. (2015). STED imaging of green fluorescent nanodiamonds containing nitrogen-vacancy-nitrogen centers. *Biomed Opt Express* 7: 34–44.

20 Yang, B., Fang, C.Y., Chang, H.C. et al. (2015). Polarization effects in lattice-STED microscopy. *Faraday Discuss* 184: 37–49.

21 Bretschneider, S., Eggeling, C., and Hell, S.W. (2007). Breaking the diffraction barrier in fluorescence microscopy by optical shelving. *Phys Rev Lett* 98: 218103.

22 Hell, S.W. and Kroug, M. (1995). Ground-state-depletion fluorescence microscopy: a concept for breaking the diffraction resolution limit. *Appl Phys B Laser Opt* 60: 495–497.

23 Chen, X., Zou, C., Gong, Z. et al. (2015). Subdiffraction optical manipulation of the charge state of nitrogen vacancy center in diamond. *Light Sci Appl* 4: e230.

24 Chen, X.D., Li, S., Shen, A. et al. (2017). Near-infrared-enhanced charge-state conversion for low-power optical nanoscopy with nitrogen-vacancy centers in diamond. *Phys Rev Appl* 7: 014008.

25 Allen, J.R., Ross, S.T., and Davidson, M.W. (2014). Structured illumination microscopy for superresolution. *ChemPhysChem* 15: 566–576.

26 Fujita, K., Kobayashi, M., Kawano, S. et al. (2007). High-resolution confocal microscopy by saturated excitation of fluorescence. *Phys Rev Lett* 99: 228105.

27 Yamanaka, M., Tzeng, Y.K., Kawano, S. et al. (2011). SAX microscopy with fluorescent nanodiamond probes for high-resolution fluorescence imaging. *Biomed Opt Express* 2: 1946–1954.

28 Dolde, F., Jakobi, I., Naydenov, B. et al. (2013). Room-temperature entanglement between single defect spins in diamond. *Nat Phys* 8: 1–5.

29 Pfender, M., Aslam, N., Waldherr, G. et al. (2014). Single-spin stochastic optical reconstruction microscopy. *Proc Natl Acad Sci USA* 111: 14669–14674.

30 Chen, E.H., Gaathon, O., Trusheim, M.E., and Englund, D. (2013). Wide-field multispectral super-resolution imaging using spin-dependent fluorescence in nanodiamonds. *Nano Lett* 13: 2073–2077.

31 Dunn, R.C. (1999). Near-field scanning optical microscopy. *Chem Rev* 99: 2891–2928.

32 Pohl, D.W., Denk, W., and Lanz, M. (1984). Optical stethoscopy: image recording with resolution λ/20. *Appl Phys Lett* 44: 651–653.

33 Betzig, E. and Trautman, J.K. (1992). Near-field optics: microscopy, spectroscopy, and surface modification beyond the diffraction limit. *Science* 257: 189–195.

34 Sanchez, E.J., Novotny, L., and Xie, X.S. (1999). Near-field fluorescence microscopy based on two-photon excitation with metal tips. *Phys Rev Lett* 82: 4014–4017.

35 Mauser, N. and Hartschuh, A. (2014). Tip-enhanced near-field optical microscopy. *Chem Soc Rev* 43: 1248–1262.

36 Gerton, J.M., Wade, L.A., Lessard, G.A. et al. (2004). Tip-enhanced fluorescence microscopy at 10 nanometer resolution. *Phys Rev Lett* 93: 180801.

37 Huang, F.M., Festy, F., and Richards, D. (2005). Tip-enhanced fluorescence imaging of quantum dots. *Appl Phys Lett* 87: 183101.

38 Hui, Y.Y., Lu, Y.C., Su, L.J. et al. (2013). Tip-enhanced sub-diffraction fluorescence imaging of nitrogen-vacancy centers in nanodiamonds. *Appl Phys Lett* 102: 013102.

39 Beams, R., Smith, D., Johnson, T.W. et al. (2013). Nanoscale fluorescence lifetime imaging of an optical antenna with a single diamond NV center. *Nano Lett* 13: 3807–3811.

40 Halliday, D., Walker, J., and Resnick, R. (2010). *Fundamentals of Physics*, 5e, 993–995. Wiley.

41 Yacobi, B.G. and Holt, D.B. (1990). *Cathodoluminescence Microscopy of Inorganic Solids*. Springer.

42 Glenn, D.R., Zhang, H., Kasthuri, N. et al. (2012). Correlative light and electron microscopy using cathodoluminescence from nanoparticles with distinguishable colours. *Sci Rep* 2: 865.

43 Tizei, L.H. and Kociak, M. (2012). Spectrally and spatially resolved cathodoluminescence of nanodiamonds: local variations of the NV^0 emission properties. *Nanotechnology* 23: 175702.

44 Tizei, L.H.G., Meuret, S., Nagarajan, S. et al. (2013). Spatially and spectrally resolved cathodoluminescence with fast electrons: a tool for background subtraction in luminescence intensity second-order correlation measurements applied to subwavelength inhomogeneous diamond nanocrystals. *Phys Status Solidi A* 210: 2060–2065.

45 Meuret, S., Tizei, L.H., Cazimajou, T. et al. (2015). Photon bunching in cathodoluminescence. *Phys Rev Lett* 114: 197401.

46 Nawa, Y., Inami, W., Lin, S. et al. (2014). Multi-color imaging of fluorescent nanodiamonds in living HeLa cells using direct electron-beam excitation. *ChemPhysChem* 15: 721–726.

47 Nishiyama, H., Suga, M., Ogura, T. et al. (2010). Atmospheric scanning electron microscope observes cells and tissues in open medium through silicon nitride film. *J Struct Biol* 169: 438–449.

48 Nagarajan, S., Pioche-Durieu, C., Tizei, L.H. et al. (2016). Simultaneous cathodoluminescence and electron microscopy cytometry of cellular vesicles labeled with fluorescent nanodiamonds. *Nanoscale* 8: 11588–11594.

49 de Boer, P., Hoogenboom, J.P., and Giepmans, B.N. (2015). Correlated light and electron microscopy: ultrastructure lights up. *Nat Methods* 12: 503–513.

50 Kukulski, W., Schorb, M., Welsch, S. et al. (2012). Precise, correlated fluorescence microscopy and electron tomography of lowicryl sections using fluorescent fiducial markers. *Meth Cell Biol* 111: 235–257.

51 Muller-Reichert, T. and Verkade, P. (ed.) (2102). *Correlative Light and Electron Microscopy*. Elsevier Science.

52 Zurbuchen, M.A., Lake, M.P., Kohan, S.A. et al. (2013). Nanodiamond landmarks for subcellular multimodal optical and electron imaging. *Sci Rep* 3: 2668.

53 Lake, M.P. and Bouchard, L.S. (2017). Targeted nanodiamonds for identification of subcellular protein assemblies in mammalian cells. *PLoS One* 12: e0179295.

54 Hemelaar, S.R., de Boer, P., Chipaux, M. et al. (2017). Nanodiamonds as multi-purpose labels for microscopy. *Sci Rep* 7: 720.

55 Prabhakar, N., Peurla, M., Koho, S. et al. (2018). STED-TEM correlative microscopy leveraging nanodiamonds as intracellular dual-contrast markers. *Small* 14: 1701807.

56 Hsieh, F.J., Chen, Y.W., Huang, Y.K. et al. (2018). Correlative light-electron microscopy of lipid-encapsulated fluorescent nanodiamonds for nanometric localization of cell surface antigens. *Anal Chem* 90: 1566–1571.

57 Sotoma, S., Hsieh, F.J., Chen, Y.W. et al. (2018). Highly stable lipid-encapsulation of fluorescent nanodiamonds for bioimaging applications. *Chem Commun* 54: 1000–1003.

11

Nanoscale Quantum Sensing

Richard Feynman predicted in his 1959 speech, *There's Plenty of Room at the Bottom* (Section 1.1), "When we get to the very, very small world – say circuits of seven atoms – we have a lot of new things that would happen that represent completely new opportunities for design. Atoms on a small scale behave like nothing on a large scale, for they satisfy the laws of quantum mechanics. ... We can use, not just circuits, but some system involving the quantized energy levels, or the interactions of quantized spins, etc." [1]. Indeed, 60 years later, Feynman's "quantized spins" is today's quantum reality.

As the name suggests, a nanoscale sensor (or nanosensor) is a device that can detect events or changes in its surrounding environment at the nanoscale or within a distance range of 1–1000 nm. The events or changes detected may be physical, chemical, or biological in nature. Examples of the use of nanosensors in the life sciences include high-sensitivity detection of DNA and measurement of local temperature changes in living cells. In order to serve as an effective nanosensor, the device must be smaller than 100 nm and the output signals should be detectable with high sensitivity, preferably, at the single molecule or particle level. Once properly conjugated with biologically important molecules or other functional nanomaterials, these devices can be designed for use in nanomedicine such as to identify and destroy cancer cells as well as repair abnormal or damaged structures in the human body. Diamondoids and nanodiamonds, in fact, have been proposed as the building blocks of nanorobots and nanomachines for medicinal purposes [2], although realization of the concept is still far in the future.

In the previous chapters, we have demonstrated that *fluorescent nanodiamonds* (FNDs) containing NV$^-$ centers fulfill most of the requirements of being an ideal nanoparticle platform for biomedical applications. The sensing capability of the NV$^-$ centers, however, has not yet been explored. In this chapter, we discuss the principle and practice of using the color centers for high-sensitivity temperature probing, along with magnetic field sensing and spin detection at the nanoscale. More applications of NV$^-$ for nanoscale quantum sensing using bulk diamonds can be found elsewhere [3] and, therefore, are not discussed here. As all these measurements are made possible by utilizing the magneto-optical properties of the NV$^-$ centers, this chapter starts with a brief review of the spin Hamiltonian associated with this unique atom-like quantum system.

Fluorescent Nanodiamonds, First Edition. Huan-Cheng Chang, Wesley Wei-Wen Hsiao and Meng-Chih Su.
© 2019 John Wiley & Sons Ltd. Published 2019 by John Wiley & Sons Ltd.

11.1 The Spin Hamiltonian

Spin is an intrinsic property of elementary particles like the electron. It does not have a counterpart in classical mechanics and, according to quantum mechanics, is present in the form of *angular momentum*. The spin can interact with the orbital angular momentum to yield a total angular momentum for a given particle. Analogous to the orbital angular momentum operator, the spin angular momentum operator is a vector operator, $S = (S_x, S_y, S_z)$, and

$$S^2 = S_x^2 + S_y^2 + S_z^2, \tag{11.1}$$

where S_x, S_y, and S_z are the components along the x, y, and z coordinates. The eigenvalues of S^2 are

$$S(S+1)\hbar^2, \quad S = 0, \frac{1}{2}, 1, \frac{3}{2}, \dots \tag{11.2}$$

and the eigenvalues of S_z are

$$m_s \hbar, \quad m_s = -S, -S+1, \dots, S-1, S, \tag{11.3}$$

where S is the *spin quantum number*, $\hbar = h/2\pi$, and h is the Planck constant [4].

In Section 3.4, we have briefly discussed how spin–spin interactions may give rise to the crystal field splitting and how applying an external magnetic field can lift the degeneracy of the $m_s = \pm 1$ sublevels of the NV⁻ center in diamond. To describe these two types of interactions quantitatively, we write the spin Hamiltonian as [5]

$$H = S \cdot D \cdot S + g_e \mu_B B \cdot S, \tag{11.4}$$

where D is the spin–spin coupling tensor, $g_e \sim 2.003$, $\mu_B = 14.00$ MHz mT⁻¹ is the Bohr magneton, and B is the applied magnetic field. For a quantum system with axial symmetry and without crystal strains (such as NV⁻ in a perfect diamond crystal), the Hamiltonian is given by

$$H = D\left[S_z^2 - \frac{1}{3}S(S+1)\right] + g_e \mu_B B \cdot S. \tag{11.5}$$

In the absence of the magnetic field, e.g. $B = 0$, the eigenvalues of the Hamiltonian of the $S = 1$ system are

$$E_{\pm 1} = \frac{1}{3}D \tag{11.6}$$

$$E_0 = -\frac{2}{3}D, \tag{11.7}$$

where D is the crystal field splitting or zero field splitting. In the cases where the magnetic field is nonzero, static, and aligned along the z-direction, the eigenvalues of the spin Hamiltonian are

$$E_{\pm 1} = \frac{1}{3}D \pm g_e \mu_B B, \tag{11.8}$$

$$E_0 = -\frac{2}{3}D. \tag{11.9}$$

The quantum mechanical description above indicates that the transitions between the electron spin states can be induced by applying microwave radiation in resonance with the crystal field splittings. When the microwave radiation is switched off, the electrons in the upper states will relax back to their lower energy states with two different decay time constants. The longer of the two time constants, T_1, is often called the *spin–lattice relaxation time* (or *longitudinal relaxation time*). The shorter time constant, T_2, is the *spin–spin relaxation time* (or *transverse relaxation time*). The T_1 relaxation arises from equilibration of the excess spin energy with its surroundings (the *lattice*) through the coupling with lattice motions that have approximately the right frequency to interact with the spins. The sharing of the excess energy between spins is the cause of the T_2 relaxation [5].

Another noted feature in the spin resonance spectroscopy is that, compared with stimulated emission, the spontaneous emission process in the microwave region is negligible at room temperature. According to the Einstein relation between the stimulated emission rate, $B\rho(\nu)$, and the spontaneous emission rate, A [6]:

$$B\rho(v) = \frac{c^3 A}{8\pi h v^3} \rho(v),$$ (11.10)

one has

$$B\rho(v) = \frac{A}{e^{hv/k_B T} - 1},$$ (11.11)

where $\rho(\nu)$ is the blackbody radiation field, c is the speed of light, h is the Planck constant, ν is the emission frequency, k_B is the Boltzmann constant, and T is the temperature. At $T = 300$ K and $\nu \sim 10^{15}$ Hz as in the visible region, $h\nu/k_B T \sim 10^2$ and thus $A \gg B\rho(\nu)$. However, the relation between these two rates is reversed in the microwave region, where $\nu \sim 10^{10}$ Hz and $A/B\rho(\nu) \sim 10^{-3}$, meaning that the stimulated emission dominates over the spontaneous emission, which may be disregarded.

This brief quantum mechanics refresher of electron spins paves the way for the following discussion of nanoscale temperature sensing.

11.2 Temperature Sensing

11.2.1 Ultrahigh Precision Temperature Measurement

The NV⁻ center in diamond is an artificial solid-state atom with optically addressable spin states. It has a crystal field splitting of $D = 2.87$ GHz between $m_s = 0$ and ± 1, which can be optically read out under ambient conditions even at the single spin level with the *optically detected magnetic resonance* (ODMR) technique [7], as described in Section 3.4. Remarkably, the spin states have an exceptionally long coherence time (e.g. T_2), capable of remaining coherent for more than 1 ms in isotope-free bulk diamond at room temperature [8]. Both the spin resonance and coherence time of the NV⁻ center are sensitive to the variations of temperature, magnetic field, electric field, and mechanical stress present in its environment [3]. This renders NV⁻ a useful tool for ultra-sensitive measurements of these physical parameters at the nanoscale, provided that the centers are implanted in nanoscale diamonds or near the surface of bulk diamonds.

Acosta et al. [9] were the first to investigate the effects of temperature on the spin resonance of the NV⁻ centers. Using ODMR, they measured the temperature-dependent shift of the zero field splitting from 280 to 330 K for an ensemble (10 ppb–15 ppm) of NV⁻ in bulk diamond. A thermal shift of $\Delta D/\Delta T = -75\,\text{kHz}\,\text{K}^{-1}$ was determined at around 300 K and the red-shift was attributed to thermal expansion of the diamond crystal lattice [9, 10]. Later studies over a wider temperature range for a single NV⁻ center found that the frequency shifting is actually nonlinear with respect to temperature; for example, it may change to $\Delta D/\Delta T = -150\,\text{kHz}\,\text{K}^{-1}$ at 600 K [11]. A polynomial that can properly describe the temperature dependence of the spin resonance frequency over 300–700 K is [11]

$$D(T) = 2.8697 + 9.7 \times 10^{-5}T - 3.7 \times 10^{-7}T^2 + 1.7 \times 10^{-10}T^3 \qquad (11.12)$$

where $D(T)$ is in unit of gigahertz and T in Kelvin. While the magnitude of the thermal shift is not substantial at room temperature, it can be measured with high precision due to the exceptional stability of the ODMR spectra. Another distinctive feature of the ODMR spectra is that the widths of the spin resonance peaks show little or almost no dependence on temperature up to 600 K due to the long spin coherence time. It has an important implication for the application of single NV⁻ centers as nanoscale thermometers over a broad temperature range, as discussed in the next section.

The concept of using FND for nanothermometry was initially disclosed in a study of thermal effects on the fluorescence lifetimes of NV⁻ centers. With 30-nm FNDs heated in an oven, Plakhotnik and Gruber [12] observed a 2.7-fold decrease in the fluorescence lifetime from 300 to 670 K. The heating–cooling cycle was reversible, suggesting the potential use of this nanomaterial as a luminescence thermometer with nanometric spatial resolution. Further studies of the thermal shifts of D for FNDs showed results in good agreement with that of bulk diamond [13]. Compared with other known luminescence thermometers composed of nanoparticles such as semiconductor quantum dots, rare earth doped oxides, and *gold nanoparticles* (GNPs) [14, 15], FNDs stand out for having a wider range of working temperature and single-particle detection sensitivity.

Kucsko et al. [16] reported the first application of FNDs as nanoscale thermometers in 2013. They obtained the ODMR spectra of single 100-nm FNDs spin-coated on a glass slide using a green laser for excitation (Figure 11.1a and b) and were able to determine the temperatures with a precision of 0.1 K or better within a measurement time of four seconds (Figure 11.1c). To induce local temperature changes, GNPs were introduced into the samples and subsequently irradiated by a separate laser to serve as the heat source (denoted in Figure 11.1b). A spatial resolution of 200 nm was achieved by measuring the separations of the nanoparticles in fluorescence images (Figure 11.1d). The results, along with other similar studies [17], showed that it is possible to detect the temperature variations at the nanoscale with a sensitivity of $0.1\,\text{K}\,\text{Hz}^{-1/2}$ if the FNDs contain up to 1000 NV⁻ centers per particle. Further improvement of the temperature sensing sensitivity to the $1\,\text{mK}\,\text{Hz}^{-1/2}$ regime is achievable by utilizing the quantum coherence of single NV⁻ spins in bulk diamonds (Figure 11.2) [16–18]. These characteristics place FND in a category of the most sensitive and stable temperature probes known to date.

In light of its excellent biocompatibility, a natural (and logical) application of the FND nanosensor is to measure local temperatures in living cells. Kucsko et al. [16]

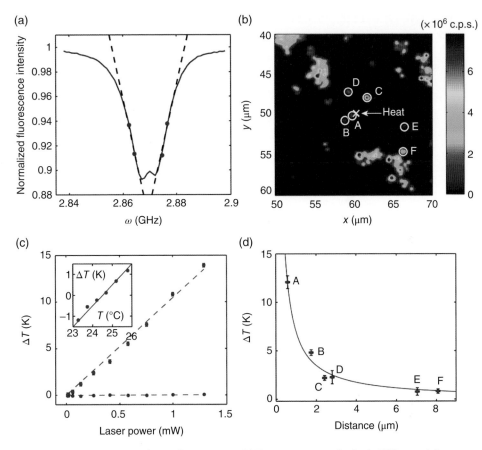

Figure 11.1 Magneto-optical nanothermometry. (a) Frequency scan of a single FND containing approximately 500 NV⁻ centers. The four red points indicate the measurement frequencies used to extract the temperature. (b) Two-dimensional confocal scan of FNDs (circles) and gold nanoparticles (cross) spin-coated onto a glass coverslip. (c) Temperature rise of a single FND as a function of laser power for two different laser-focus locations with (red) or without (blue) laser illumination on a nearby gold nanoparticle. (d) Temperature changes measured at six FND locations, indicated by circles in (b), as a function of the distance from the illuminated gold nanoparticle indicated by the cross. The blue curve represents the best-fitting of the temperature profile to a steady-state solution of the heat conduction equation. *Source:* Adapted with permission from Ref. [16]. Reproduced with permission of Nature Publishing Group.

demonstrated this applicability by introducing FNDs and GNPs (both ~100 nm in diameter) into human embryonic fibroblast cells through nanowire-assisted delivery. They then monitored the temperature changes of a FND particle, while locally heating a nearby GNP by a separate laser. At the laser power of 12 μW, a temperature increase of 0.5 K was measured at the FND location, corresponding to a change of approximately 10 K at the GNP location, and the cell was still alive. However, cell death occurred when the laser power was increased to 120 μW, which resulted in a temperature change of 3.9 K for the FND and 80 K at the GNP position. The results demonstrated that the technique could actively control cell viability at the nanoscale, enabling the optimization of nanoparticle-based photothermal therapy at the single-cell or subcellular level (cf., Section 12.2 for further discussion).

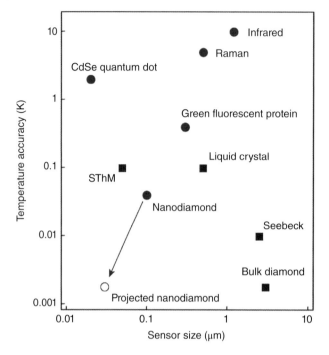

Figure 11.2 Comparison of sensor sizes and temperature accuracies for the NV-based quantum thermometer and other reported techniques. Biocompatible methods are labeled in red. The open red circle indicates the ultimate expected accuracy for FNDs. SthM, scanning thermal microscopy. *Source:* Reprinted with permission from Ref. [16]. Reproduced with permission of Nature Publishing Group.

11.2.2 Time-Resolved Nanothermometry

The plot in Figure 11.2 compares sensor sizes and temperature accuracies between the NV-based quantum thermometer and other techniques reported in the literature [16]. The comparison clearly shows that FND outperforms other nanothermometers (such as quantum dots and green fluorescent proteins) in terms of its precision in the temperature measurement. However, a nanothermometer is far from ideal if it cannot provide information on the dynamics of underlying heat transfer phenomena. The commonly used luminescence nanothermometers, such as the lanthanide complexes [19], typically have a response time in the range of seconds. This temporal resolution by no means is sufficient to follow the time evolution of any systems under investigation at the nanoscale. Similarly for FNDs, it is a time-consuming process (in minutes) to acquire the whole ODMR spectrum of the NV⁻ centers. Moreover, additional data processing after the measurements is required to determine the peak positions and thus the temperature changes.

To speed up the process, Tzeng et al. [20] have developed a three-point sampling method to determine the temperature of a system, which may or may not be necessarily in equilibrium with its surroundings, with a temporal resolution better than 10 μs. The method is based on the experimental finding that the widths of the ODMR peaks of the NV⁻ centers are nearly invariant with temperature changes. For 100-nm FNDs, the *full width at*

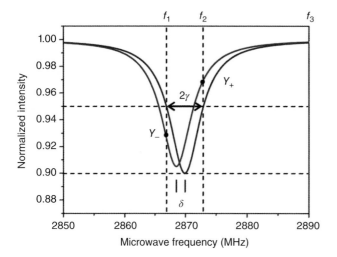

Figure 11.3 Pictorial presentation of the three-point method based on an ODMR spectrum consisting of only one peak. The peaks before (blue) and after (red) temperature change are both Lorentzian and have the same width, although their heights may vary. Overlaid on the spectra are three frequencies (f_1, f_2, f_3) chosen for the intensity measurement. *Source:* Reprinted with permission from Ref. [20]. Reproduced with permission of American Chemical Society.

half maximum (FWHM) of the ODMR peaks is typically 20 MHz in observation. Over the temperature range of 300–700 K, the width changes less than 10%, while the peak positions are shifted by more than 20 MHz [11]. By assuming a Lorentzian profile of constant width (γ) and measuring the intensity changes of the fluorescence dip at three preselected frequencies (Figure 11.3), the researchers have been able to determine the temperature shift (δ) without scanning the whole ODMR spectrum according to the following equations [20]:

$$\delta = \gamma \cdot \frac{1-\sqrt{1-2R^2}}{R}, \tag{11.13}$$

where

$$R = \frac{I_2 - I_1}{2I_3 - I_2 - I_1}, \tag{11.14}$$

and I_1, I_2, and I_3 are the fluorescence intensities measured at points f_1, f_2, and f_3, respectively. The method enables real-time measurement of the temperature changes over ±100 K and, more importantly, the studies of nanoscale heat transfer dynamics in a pump-probe configuration with a pulsed heating source.

Exploiting this unique spin resonance feature, Chang and coworkers [20] conducted time-resolved temperature measurement for a gold nanorod solution heated by a tightly focused 808 nm laser using 100-nm FNDs as the single-particle temperature sensors (Figure 11.4a). With the FNDs submerged in the medium and positioned near the 808 nm laser focus, they first observed superheating of the aqueous solution near the water–glass interface from the measured ODMR spectra (Figure 11.4b) and demonstrated a measurement precision of better than ±1 K over a temperature variation range of 100 K with a

(a)

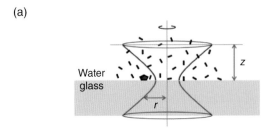

(b)

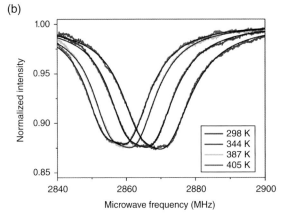

Figure 11.4 Time-resolved nanothermometry. (a) Experimental scheme for the time-resolved temperature measurement with a 100-nm FND particle submerged in aqueous solution containing 10-nm × 41-nm GNRs (black rods) heated by an 808 nm laser (red hyperboloid). (b) ODMR spectra of the GNR solution heated by the near-infrared laser with its power varying from 0 to 10 mW. (c) Time sequences of the laser, microwave, and detection pulses (all in μs) used in the time-resolved nanothermometry with a three-point method. (d) Time evolution of the heat dissipation of the GNR solution at the radial positions of $r = 1.0$ and $1.5\,\mu m$, as indicated in (a). *Source:* Adapted with permission from Ref. [20]. Reproduced with permission of American Chemical Society.

(c)

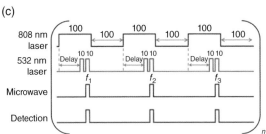

(d)

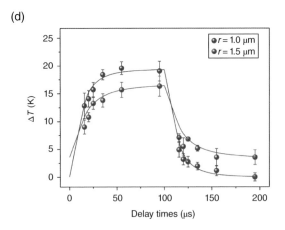

spatial resolution of approximately 0.5 μm. Later, by pulsing the heating laser (808 nm), the probing laser (532 nm), as well as the microwave radiation (Figure 11.4c), a temporal resolution of 10 μs was achieved through a pump-probe type measurement using the three-point method (Figure 11.4d). The validity of the measurements was finally verified with finite-element numerical simulations in this proof-of-principle experiment. Compared with the time-resolved temperature measurement with dye molecules in water [21], the results represent a significant improvement of both spatial and temporal resolution by 1 and 3 orders of magnitude, respectively.

11.2.3 All-Optical Luminescence Nanothermometry

The thermal shifts of the spin resonances of NV$^-$ centers, as illustrated above, provide an effective means for nanoscale temperature sensing with both high temporal and spatial resolutions. However, the method is technically demanding because microwave excitation is required to deplete the fluorescence emission. Additionally, the microwave has to be delivered through a metal wire or nanostructure placed close to the systems under investigation (typically within 10 μm). The requirement of such close proximity severely limits the practical use of the OMDR technique in the life science research. Alternatively, one may focus on the thermal shift of the *zero-phonon line* (ZPL) of the NV$^-$ centers. Chen et al. [10] have reported that the trend of the thermal shifting of the ZPL at 638 nm matches closely with that of the spin transitions at 2.87 GHz. At room temperature, the magnitude of the thermal shift is ΔZPL$/\Delta T = +0.015$ nm K^{-1} with a FWHM approximately 5 nm for the ZPL [10, 22]. Small as it may seem, the shift is sufficient for practical temperature measurement because both the position and width of the ZPL in the fluorescence spectra are highly stable. In comparison to the temperature sensing by ODMR, this all-optical method has the advantages of being simple, straightforward, and readily implementable by any confocal microscope equipped with a CCD-based monochromator.

Tsai et al. [23] tried out the above idea with 100-nm FNDs containing about 900 NV$^-$ centers per particle. Specifically, they used a 594 nm laser as the light source to avoid exciting the NV^0 centers whose fluorescence emission band is partially overlapped with that of NV$^-$ at 638 nm and thus can complicate the spectral analysis. Using FNDs dispersed in water as an example, while the overall feature of the emission band does not change much with temperature over 28–75 °C, its ZPL is significantly broadened and shifted to the red as the solution temperature increases (Figure 11.5a). To determine the sensitivity of this all-optical method, the research team spin-coated 100-nm FNDs on a glass coverslip and obtained their fluorescence spectra at 25 °C for the individual particles (Figure 11.5b). The ZPL centers (λ_0) were then determined by fitting the spectra over 610–660 nm to a Lorentzian function along with an exponential function for the baseline correction as [24]

$$I(\lambda) = B\exp(b\lambda) + \frac{A\Gamma^2}{\Gamma^2 + (\lambda - \lambda_0)^2},\tag{11.15}$$

where $I(\lambda)$ is the fluorescence intensity, λ is the wavelength, Γ is the Lorentzian half-width, λ_0 is the ZPL center, and B, b, and A are constants. Signal averaging over 6 seconds by curve-fitting of 60 independent spectra yielded $\lambda_0 = 638.519 \pm 0.013$ nm at a 95%

(a)

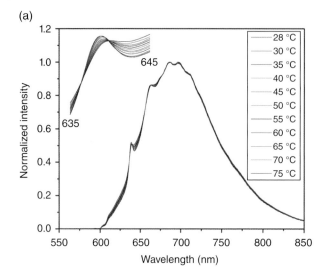

(b)

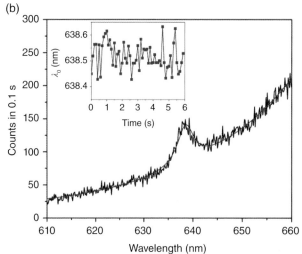

Figure 11.5 All-optical nanothermometry. (a) Area-normalized temperature-dependent fluorescence spectra of 100-nm FNDs illuminated by a 594 nm laser in solution. Inset: Enlarged view of the temperature-induced shift of the ZPLs for spectra acquired at 28–75 °C. (b) Typical fluorescence spectrum of a single 100-nm FND spin-coated on a glass coverslip at room temperature over 610–660 nm. The exposure time of the sample to the 594 nm laser with a power of 30 μW is 0.1 second. Inset: Changes of λ_0 with time over six seconds, with a mean value of 638.519±0.013 nm for a 95% confidence interval. *Source:* Reprinted with permission from Ref. [23]. John Wiley & Sons.

confidence interval (inset in Figure 11.5b), suggesting a temperature measurement sensitivity of ±2 K Hz$^{-1/2}$. Although the effective working range of this method is limited to be less than 100 °C due to the significant band broadening as temperature increases, it is well suitable for biological applications (cf., Section 12.2).

In a separate study, Plakhotnik et al. [24] demonstrated an all-optical ratiometric temperature sensor consisting of a 50-nm FND hosting more than 100 NV⁻ centers in the crystal matrix. Instead of detecting the thermal shift, they characterized the amplitude of the ZPL by fitting the fluorescence spectra with Eq. (11.15) and achieved a temperature measurement sensitivity of $0.3\,\mathrm{K\,Hz^{-1/2}}$ using a 590 nm laser for the excitation. In addition to that, a long-term stability was attained within 0.6 K (peak-to-peak value) for a single 50-nm FND spin-coated on a glass substrate and imaged with a home-built microscope system. Although the method is intrinsically more sensitive than the detection of the ZPL shift, it is more susceptible to the interference from background (or baseline) signals. Fortunately, the background interference can be properly removed by conducting either fluorescence lifetime gating or magnetic modulation as discussed in Section 6.2.

11.2.4   Scanning Thermal Imaging

Thermal mapping is a process by which the spatial variation of the temperature of an object is measured. While feasible, it remains a challenge to apply the single-particle FND nanothermometry for any practical thermal mapping at the nanoscale. In particular, to conduct the mapping, the FND probe must be scanned through the sample in order to obtain information about the spatial temperature distribution. To meet the challenge, Tetienne et al. [25] grafted a single FND particle (~100 nm in diameter) as a thermal probe onto the tip of an *atomic force microscope* (AFM). Hosting an ensemble of spins (~100 NV⁻ centers), the FND provided a significantly improved acquisition time by as much as an order of magnitude. Additionally, the combination of these two techniques preserved the sub-100-nm spatial resolution inherited from the scanning probe microscope for both temperature sensing and topographic imaging. The researchers applied this method, in conjunction with ODMR, to map the local temperature rise due to laser irradiation of a single 40-nm GNP submerged in aqueous solution (cf., captions in Figure 11.6 for details). The results pave the way toward new applications of the FND nanothermometry in areas such as thermal imaging of material processing and microelectronic devices in operation.

Besides mapping nanoscale temperature changes, the FND nanothermometry is also capable of imaging the thermal conductivity (k) of a nanostructure. With this perspective, Laraoui et al. [26] attached a FND to the apex of a silicon tip of an AFM as a local temperature sensor. They then applied an electrical current to heat up the tip and brought the FND in contact with a surface of varying thermal conductivities with nanometer precision. The fast response time ($<200\,\mu s$) of the sensor, due to its small size and high thermal conductivity, made it possible to study the nanoscale heat transfer dynamics triggered by an electrical pulse. By combining AFM and confocal fluorescence microscopy, the researchers attained excellent matching between thermal conductivity and topographic maps for a phantom microstructure made of a gold grid ($k = 314\,\mathrm{W\,m^{-1}K^{-1}}$) on a sapphire substrate ($k = 30\,\mathrm{W\,m^{-1}K^{-1}}$) with distinct thermal conductivities between these two materials. Such an integrated method promises multiple applications, ranging from the investigation of phonon dynamics in nanostructures to the characterization of heterogeneous phase transitions and chemical reactions in various solid-state systems.

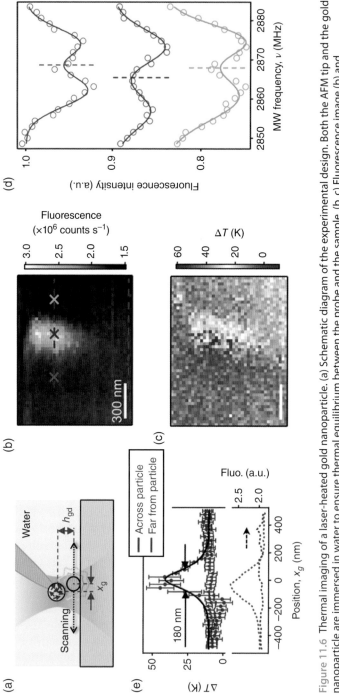

Figure 11.6 Thermal imaging of a laser-heated gold nanoparticle. (a) Schematic diagram of the experimental design. Both the AFM tip and the gold nanoparticle are immersed in water to ensure thermal equilibrium between the probe and the sample. (b, c) Fluorescence image (b) and temperature map (c) obtained simultaneously by scanning a 40-nm gold nanoparticle relative to the FND probe and its excitation laser. (d) ODMR spectra corresponding to three different pixels of the scan, located as indicated by the crosses in (b) with matching colors. (e) Line cuts extracted from (b) and (c) taken along the dashed lines shown in (b) with matching colors. The black solid curve is a fit of the experimental data to a theoretical model. *Source:* Adapted with permission from Ref. [25]. Reproduced with permission of American Chemical Society.

11.3   Magnetic Sensing

11.3.1   Continuous-Wave Detection

In 1997, Wrachtrup and coworkers [7] demonstrated for the first time that the electron spin resonance spectra of a single NV⁻ center in a bulk diamond could be optically read out at the single-molecule level by using the ODMR technique. In the absence of crystal strain, the spin resonance spectrum showed a single peak at 2.87 GHz, which would split into two if an external magnetic field was applied to the NV⁻ center, as verified by experiments (Figure 3.9) [27]. Mathematically, it has been shown that at $B << D/\gamma_e$, the two components of the splitting have the frequencies of [28]

$$f_{\pm} = D + \frac{3\gamma_e^2 B^2}{2D}\sin^2\theta_B \pm \gamma_e B\cos\theta_B\sqrt{1+\frac{\gamma_e^2 B^2}{4D^2}\tan^2\theta_B\sin^2\theta_B}, \tag{11.16}$$

where $\gamma_e = g_e\mu_B/h = 28.03\,\text{MHz}\,\text{mT}^{-1}$ and θ_B is the angle between the magnetic field and the NV⁻ center's major symmetry axis. Using this equation, one can accurately determine the strength of the external magnetic field based on the observed Zeeman splittings with the NV⁻ center as a highly sensitive magnetic sensor. However, as pointed out by Schirhagl et al. [3], while the method is direct, the detection sensitivity is compromised by the significant broadening of the spin resonance bands. Much enhancement of the sensitivity can be gained by changing the detection to the pulsed modes, which take advantage of the NV⁻ center's long coherence time in bulk diamond as well as the many refined microwave sequences that have been developed over the past few decades for magnetic resonance spectroscopy [29]. The combination of this unique spin system and the ODMR technique has enabled ultrasensitive and rapid detection of single electronic spin states under ambient conditions for nanoscale magnetic resonance imaging [30].

Some research groups have applied FNDs as a nanoscale imaging magnetometer to detect weak magnetic fields using the continuous-wave approach [27, 31]. Specifically, Balasubramanian et al. [27] incorporated a 40-nm FND hosting a single NV⁻ center into the cantilever of an AFM (Figure 11.7a and b) and applied the cantilever with the ODMR setup as a scanning probe microscope to achieve subwavelength imaging resolution of a nickel nanostructure. Additionally, by applying microwaves resonant at 2.750 GHz to the NV⁻ center, they observed a narrow dark line close to the corner of the triangular structure (Figure 11.7c). The width of the line was about 20 nm, corresponding to a measurement resolution of 0.5 mT.

Using a similar setup, Rondin et al. [31] conducted nanoscale magnetic imaging with a FND-attached AFM tip to measure large off-axis magnetic fields, which could induce spin level mixing and thus change the fluorescence intensity (Section 3.4). The method was all-optical, requiring no microwave control, and hence expanding the operation range of the NV-based magnetometry. The performance of the FND-scanning probe magnetometer was demonstrated by mapping the magnetic field distribution created by a commercial magnetic hard disk. A sensitivity of $10\,\mu\text{T}\,\text{Hz}^{-1/2}$ was reported by the researchers in this study. The ultimate accuracy of these magnetic field measurements is about $20\,\mu\text{T}$, limited by the couplings of the NV⁻ centers with paramagnetic entities such as other NV centers, ^{13}C, and ^{14}N nuclei in the FNDs [32].

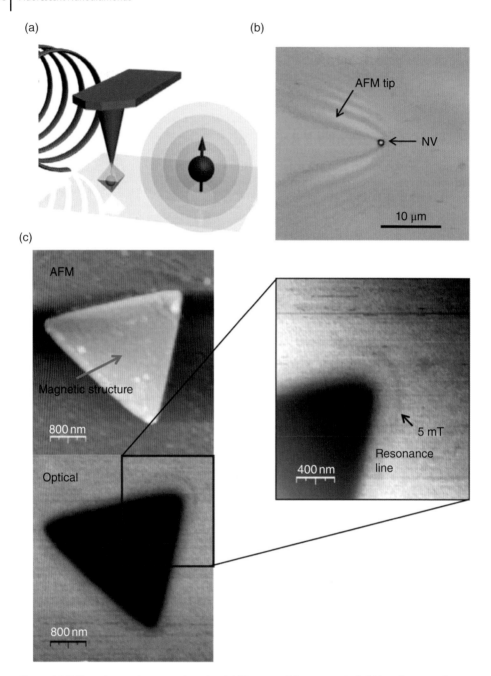

Figure 11.7 Scanning probe magnetometry. (a) Diagram of the magnetic field imaging experiment. A magnetic nanoparticle (red) is imaged with a single NV⁻ in FND (green dot within the blue nanocrystal) fixed at the scanning probe tip (black). (b) Optical image of a FND attached to an AFM tip. The scattered light image of the tip is overlapped with the fluorescence image of the nanocrystal. The bright red spot (arrowed) represents fluorescence of the single NV⁻. (c) Field reconstruction using the scanning probe single spin magnetometer, showing an AFM image of a nickel magnetic nanostructure (top left), a magneto-optical image of the same structure (bottom left), and the fluorescence signal when resonant microwaves at 2.750 GHz are applied to the NV⁻ center. *Source:* Reprinted with permission from Ref. [27]. Reproduced with permission of Nature Publishing Group.

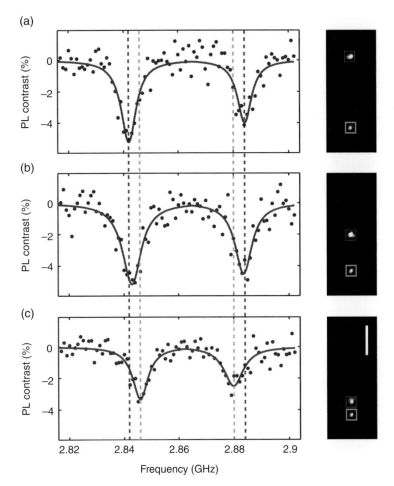

Figure 11.8 ODMR spectra (left) and camera images (right) of a magnetic nanoparticle (red square) and a FND (green square) separated by a distance of (a) 7.25 μm, (b) 3.62 μm, and (c) 1.50 μm. Red curves in the spectra are Lorentzian fits to the measured data and dashed lines in purple and orange are the center frequencies of the Lorentzian fits in (a) and (c), respectively. PL, photoluminescence. Scale bar: 3 μm. *Source:* Reprinted with permission from Ref. [33]. Reproduced with permission of American Chemical Society.

To demonstrate the chemical sensing capability of the FND-based luminescence magnetometry, Lim et al. [33] integrated the NV-based magnetic sensor with a microfluidic system. They manipulated a single magnetic particle in three dimensions in a microfluidic device with a combination of planar electro-osmotic flow control and vertical magnetic actuation, and then applied a fixed FND hosting a single NV⁻ center to measure the strength of the magnetic field generated by the magnetic particle at a separation distance of 7.25 μm (Figure 11.8a), 3.62 μm (Figure 11.8b), and 1.50 μm (Figure 11.8c). From the peak shifts in the ODMR spectra, the magnetic field distribution was mapped out with nanometer accuracy. Their result compared favorably with the calculated magnetic field intensity of a model magnetic dipole as a function of distance from the FNDs.

Taking a different approach, Horowitz et al. [34] applied an infrared laser as an optical tweezer for the three-dimensional control of a single 100-nm FND particle in solution.

Despite the motion and random orientation of the FND in the trap, they were able to observe distinct peaks in the ODMR spectra of an ensemble of NV^- centers in the particle. Similarly, Kayci et al. [35] employed an anti-Brownian electro-kinetic trap to manipulate a single FND particle under ambient and aqueous conditions for magneto-optical spin detection. Furthermore, Geiselmann et al. [36] developed a deterministic optical trapping technique and demonstrated the three-dimensional manipulation of individual FNDs hosting multiple NV^- centers to measure the local magnetic field in a glycerol–water mixture. The development of all these techniques has opened new opportunities to probe the magnetic fields in complex environments such as the interiors of microfluidic channels and live cells that are inaccessible by the scanning probe techniques existing today.

Developed originally to study condensed-phase phenomena, the magnetic sensing techniques discussed above have also been applied to using the NV^- centers as nanoprobes in the gas phase [37–39]. For example, Hoang et al. [38] optically levitated a 100-nm FND particle for electron spin control of its built-in NV^- centers in low vacuum. Measuring the absolute internal temperature of the levitated FNDs by ODMR, the researchers observed an enhancement in the strength of the spin resonance signal as the air pressure was reduced, suggesting that oxygen gases could exert an effect on both the fluorescence intensity and the ODMR contrast. The study implies a potential application of NV^- for sensing gaseous oxygen, which has two unpaired electrons and is paramagnetic.

11.3.2 Relaxometry

Now we turn to the pulsed operations. As discussed in the previous section, the pulsed operations may provide a more sensitive means for magnetic sensing. *Relaxometry* is one of such techniques, which are traditionally used in nuclear magnetic resonance spectroscopy to measure nuclear spin relaxation rates that depend strongly on the fluctuations of the magnetic fields in microscopic environments. The technique is adopted here to monitor the relaxation of NV^- centers by optically polarizing its $m_s = 0$ state and then measuring the probability of finding the spins recovered in the same state at a later time, T_1, i.e. the spin–lattice relaxation time [40–44]. Time-resolved measurements with a temporal resolution in the range of one minute are readily achievable with this technique [44].

With the relaxometry, Ermakova et al. [42] demonstrated the detection of a few ferritin molecules loaded on a single FND particle hosting multiple NV^- centers. Ferritin is an iron storage protein capable of transporting as many as 4500 atoms of Fe with a magnetic moment of $300\,\mu_B$. The researchers first attached ferritin to the surface of a 40-nm FND by noncovalent conjugation to form ferritin–FND complexes. From a comparison of the measured T_1 relaxation times between free FNDs and the ferritin–FND complexes for more than 30 particles in each group, a significant reduction of the T_1 time (from $41.8\,\mu s$ of free FNDs to $5.7\,\mu s$ of the ferritin–FND complexes) due to the protein conjugation was found. Compared with existing magnetic force microscopy techniques, the method has an unmatched advantage in that it does not require cryogenic temperature or vacuum, a major improvement in a practical sense.

In a proof-of-principle experiment for biological sensing with the relaxometry, Kaufmann et al. [43] applied the single NV^- centers in FNDs to detect gadolinium (Gd) spin labels in an artificial cell membrane under ambient conditions (cf., captions in

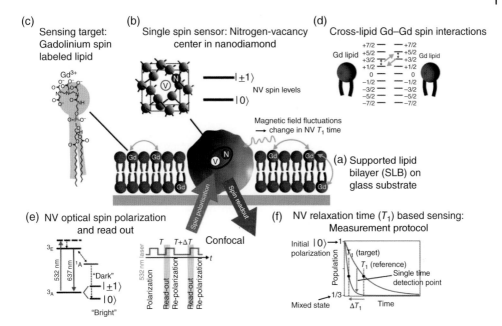

Figure 11.9 Schematic of the nanoscopic detection of spin labels in an artificial cell membrane using a single-spin FND sensor. (a) A supported lipid bilayer (SLB) formed around a FND immobilized on a glass substrate. (b) A FND hosting a single NV^- center acting as a single-spin sensor by virtue of the magnetic levels in the ground state. (c) A Gd^{3+} ion binding with a lipid molecule in the SLB. (d) Gd spin labels causing magnetic field fluctuations that affect the quantum states of the NV^- centers in FNDs, measured through its spin–lattice relaxation time, T_1. (e) Electronic energy structure of the NV^- center, showing the fluorescence cycle and optical spin readout of the spin states, $m_s = 0$ and ± 1, and the protocol for the T_1 measurement. (f) Schematic illustration of the T_1 measurement. The relaxation of the NV spin in the target environment is compared with that in the reference environment. Measurement at a single time point in the evolution allows faster detection. *Source:* Reprinted with permission from Ref. [43].

Figure 11.9 for details). The Gd^{3+} ion is a common magnetic resonance imaging contrast agent with spin 7/2 and it can bind with the negatively charged head of the lipid molecule in a supported lipid bilayer by electrostatic forces. The Gd-labeled lipids can thus produce considerable magnetic fluctuations in the artificial membrane by a combined effect of Gd–Gd spin dipole interactions, motional diffusion, and intrinsic Gd spin relaxation. The researchers found that the T_1 relaxation times of the membrane-bound FNDs were significantly shortened when the artificial membrane was labeled with 10% Gd spin-labeled lipids, suggesting a sensitivity of approximately $5\,Gd\ spins\,Hz^{-1/2}$. The ability to detect such a small number of spins in a model biological setting has established a new avenue for in situ nanoscale investigation of the dynamical processes in a single living cell.

References

1 Feynman, R.P. (1959). There's plenty of room at the bottom. http://calteches.library. caltech.edu/1976/1/1960Bottom.pdf (accessed 16 April 2018).

2 Freitas, R.A. Jr. (2003). *Nanomedicine (Vol IIA): Biocompatibility*. Landes Bioscience.

3 Schirhagl, R., Chang, K., Loretz, M., and Degen, C.L. (2014). Nitrogen-vacancy centers in diamond: nanoscale sensors for physics and biology. *Annu Rev Phys Chem* 65: 83–105.

4 Levine, I.N. (1983). *Quantum Chemistry*, 3e. Allyn and Bacon.

5 Atherton, N.M. (1993). *Principles of Electron Spin Resonance*. Ellis Horwood, PTR Prentice Hall.

6 Atkins, P. and de Paula, J. (2006). *Atkins' Physical Chemistry*. Oxford: Oxford University Press.

7 Gruber, A., Drabenstedt, A., Tietz, C. et al. (1997). Scanning confocal optical microscopy and magnetic resonance on single defect centers. *Science* 276: 2012–2014.

8 Balasubramanian, G., Neumann, P., Twitchen, D. et al. (2009). Ultralong spin coherence time in isotopically engineered diamond. *Nat Mater* 8: 383–387.

9 Acosta, V.M., Bauch, E., Ledbetter, M.P. et al. (2010). Temperature dependence of the nitrogen-vacancy magnetic resonance in diamond. *Phys Rev Lett* 104: 070801.

10 Chen, X.D., Dong, C.H., Sun, F.W. et al. (2011). Temperature dependent energy level shifts of nitrogen-vacancy centers in diamond. *Appl Phys Lett* 99: 161903.

11 Toyli, D.M., Christle, D.J., Alkauskas, A. et al. (2012). Measurement and control of single nitrogen-vacancy center spins above 600 K. *Phys Rev X* 2: 031001.

12 Plakhotnik, T. and Gruber, D. (2010). Luminescence of nitrogen-vacancy centers in nanodiamonds at temperatures between 300 and 700 K: perspectives on nanothermometry. *Phys Chem Chem Phys* 12: 9751–9756.

13 Plakhotnik, T., Doherty, M.W., Cole, J.H. et al. (2014). All-optical thermometry and thermal properties of the optically detected spin resonances of the NV⁻ center in nanodiamond. *Nano Lett* 14: 4989–4996.

14 Brites, C.D., Lima, P.P., Silva, N.J. et al. (2012). Thermometry at the nanoscale. *Nanoscale* 4: 4799–4829.

15 Jaque, D. and Vetrone, F. (2012). Luminescence nanothermometry. *Nanoscale* 4: 4301–4326.

16 Kucsko, G., Maurer, P.C., Yao, N.Y. et al. (2013). Nanometer-scale thermometry in a living cell. *Nature* 500: 54–58.

17 Neumann, P., Jakobi, I., Dolde, F. et al. (2013). High-precision nanoscale temperature sensing using single defects in diamond. *Nano Lett* 13: 2738–2742.

18 Toyli, D.M., de las Casas, C.F., Christle, D.J. et al. (2013). Fluorescence thermometry enhanced by the quantum coherence of single spins in diamond. *Proc Natl Acad Sci USA* 110: 8417–8421.

19 Piñol, R., Brites, C.D., Bustamante, R. et al. (2015). Joining time-resolved thermometry and magnetic-induced heating in a single nanoparticle unveils intriguing thermal properties. *ACS Nano* 9: 3134–3142.

20 Tzeng, Y.K., Tsai, P.C., Liu, H.Y. et al. (2015). Time-resolved luminescence nanothermometry with nitrogen-vacancy centers in nanodiamonds. *Nano Lett* 15: 3945–3952.

21 Cordero, M.L., Verneuil, E., Gallaire, F., and Baroud, C.N. (2009). Time-resolved temperature rise in a thin liquid film due to laser absorption. *Phys Rev E* 79: 011201.

22 Doherty, M.W., Acosta, V.M., Jarmola, A. et al. (2014). Temperature shifts of the resonances of the NV⁻ center in diamond. *Phys Rev B* 90: 041201(R).

23 Tsai, P.C., Epperla, C.P., Huang, J.S. et al. (2017). Measuring nanoscale thermostability of cell membranes with single gold-diamond nanohybrids. *Angew Chem Int Ed* 56: 3025–3030.

24 Plakhotnik, T., Aman, H., and Chang, H.C. (2015). All-optical single-nanoparticle ratiometric thermometry with a noise floor of 0.3 K·Hz$^{-1/2}$. *Nanotechnology* 26: 245501.

25 Tetienne, J.P., Lombard, A., Simpson, D.A. et al. (2016). Scanning nanospin ensemble microscope for nanoscale magnetic and thermal imaging. *Nano Lett* 16: 326–333.

26 Laraoui, A., Aycock-Rizzo, H., Gao, Y. et al. (2015). Imaging thermal conductivity with nanoscale resolution using a scanning spin probe. *Nat Commun* 6: 8954.

27 Balasubramanian, G., Chan, I.Y., Kolesov, R. et al. (2008). Nanoscale imaging magnetometry with diamond spins under ambient conditions. *Nature* 455: 648–651.

28 Doherty, M.W., Michl, J., Dolde, F. et al. (2014). Measuring the defect structure orientation of a single NV$^-$ centre in diamond. *New J Phys* 16: 063067.

29 Slichter, C.P. (1990). *Principles of Magnetic Resonance*, 3e. Springer.

30 Rugar, D., Mamin, H.J., Sherwood, M.H. et al. (2015). Proton magnetic resonance imaging using a nitrogen-vacancy spin sensor. *Nat Nanotechnol* 10: 120–124.

31 Rondin, L., Tetienne, J.P., Spinicelli, P. et al. (2012). Nanoscale magnetic field mapping with a single spin scanning probe magnetometer. *Appl Phys Lett* 100: 153118.

32 Aman, H. and Plakhotnik, T. (2016). Accuracy in the measurement of magnetic fields using nitrogen-vacancy centers in nanodiamonds. *J Opt Soc Am B* 33: B19–B27.

33 Lim, K., Ropp, C., Shapiro, B. et al. (2015). Scanning localized magnetic fields in a microfluidic device with a single nitrogen vacancy center. *Nano Lett* 15: 1481–1486.

34 Horowitz, V.R., Alemán, B.J., Christle, D.J. et al. (2012). Electron spin resonance of nitrogen-vacancy centers in optically trapped nanodiamonds. *Proc Natl Acad Sci USA* 109: 13493–13497.

35 Kayci, M., Chang, H.C., and Radenovic, A. (2014). Electron spin resonance of nitrogen-vacancy defects embedded in single nanodiamonds in an ABEL trap. *Nano Lett* 14: 5335–5341.

36 Geiselmann, M., Juan, M.L., Renger, J. et al. (2013). Three-dimensional optical manipulation of a single electron spin. *Nat Nanotechnol* 8: 175–179.

37 Neukirch, L.P., von Haartman, E., Rosenholm, J.M., and Vamivakas, A.N. (2015). Multi-dimensional single-spin nano-optomechanics with a levitated nanodiamond. *Nat Photon* 9: 653–657.

38 Hoang, T.M., Ahn, J., Bang, J., and Li, T. (2016). Electron spin control of optically levitated nanodiamonds in vacuum. *Nat Commun* 7: 12250.

39 Kumar, P. and Bhattacharya, M. (2017). Magnetometry via spin-mechanical coupling in levitated optomechanics. *Opt Express* 25: 19568–19582.

40 Ziem, F.C., Götz, N.S., Zappe, A. et al. (2013). Highly sensitive detection of physiological spins in a microfluidic device. *Nano Lett* 13: 4093–4098.

41 Steinert, S., Ziem, F., Hall, L.T. et al. (2013). Magnetic spin imaging under ambient conditions with sub-cellular resolution. *Nat Commun* 4: 1607.

42 Ermakova, A., Pramanik, G., Cai, J.M. et al. (2013). Detection of a few metallo-protein molecules using color centers in nanodiamonds. *Nano Lett* 13: 3305–3309.

43 Kaufmann, S., Simpson, D.A., Hall, L.T. et al. (2013). Detection of atomic spin labels in a lipid bilayer using a single-spin nanodiamond probe. *Proc Natl Acad Sci USA* 110: 10894–10898.

44 Rendler, T., Neburkova, J., Zemek, O. et al. (2017). Optical imaging of localized chemical events using programmable diamond quantum nanosensors. *Nat Commun* 8: 14701.

12

Hybrid Fluorescent Nanodiamonds

Many natural materials such as wood and bones are made of two or more materials, known as *composites*. If the materials are mixed at the nanometer or molecular level, these composites are called *hybrids* [1]. Depending on how they are made, the mixing may lead to the formation of new materials with novel properties and/or multiple functions. For example, the hydroxyapatite (a calcium phosphate) in the bone provides mechanical strength and the collagen (a protein) in the bone promotes the bonding between the inorganic building blocks and the soft tissue.

In the previous chapters, we have discussed the unique chemical and physical properties of *fluorescent nanodiamonds* (FNDs) alone. Here, we extend the discussion to hybrid FNDs, e.g. FNDs either covalently or noncovalently conjugated with other inorganic nanomaterials such as silica, gold, silver, or iron oxide nanoparticles, all of which have found practical applications for drug delivery and theranostics [2]. In preparing these nanohybrids or nanodevices, the surface chemistry of nanodiamonds, such as that illustrated in Chapter 4, clearly plays a central role. Research and development along this line are expected to add brand new dimensions to the use of FNDs in the life sciences and other research areas. Despite the fact that the field is still at its infant stage, various strategies have been developed and implemented to overcome the obstacles and challenges encountered in sample preparation. In this chapter, we discuss the synthesis and characterization of these nanohybrids, together with their applications in different facets of science and technology.

12.1   Silica/Diamond Nanohybrids

So, where do we begin to find a good hybrid partner for diamond? It all, intuitively at the least, seems that silica ought to be the first choice because both silicon and carbon are in the same family on the Periodic Table, sharing similar bonding configuration and chemistry. It is true that silica and diamond are compatible chemically, able to form stable hybrids in more ways than just one. We will examine two examples of silica/diamond nanohybrids here as a general introduction.

In Section 4.3.2, we have introduced how FNDs can be encapsulated in a silica shell, where two distinct types of nanomaterials are mixed at the molecular level to form a hybrid. Unlike other nanohybrids discussed in the following sections, the encapsulation transforms the irregular shape of the FND into a spheroid (Figure 4.10), a unique aspect of the

Fluorescent Nanodiamonds, First Edition. Huan-Cheng Chang, Wesley Wei-Wen Hsiao and Meng-Chih Su.
© 2019 John Wiley & Sons Ltd. Published 2019 by John Wiley & Sons Ltd.

(a)

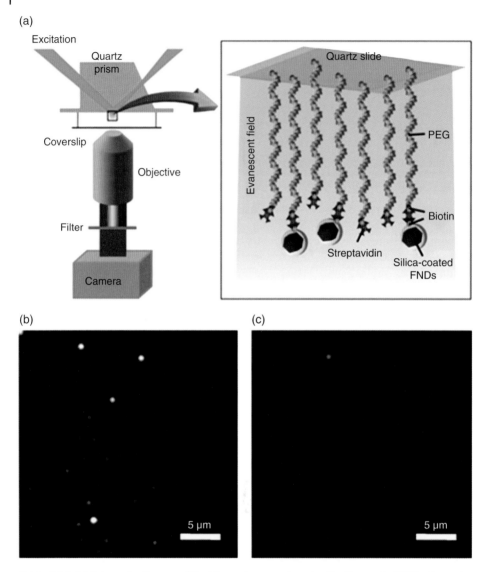

(b)   (c)

Figure 12.1 (a) Schematic diagram of the biosensing experiment with silica-coated FNDs. A quartz slide passivated with biotinylated PEG was first saturated with streptavidin. The biotinylated FNDs were then flowed into a microfluidic cell and specific attachment to streptavidin–biotin-PEG was probed by comparing the number of bound particles before and after washing to remove nonspecifically bound particles. FNDs were excited by the evanescent field (green) in a prism-type TIRFM. (b, c) Images showing that (b) 60% of the biotinylated silica-coated FNDs and (c) 4% of the nonbiotinylated silica-coated FNDs remained attached to the surface after acid–base washes. *Source:* Reprinted with permission from Ref. [4]. Reproduced with permission of American Chemical Society.

technique [3]. Another benefit of the encapsulation method is that it renders the FND's surface readily functionalizable through routine crosslinking chemistry. As an example, Bumb et al. [4] attached biotin to silica-coated FNDs and quantified their specific binding to a surface passivated with streptavidin-labeled poly-ethylene glycol (PEG) using *total internal reflection fluorescence* microscopy (TIRFM) (Section 8.2.2 and Figure 12.1a). It was found that the binding between the silica-coated FNDs and the streptavidin-passivated surface was approximately 15-fold greater with biotin than without one (Figure 12.1b and c), showing the potential use of these nanohybrids for biosensing applications. Additionally, the researchers applied the method to track the three-dimensional motion of a biotinylated silica-FND tethered by a single DNA molecule on a glass slide with both high spatial and temporal resolution. They were able to measure the persistence length of a single DNA molecule after photobleaching the observation area to reduce background fluorescence signals prior to tracking the single DNA-tethered FND particle. Finally, an application of the nanohybrids was made by the research team to map out the lymph modes in a mouse by background-free fluorescence imaging of FNDs through magnetic modulation of the emission (Section 9.3.4), demonstrating the versatility of this method [5].

In another case, Rosenholm and coworkers [6, 7] synthesized silica-FND nanohybrids for use in both bioimaging and drug delivery. Instead of making a thin shell, they coated the FND core with *mesoporous silica nanoparticles* (MSNs), which have high porosity and tunable pore sizes, achieved by using a supramolecular templating method. Initially, the FND had an average size of about 50 nm (Figure 12.2a), which increased to more than 300 nm after silica coating (Figure 12.2b–d). *Dynamic light scattering* (DLS) and *transmission electron microscopy* (TEM) showed homogeneous coating of the individual FND cores, which in effect removed the undesirable shape irregularity from the primary diamond nanoparticles. The nanohybrid is novel in the sense that it integrates the superb optical properties of FNDs with the high drug delivery capability of MSNs. By conducting cell viability, cargo delivery assays, and live cell imaging, the studies have provided a valuable demonstration of how this novel nanohybrid can be exploited for future theranostic applications in nanoscale biology and medicine.

12.2 Gold/Diamond Nanohybrids

12.2.1 Photoluminescence Enhancement

While silica may have been a logic pick for hybridization, it was gold/diamond that emerged among the first nanohybrids reported in the literature. The development of such nanohybrids was aimed at investigating the plasmonic effect of *gold nanoparticles* (GNPs) on the optical properties of the nitrogen-vacancy (NV) centers in FNDs (Section 10.2.4). An early plasmonic study was made by Liu and Sun [8] to couple different-sized GNPs with FNDs using two complementary DNA strands for physical adsorption. A significant enhancement in the photoluminescence signals was observed when the particles were excited at 488 nm. The enhancement was envisioned useful for biosensing applications.

At a technically more sophisticated level, Schietinger et al. [9] investigated the plasmon-enhanced effect employing an *atomic force microscope* (AFM) to achieve controlled coupling of GNPs with a FND that contained only a single NV$^-$ center. Either one or two GNPs were assembled with the FND (~30 nm in diameter) in a step-by-step manner (Figure 12.3a). The NV$^-$ center was observed to have an increase of nearly 10-fold in both excitation rate and radiative decay rate, a result of plasmonic enhancement as supported by numerical simulations (Figure 12.3b). Following this study, Shen et al. [10] showed that unidirectional emission from a single FND particle could be generated if it was coupled with a single *gold nanorod* (GNR). The emission patterns were controlled

(a)

(b)

(c)

(d)

Figure 12.2 TEM images of FNDs and MSN-FNDs. (a) Pure FND cores as imaged by TEM and (b) individual MSN-FNDs imaged by high-resolution TEM. (c, d) Corresponding (c) bright field and (d) dark field images revealing the presence of a crystalline diamond core. *Source:* Reprinted with permission from Ref. [7]. Reproduced with permission of Royal Society of Chemistry.

by adjusting the GNR orientation and position with respect to the FND. The emission remained highly unidirectional even when the emitter was positioned away from the GNR antenna by a distance of up to 50 nm. The result was attributed to the interference between the electromagnetic fields produced by the dipole-like source and the out-of-phase dipole induced in the GNR. This hybrid GNR-FND system, together with GNP-FND, may serve as an important building block for novel nanophotonic light sources in the advanced plasmonic devices that are stable even at room temperature.

12.2.2 Dual-Modality Imaging

Zhang et al. [11] first used gold/diamond nanohybrids as a dual-modality contrast agent for combined fluorescence and photoacoustic imaging. *Photoacoustic effect* is the formation of sound waves upon absorption of light in a sample [12]. Typically, a pulsed

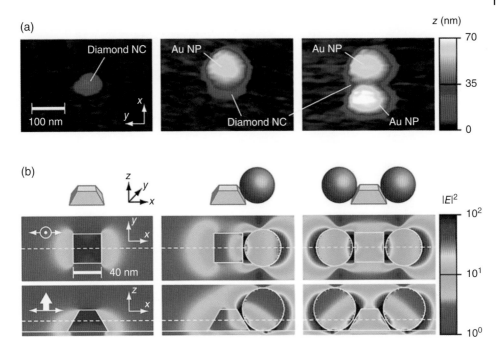

Figure 12.3 Experimental realization and numerical simulation of a GNP-FND hybrid. (a) AFM images of a single diamond nanocrystal (left), to which one (middle) or two (right) GNPs (or Au NPs) are coupled. (b) Corresponding numerical simulations of the intensity enhancement of the excitation light linearly polarized along the *x* axis. Upper row: schematic representation of the particle configuration. Middle row: *x–y* cross section. Lower row: *x–z* cross section. The field intensity is normalized to the value at the center of the bare FND. *Source:* Reprinted with permission from Ref. [9]. Reproduced with permission of American Chemical Society.

nanosecond laser is used for the excitation and a transducer is used to detect the sound wave signals. Compared with all optical methods, the technique allows detection of light-absorbing chromophores with a greater penetration depth in tissue (Figure 12.4) [13]. Employing a nanosecond optical parametric oscillator operating at 532 nm as the light source, the researchers found that the photoacoustic signals of FNDs (~100 nm in diameter) alone were weak but could be greatly enhanced by a factor of 30 after conjugation with GNPs (~20 nm in diameter). The GNPs were selected here because their *surface plasmon resonance* (SPR) matches closely with the absorption maximum of FNDs at 560 nm. Moreover, the GNPs were amine-terminated and thus could be covalently conjugated with carboxylated FNDs through carbodiimide chemistry (Section 4.2.2) to form stable nanohybrids. Furthermore, the red emission of the NV⁻ centers could also be detected when the nanoparticles were excited by green-yellow light, indicating that the GNP-attached FNDs were still useful as a fluorescent contrast agent. The method is applicable to conjugate FNDs with GNPs of various sizes and shapes to improve the photoacoustic detection sensitivity.

In a recent study, Liu et al. [14] prepared GNP-FND nanohybrids for combined fluorescence and electron microscopy imaging. The nanohybrids were synthesized via the adsorption of GNPs (~10 nm in diameter) and human serum albumin (HSA) to FNDs (~33 nm in diameter) after careful adjustments of the relative ratios of FND, GNP, and

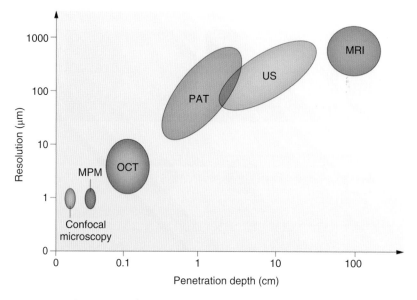

Figure 12.4 Comparison of resolution and penetration depth between different imaging modalities, including confocal microscopy, multiphoton microscopy (MPM), optical coherence tomography (OCT), photoacoustic tomography (PAT), ultrasonography (US), and magnetic resonance imaging (MRI). *Source:* Reprinted with permission from Ref. [13]. Reproduced with permission of Nature Publishing Group.

HSA in aqueous solution. To ensure good stability of the GNP-FND nanohybrids, the carboxyl groups on the surfaces of both FNDs and GNPs were crosslinked with the amine groups of HSA via amide bond formation. TEM revealed that about 1–2 GNPs could be anchored on each FND (Figure 12.5a). The separation between these two types of nanoparticles was approximately 1 nm (Figure 12.5b). An analytical ultracentrifugation method was applied to isolate the GNP-FNDs in the desired size range. The research team used this nanohybrid as a new class of "all-in-one" nanodevice for multimodal applications, including optical microscopy, electron microscopy, and potentially quantum sensing.

12.2.3 Hyperlocalized Hyperthermia

Gold particles are one of the most frequently used heating agents in plasmon-based photothermal therapy [15, 16]. A member of particular interest in this family is the GNR, because its *longitudinal surface plasmon resonance* (LSPR) band can be conveniently tuned by increasing the aspect ratio of the particle to the near-infrared range (cf., captions in Figure 12.6) [17], where light has a relatively long penetration depth through tissue (Section 9.3). It is therefore more suitable than spherical GNPs for *in vivo* photoacoustic imaging and hyperthermia applications. Another outstanding feature of GNR is that its LSPR band has an exceptionally large molar extinction coefficient on the order of $10^9\,\mathrm{M}^{-1}\,\mathrm{cm}^{-1}$ at 800 nm [18]. Numerical simulations have shown that the light absorbed by the nanoparticles will be effectively converted into heat and thermal equilibration at the particle/water interface can be rapidly reached within 1 ns due to their small size [19].

(a)

(b)

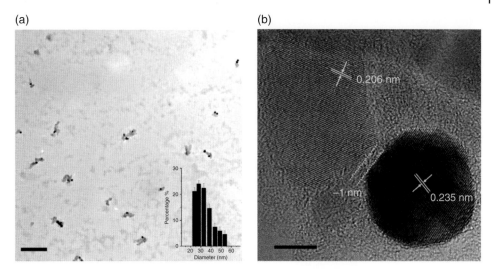

Figure 12.5 Characterization of GNP-FND nanohybrids. (a) TEM image of GNP-FNDs. The inset depicts the size distribution of the particles with a mean diameter of 33 nm. (b) High-resolution TEM image of the GNP-FND on graphene as a supporting film. The lattice fringes of FND (0.206 nm) and GNP (0.235 nm) are clearly resolved. The separation width between these two nanoparticles is approximately 1 nm. Scale bars: 200 nm in (a) and 5 nm in (b). *Source:* Reprinted with permission from Ref. [14]. Reproduced with permission of American Chemical Society.

A number of studies have applied GNPs to nanoplasmonic heating of synthetic phospholipid membrane [20] and human cell membrane [21, 22]. The technique, also known as *hyperlocalized hyperthermia* [23, 24], surpasses conventional thermal therapy by reducing the heating zone down to the nanometer size. When applying the hyperlocalized hyperthermia technique to plasma membrane, for example, only a small heating zone is created on the cell surface. The spatial restriction of the heating zone greatly reduces the energy input required to reach the critical temperature that leads to cell death. However, in order to visualize the particles, which are only weakly fluorescent, either dark-field microscopy [20] or two-photon excitation microscopy [21] is necessary, which hinders popular uses of the methods. A way to circumvent this problem is to conjugate GNRs with FNDs, which are both photophysically and thermally stable. To demonstrate the concept, Chang and coworkers [25] prepared GNR-FND nanohybrids that could be individually imaged and heated at the same time with a single laser. In this experiment, carboxylated FNDs were first covalently linked with polyethylenimine (PEI) through carbodiimide chemistry, followed by decorating the cationic FND surface with citrate-capped GNRs through electrostatic interactions to form dual-functional nanodevices. The advantage of this noncovalent conjugation was that it allowed multiple GNRs to be attached on a FND, as demonstrated for GNPs on silica spheres [26, 27]. For GNRs with dimensions of 10 nm × 16 nm (diameter × length) as an example, nanohydrids consisting of more than 50 GNRs on a single 100-nm FND could be readily synthesized (inset in Figure 12.7a). This large loading made the nanohybrid an efficient light absorber and thus an effective nanoheater, suitable for a wide range of hyperthermia applications.

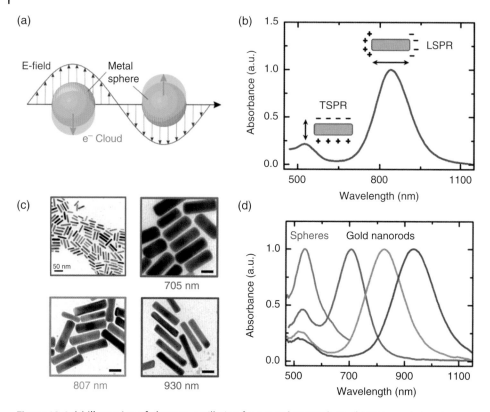

Figure 12.6 (a) Illustration of plasmon oscillation for a metal nanosphere. (b) Measured absorbance spectrum of a GNR solution. The insets show the schematic of the transverse and longitudinal SPR modes responsible for the two absorption bands observed in the spectrum. (c) TEM images of as-synthesized GNRs with different longitudinal SPR wavelengths as noted. Unlabeled scale bars: 20 nm. (d) Measured absorbance spectra of gold nanospheres and GNRs, whose TEM images are shown in (c). *Source:* Reprinted with permission from Ref. [17].

Figure 12.7a shows an absorption spectrum of 10-nm × 16-nm GNRs, which exhibit a LSPR band peaking at approximately 590 nm. The band matches closely with the laser excitation wavelength of 594 nm but does not significantly quench the NV⁻ fluorescence. The nanohybrids can thus be applied as a two-in-one nanodevice using a single laser equipped in a fluorescence microscope for imaging and heating at the same time. An example of the application is given in Figure 12.7b, where the GNR-FNDs were sparsely anchored on the plasma membrane of a cell cluster derived from the human embryonic kidney cell line (HEK293T). These nanohybrids could be readily identified individually by confocal fluorescence microscopy at a laser power as low as 30 μW without damaging the cell's membrane. To access a viable thermal range, Tsai et al. [25] increased the laser power by a factor of 10 and observed severe membrane blebbing within merely an irradiating time of six seconds. The blebbing was a sign of cell death [28], caused by laser heating of the membrane-attached GNR-FNDs. To examine the events more closely, the research team imaged the cells in the presence of propidium iodide (PI) before and after laser irradiation of the nanohybrids. PI is a membrane-impermeable dye, which increases its fluorescence intensity by more than 10-fold

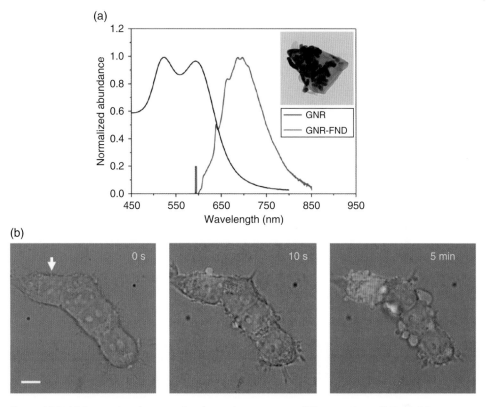

Figure 12.7 (a) Comparison between the absorption spectrum of 10-nm × 16-nm GNRs and the emission spectrum of 100-nm GNR-FNDs. The thick green bar denotes the excitation wavelength at 594 nm. Inset: TEM image of a GNR-decorated FND having more than 30 GNRs on the observable hemisphere or a surface coverage of greater than 50%. (b) Merged bright-field/fluorescence images of an HEK293T cell cluster with GNR-FNDs attached to the plasma membrane after exposure to the 594 nm laser in the presence of PI at different time points. Intracellular PI was observed almost immediately after laser irradiation of the particle indicated by the white arrow. Many blebs appeared on the cells not being irradiated at five minute after the hyperlocalized hyperthermia treatment. Scale bar: 10 μm. *Source:* Adapted with permission from Ref. [25]. Reproduced with permission of John Wiley & Sons.

once bound to nucleic acids in cells. The labeling enabled the observation of the cell membrane damage by detecting the red emission of PI inside the cytoplasm in three dimensions and real time (e.g. zero to five minutes in this case).

While the noncovalently bound GNR-FND complexes have been demonstrated to be a powerful nanodevice for both hyperlocalized hyperthermia and nanoscale thermometry applications, their chemical stability is a concern. Covalent conjugation of GNPs or GNRs with FNDs to ensure long-term stability is desired. More preferably, a uniform thin shell of gold should form on the FND surface, similar to that of silica-encapsulated FNDs. Experiments to realize such nanostructures have been undertaken with promising results so far [29, 30]. Further development of the technology, along with the incorporation of the specific cell targeting functionality [31–33], is expected to make the FND-assisted photothermal therapy a powerful new tool for cancer treatment.

12.2.4 NV-Based Nanothermometry

Hyperlocalized hyperthermia differs from conventional local hyperthermia in that it creates a much larger temperature gradient in cells. It is well known in materials science and engineering that a high spatial temperature gradient can cause structural stress detrimental to fragile solids. Hence, the true culprit of the cell killing effect as observed in Figure 12.7b might have been the large change in temperature difference within a short distance, rather than the absolute temperature. With the GNR-FND nanohybrids becoming available commercially, it is possible to explore the mechanism of the hyperlocalized hyperthermia by conducting nanothermometric measurements within a single cell. The combination of the FND's temperature sensing ability and the GNR's heating capability allows in situ temperature tuning while at the same time reading out of the local temperatures of the cell.

Using FNDs, we now have two ways of measuring nanoscale temperatures: One is through *optically detected magnetic resonance* (ODMR) and the other is by all-optical detection, as illustrated in Section 11.2. Both methods have been applied to GNR-FNDs as two-in-one nanodevices [23, 25]. Specifically, Tsai et al. [23] conjugated FNDs with GNRs (aspect ratio = 4 : 1) for heating by a near-infrared laser (808 nm) and measured the resulting temperature rise with the ODMR technique. With less than five GNRs attached to each FND, a temperature rise of 10 K could be readily achieved at a laser power of 0.4 mW in cells. To elucidate the nanoscale heating process, the research team conducted numerical simulations with finite element analysis. Their results showed that the temperature would be distributed almost uniformly over the entire nanohybrid as the number of GNRs on a FND reached 10 or greater. A large thermal gradient would develop right outside the gold shell in water, where the thermal conductivity is 2–3 orders of magnitude lower than that of gold and diamond. By treating the GNR-FND nanohybrid as a spherical core-shell structure, the temperature rise relative to ambient temperature around the nanohybrid (i.e. $r \geq r_0$) could be approximated by

$$\Delta T(r) = \frac{\dot{Q}}{4\pi k r} \tag{12.1}$$

where $\dot{Q}$ is the heat generation rate, k is the thermal conductivity of water, r is the distance from the FND center, and r_0 is the outer radius of the gold shell [25].

In a later study, Tsai et al. [25] employed the all-optical method for nanothermometric measurements using GNR-FND nanohybrids composed of 10-nm × 16-nm GNRs as the bifunctional devices (Figure 12.7a). In addition to the hyperthermia experiment as illustrated in Figure 12.7b, the local thermostability (or heat tolerance) of subcellular components was probed at the nanoscale. The researchers chose to study the thermally induced rupture of the membrane *tunneling nanotubes* (TNTs) of HEK293T cells (Section 7.2.2 and Figure 7.8), because the events of membrane disruption and retraction could be directly visualized with optical microscopy. By using a 594 nm laser for both heating and probing of single GNR-FNDs confined in the TNTs, they measured the rupture temperatures of the nanotubes one by one (cf., captions in Figure 12.8 for details). A temperature difference as large as 10 °C was found between the results of the local heating by lasers and the global heating with an on-stage incubator in separate

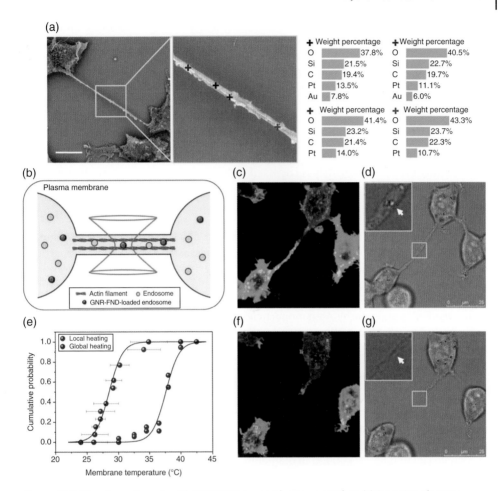

Figure 12.8 Nanohyperthermia with GNR-FNDs in membrane nanotubes. (a) Scanning electron microscopy (SEM) image of a membrane nanotube formed between two HEK293T cells. Elemental analysis by energy-dispersive X-ray spectroscopy (EDS) shown to the right reveals the presence of GNR-FNDs (denoted by black and red crosses) in the TNT. Scale bar: 10 μm. (b) Pictorial presentation of the experiment using a tightly focused 594 nm laser beam for both heating and temperature sensing of the GNR-FNDs trapped in the endosomes of a membrane nanotube. (c, d) Fluorescence (c) and merged bright-field/fluorescence (d) images of HEK293T cells transduced with actin-GFP fusion proteins (green) and labeled with GNR-FNDs (red). (e) Empirical cumulative distribution plot of the membrane temperatures at which TNTs are ruptured by local heating and global heating of GFP-transduced HEK293T cells. Data of two independent global heating experiments (blue dots) are shown in the plot. (f, g) Fluorescence (f) and merged bright-field/fluorescence (g) images of GNR-FND-labeled, GFP-transduced HEK293T cells after exposure to the 594 nm laser with a power of 330 μW for six seconds. TNT breaking and retraction could be clearly observed after laser irradiation. Insets of (d) and (g): Enlarged views of the regions with the particle irradiated by the 594 nm laser in the yellow boxes. White arrows indicate the particles being irradiated. GFP, green fluorescent protein. *Source:* Reprinted with permission from Ref. [25]. Reproduced with permission of John Wiley & Sons.

experiments. The local temperature rise leading to cell death could also be measured with this technique. The experiment demonstrated a new paradigm for hyperthermia research and application. It is anticipated that further advancement of the technologies will lead to the production of multifunctional nanoparticles for all-in-one hyperthermia therapy with tracking, heating, and sensing capabilities.

12.3   Silver/Diamond Nanohybrids

Silver nanostructures are widely recognized to have a profound effect on Raman scattering, known as *surface-enhanced Raman scattering* (SERS) [34]. The nanostructures also help increase the photoluminescence intensities of fluorophores in the vicinity [35]. A study in 2009 first reported that the fluorescence intensity of FNDs (35 or 140 nm in size) could be significantly enhanced after deposition on nanocrystalline Ag films without buffer layers [36]. The intrinsic photostability of the FND nanoparticles was preserved even in direct contact with the Ag film. However, the fluorescence enhancement effect diminished when the individual FND particles were wrapped around by DNA molecules, which increased the distance between the NV$^-$ centers inside the FNDs and nearby Ag nanoparticles. An enhancement factor of up to 10 was measured for 35-nm FNDs, confirmed by time-gated fluorescence imaging (Section 8.2.4) and histogram analysis.

The Ag film mentioned above consisted of Ag nanoparticles in various sizes and shapes, making it difficult to understand the underlying physics of the plasmon-enhanced effect. To overcome this difficulty, Andersen et al. [37] studied both experimentally and theoretically the spontaneous emission of the NV$^-$ centers in a FND placed near a silver nanocube with an AFM. The controlled assembly of the Ag-FND system allowed the researchers to compare directly its fluorescence properties with that of a single isolated FND (~40 nm in diameter). It was found that the cube-coupled NV$^-$ centers emitted strongly polarized fluorescence. At the optimal pump laser polarization, the rate of photons emitted by the FNDs in the nanohybrids was increased by a factor of 4 due to the local field enhancement. A similar result was obtained by Lourenco-Martins et al. [38] in a cathodoluminescence study to probe the plasmonic coupling of the NV^0 centers in FND aggregates with nearby Ag nanocubes on the nanometer scale using both photons and fast electrons equipped in a scanning transmission electron microscope (Section 10.3).

The Ag-FND nanohybrids discussed so far are all formed by physical contacts among the individual particles and therefore may not be stable enough for practical applications. To strengthen the bonding, Gong et al. [39] developed a general bottom-up approach to fabricate freestanding FND-based hybrid nanostructures. Figure 12.9a displays a schematic diagram of the Ag-FND synthesis, which started with size-selected FNDs after extensive acid treatment to passivate their surface with carboxyl groups for good dispersion of the FND particles in water. The tradeoff was that the surface became relatively inert, impeding direct growth of Ag nanoparticles. To enable the nucleation and good control over the coverage of the externally coupled units, the FND surface was first functionalized with poly(vinylpyridine) (PVP) molecules, of which the pyridyl

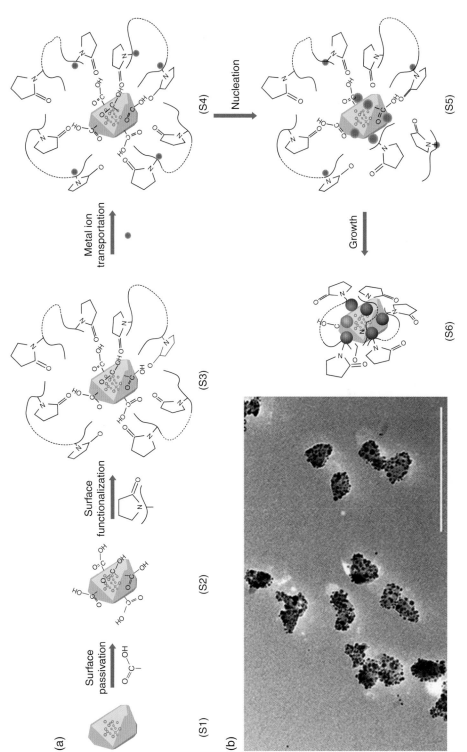

Figure 12.9 (a) Schematic synthetic paradigm illustrating different growth stages (S1–S6). (b) A typical large-scale TEM image showing excellent dispersion and uniformity of hybrid Ag-FND nanostructures made by following synthetic scheme in (a). Scale bar: 200 nm. *Source:* Reprinted with permission from Ref. [39].

groups could interact with the nonmetallic polar surface containing carboxyl groups through hydrogen bonding. The unbound nitrogen atoms at the end of the pyridyl groups were then utilized to attract and transport Ag^+ ions that eventually led to the nucleation and growth of Ag nanoparticles on the FND surface. TEM images confirmed the uniformity of the as-grown Ag-FND nanohybrids with the size of the constituent Ag nanoparticles in the range of 10 nm (Figure 12.9b). The research team also investigated in detail the coupling between the NV^- centers in FNDs and the SPRs of the Ag subunits by conducting size- and coverage-dependent ODMR, in addition to fluorescence lifetime measurements for the hybrid nanostructures. The strategy, in principle, can be applied to studying the interactions of NV^- and other nanostructures (such as quantum dots) as well as the synthesis of a wide range of metal-FND nanohybrids [40] with improved quantum sensing capabilities (Chapter 11).

With regard to biological applications of Ag-FNDs, a potential use of the nanohybrids is to improve the potency of silver nanoparticles in clinical medicine. Silver is a natural antibiotic material because it kills bacteria. People have used silver products for thousands of years and the material continues to play an important role in hospital as medical and health care appliance nowadays [41]. However, concerns about the possible harm of the heavy metal to human health exist. "Blue blood" is a description for people who are exposed to an excessive amount of silver and suffered from such conditions. The most dramatic symptom is that the skin of the patients turns purple or blue, known as "Argyria" [42]. A solution for *Argyria* is to use silver nanoparticles, which are more active than bulk silver due to their increased surface-to-volume ratios and thus a lower dose is required to achieve the same antimicrobial effects. Studies have shown that some bacteria (such as antibiotic-resistant bacteria) are unable to develop a resistance to silver nanoparticles even after repeated exposure [43, 44]. The conjugation of FND (or diamond in general) with silver is expected to reduce the toxicity of the metal particles while maintaining their antibacterial activities both *in vitro* and *in vivo*.

12.4   Iron Oxide/Diamond Nanohybrids

12.4.1   Single-Domain Magnetization

Superparamagnetic iron oxides (SPIOs) are ferrimagnetic nanoparticles composed of single magnetic domains [45]. This type of magnetism occurs in small particles (typical diameter < 50 nm), where thermal fluctuations randomly flip their magnetization directions and the average magnetization is zero. However, in the presence of an external magnetic field, the nanoparticles can be magnetized in the superparamagnetic states, developing strong internal magnetization. Unlike ferromagnetic substances, superparamagnetic materials do not retain any net magnetization once the external field is removed and thus have no magnetic memory. Composed of Fe_3O_4, which is easily oxidized to Fe_2O_3 in air, SPIOs have found wide applications as contrast agents for *magnetic resonance imaging* (MRI) [46] and heating agents for hyperthermia therapy [47]. Compared with other nanoparticles made of silica, gold, and silver, SPIOs have a higher level of biocompatibility, posing as the most promising candidate for clinical use. Therefore, when SPIOs are integrated with FNDs, it is anticipated to open up an exciting new horizon for the application of the nanohydrids in biology and medicine.

Plakhotnik et al. [48] presented a method to synthesize this new class of nanohybrids by first modifying the surface of FNDs (~40 nm in diameter) with amino groups, followed by coating of the cationic FNDs with SPIOs (~12 nm in diameter) through electrostatic self-assembly. TEM imaging showed a high-density layer of SPIOs on the FND surface. To characterize the magnetic properties of the surface-bound SPIOs, the researchers utilized the NV⁻ centers embedded in the FNDs for nanomagnetometric measurements. By using the ODMR technique as described in Section 3.4, saturation of the average magnetization was observed for the individual nanomagnets as the external magnetic field strength increased from 10 to 40 mT. The result was in agreement with a simple single-domain magnetization model, demonstrating the feasibility of forming this novel nanohybrid with the noncovalent conjugation method.

12.4.2 Magnetic Resonance Imaging

MRI is a powerful tool that has revolutionized the practice of medical research since its inception in the early 1970s [49]. The technique is noninvasive and yet able to produce three-dimensional anatomical and functional images of the human body without the use of harmful ionizing radiation [50]. Compared with other imaging modalities as listed in Figure 12.4, MRI provides a considerably higher penetration depth in tissue but suffers from poor sensitivity and low spatial resolution. To improve the sensitivity, SPIOs and *ultrasmall superparamagnetic iron oxides* (USPIOs) have been harnessed as MRI contrast agents for their excellent biocompatibility, high biodegradability, and large magnetic susceptibility. By properly conjugating SPIOs or USPIOs with FNDs, high-sensitivity imaging *in vitro* and *in vivo* is possible with these fluorescent magnetic nanohybrids.

In a contemporary work aiming at developing the dual-modality imaging agents, we covalently linked carboxylated FNDs and aminated USPIOs together by carbodiimide chemistry to form nanohybrids. DLS showed a number-averaged hydrodynamic diameter of 10.6, 84.9, and 96.4 nm for USPIOs, FNDs, and USPIO-FNDs, respectively (Figure 12.10a). TEM imaging of USPIO-FNDs revealed that the surface of FND could be covered by a large number (>100) of USPIO particles with a surface coverage of more than 50% (Figure 12.10b). The content of Fe atoms in the USPIO-FND nanohybrids was approximately 7% (w/w), as determined by *inductively coupled plasma* (ICP) mass spectrometry. We then measured the saturation magnetization (M_s) of USPIOs and USPIO-FNDs separately by using a *superconducting quantum interference device* (SQUID), and obtained a value of $M_s = 63$ emu g^{-1} for USPIOs in the nanohybrids (Figure 12.10c), in agreement with $M_s = 60$ emu g^{-1} of free USPIOs. Additionally, the hysteresis curves of USPIO-FNDs and USPIOs matched closely with each other, indicating that there was no significant difference in the inherent super paramagnetic properties of USPIOs before and after covalent conjugation with FNDs. Further investigations of the magnetic properties of USPIO-FNDs were carried out by measuring the transverse (spin–spin) relaxation times of water protons with (T_2) and without ($T_{2,w}$) the nanoparticles (Section 11.1). The equation used to quantify the relaxation times was

$$\frac{1}{T_2} = \frac{1}{T_{2,w}} + r_2 C_{Fe},\tag{12.2}$$

where r_2 is the relaxivity and C_{Fe} is the Fe concentration calculated from the weight of USPIOs. Relaxation time measurements against C_{Fe} yielded $r_2 = 149$ mM^{-1}s^{-1} for

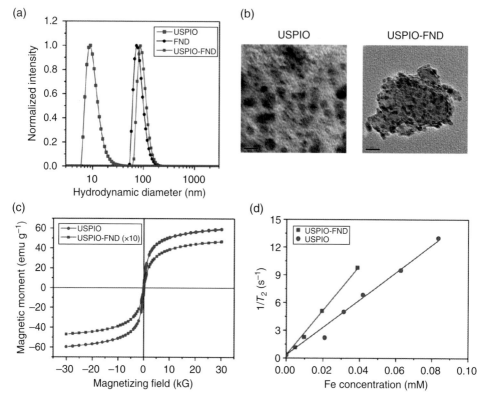

Figure 12.10 Characterization of USPIOs, FNDs, and USPIO-FNDs. (a) Size distributions, (b) TEM images, (c) magnetizations, and (d) transverse relaxivities (r_2) of USPIOs before and after conjugation with FND. Scale bars in (b): 10 nm (USPIO) and 20 nm (USPIO-FND). The values of the transverse relaxivity (r_2) are given by the slopes of the linear fitting to the experimental data in (d).

USPIOs and $r_2 = 240$ mM^{-1}s^{-1} for USPIO-FNDs (Figure 12.10d). The increase of r_2 by approximately 60% upon conjugation with FNDs suggested that USPIO-FNDs may serve as a better contrast agent than USPIOs alone for MRI imaging.

An application of SPIO-FNDs with a substantial impact is the tracking of tumor cells *in vivo* for cancer detection. To demonstrate the feasibility, we labeled human brain tumor cells (U-87 MG) with USPIO-FNDs in a culture medium. The USPIO-FND nanohybrids could be avidly taken up by the cells under serum-free conditions. To quantify the USPIO-FND uptake and the Fe loading capacity per cell, we utilized the magnetically modulated fluorescence technique for FNDs as described in Section 9.3.4. For cells labeled at the USPIO-FND concentration of 200 µg ml^{-1}, the amounts determined were 2.4×10^4 FND particles per cell and 3.15 pg of Fe per cell. The high Fe loading capacity allowed us to conduct *in vivo* MRI of these labeled cells transplanted through subcutaneous injection into mice. An example of such results is given in Figure 12.11a, where coronal slices (three each) of the T_2-weighted images were acquired at different time points. The sizes of both the unlabeled (right thigh) and USPIO-FND-labeled (left thigh) brain tumors were monitored in real time without having to sacrifice the animals. No adverse effects were found in the tumor growth due to the USPIO-FND labeling. The results proved that the nanohybrid-enabled MRI was

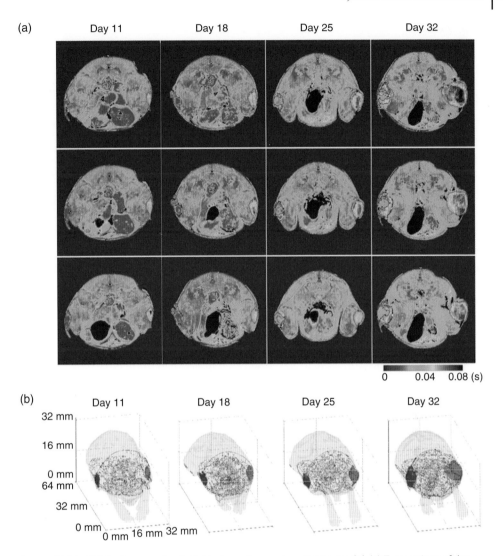

Figure 12.11 MRI for the growth of glioblastoma tumors in a mouse model. (a) T_2 mappings of three coronal slices on day 11, 18, 25, and 32. The unlabeled glioblastoma tumors (right thigh) were circled by thin red lines and the USPIO-FND-labeled glioblastoma tumors were circled by thin black lines (left thigh). Given shorter T_2 times, the contrast-enhanced regions within the labeled glioblastoma tumors (left thigh) are in dark blue due to the presence of USPIO-FNDs. (b) Three-dimensional images of the unlabeled glioblastoma tumor (right thigh, red area) and the USPIO-FND-labeled glioblastoma tumor (left thigh, green area) on day 11, 18, 25, and 32.

capable of three-dimensional tracking noninvasive to the growth of glioblastoma tumors (Figure 12.11b). Furthermore, detailed histological analysis of the USPIO-FND-labeled glioblastoma tissue sections could be performed by hematoxylin and eosin staining, immunofluorescence, and time-gated fluorescence imaging of FNDs (Section 9.3.3). Altogether, it is feasible now to extend the spatial resolution of the imaging from *in vivo* MRI on the centimeter scale to *in vitro* fluorescence microscopy on the micrometer or even nanometer scale, approaching true nanomedicine.

References

1 Kickelbick, G. (2007). Introduction to hybrid materials. In: *Hybrid Materials: Synthesis, Characterization, and Applications* (ed. G. Kickelbick), 1–48. Wiley.

2 Menon, J.U., Jadeja, P., Tambe, P. et al. (2013). Nanomaterials for photo-based diagnostic and therapeutic applications. *Theranostics* 3: 152–166.

3 Rehor, I., Slegerova, J., Kucka, J. et al. (2014). Fluorescent nanodiamonds embedded in biocompatible translucent shells. *Small* 10: 1106–1115.

4 Bumb, A., Sarkar, S.K., Billington, N. et al. (2013). Silica encapsulation of fluorescent nanodiamonds for colloidal stability and facile surface functionalization. *J Am Chem Soc* 135: 7815–7818.

5 Sarkar, S.K., Bumb, A., Wu, X. et al. (2014). Wide-field *in vivo* background free imaging by selective magnetic modulation of nanodiamond fluorescence. *Biomed Opt Express* 5: 1190–1202.

6 von Haartman, E., Jiang, H., Khomich, A.A. et al. (2013). Core-shell designs of photoluminescent nanodiamonds with porous silica coatings for bioimaging and drug delivery I: fabrication. *J Mater Chem B* 1: 2358–2366.

7 Prabhakar, N., Nareoja, T., von Haartman, E. et al. (2013). Core-shell designs of photoluminescent nanodiamonds with porous silica coatings for bioimaging and drug delivery II: application. *Nanoscale* 5: 3713–3722.

8 Liu, Y.L. and Sun, K.W. (2011). Plasmon-enhanced photoluminescence from bioconjugated gold nanoparticle and nanodiamond assembly. *Appl Phys Lett* 98: 153702.

9 Schietinger, S., Barth, M., Aichele, T., and Benson, O. (2009). Plasmon-enhanced single photon emission from a nanoassembled metal-diamond hybrid structure at room temperature. *Nano Lett* 9: 1694–1698.

10 Shen, H., Chou, R.Y., Hui, Y.Y. et al. (2016). Directional fluorescence emission from a compact plasmonic-diamond hybrid nanostructure. *Laser Photon Rev* 10: 647–655.

11 Zhang, B., Fang, C.Y., Chang, C.C. et al. (2012). Photoacoustic emission from fluorescent nanodiamonds enhanced with gold nanoparticles. *Biomed Opt Express* 3: 1662–1669.

12 Tam, A. (1986). Applications of photoacoustic sensing techniques. *Rev Mod Phys* 58: 381–431.

13 Fried, N.M. and Burnett, A.L. (2015). Novel methods for mapping the cavernous nerves during radical prostatectomy. *Nat Rev Urol* 12: 451–460.

14 Liu, W., Naydenov, B., Chakrabortty, S. et al. (2016). Fluorescent nanodiamond-gold hybrid particles for multimodal optical and electron microscopy cellular imaging. *Nano Lett* 16: 6236–6244.

15 Huang, X., Jain, P.K., El-Sayed, I.H., and El-Sayed, M.A. (2008). Plasmonic photothermal therapy (PPTT) using gold nanoparticles. *Lasers Med Sci* 23: 217–228.

16 Jaque, D., Martínez Maestro, L., del Rosal, B. et al. (2014). Nanoparticles for photothermal therapies. *Nanoscale* 6: 9494–9530.

17 Li, J., Guo, H., and Li, Z.Y. (2013). Microscopic and macroscopic manipulation of gold nanorod and its hybrid nanostructures. *Photon Res* 1: 28–41.

18 Park, K., Biswas, S., Kanel, S. et al. (2014). Engineering the optical properties of gold nanorods: independent tuning of surface plasmon energy, extinction coefficient and scattering cross section. *J Phys Chem C* 118: 5918–5926.

19 Ekici, O., Harrison, R.K., Durr, N.J. et al. (2008). Thermal analysis of gold nanorods heated with femtosecond laser pulses. *J Phys D Appl Phys* 41: 185501.

20 Urban, A.S., Fedoruk, M., Horton, M.R. et al. (2009). Controlled nanometric phase transitions of phospholipid membranes by plasmonic heating of single gold nanoparticles. *Nano Lett* 9: 2903–2908.

21 Tong, L., Zhao, Y., Huff, T.B. et al. (2007). Gold nanorods mediate tumor cell death by compromising membrane integrity. *Adv Mater* 19: 3136–3141.

22 Delcea, M., Sternberg, N., Yashchenok, A.M. et al. (2012). Nanoplasmonics for dual-molecule release through nanopores in the membrane of red blood cells. *ACS Nano* 6: 4169–4180.

23 Tsai, P.C., Chen, O.Y., Tzeng, Y.K. et al. (2015). Gold/diamond nanohybrids for quantum sensing applications. *EPJ Quantum Technol* 2: 19.

24 Epperla, C.P., Chen, O.Y., and Chang, H.C. (2016). Gold/diamond nanohybrids may reveal how hyperlocalized hyperthermia kills cancer cells. *Nanomedicine (Lond)* 11: 443–445.

25 Tsai, P.C., Epperla, C.P., Huang, J.S. et al. (2017). Measuring nanoscale thermostability of cell membranes with single gold-diamond nanohybrids. *Angew Chem Int Ed* 56: 3025–3030.

26 Sadtler, B. and Wei, A. (2002). Spherical ensembles of gold nanoparticles on silica: electrostatic and size effects. *Chem Commun* 15: 1604–1605.

27 Pastoriza-Santos, I., Gomez, D., Pérez-Juste, J. et al. (2004). Optical properties of metal nanoparticle coated silica spheres: a simple effective medium approach. *Phys Chem Chem Phys* 6: 5056–5060.

28 Charras, G.T. (2008). A short history of blebbing. *J Microsc* 231: 466–478.

29 Minati, L., Cheng, L.C., Lin, C.Y. et al. (2015). Synthesis of novel nanodiamonds-gold core shell nanoparticles. *Diam Relat Mater* 53: 23–28.

30 Matassa, R., Orlanducci, S., Reina, G. et al. (2016). Structural and morphological peculiarities of hybrid Au/nanodiamond engineered nanostructures. *Sci Rep* 6: 31163.

31 Chang, B.M., Lin, H.H., Su, L.J. et al. (2013). Highly fluorescent nanodiamonds protein-functionalized for cell labeling and targeting. *Adv Funct Mater* 23: 5737–5745.

32 Cheng, L.C., Chen, H.M., Lai, T.C. et al. (2013). Targeting polymeric fluorescent nanodiamond-gold/silver multi-functional nanoparticles as a light-transforming hyperthermia reagent for cancer cells. *Nanoscale* 5: 3931–3940.

33 Rehor, I., Lee, K.L., Chen, K. et al. (2015). Plasmonic nanodiamonds: targeted core-shell type nanoparticles for cancer cell thermoablation. *Adv Healthc Mater* 4: 460–468.

34 Schlücker, S. (2014). Surface-enhanced Raman spectroscopy: concepts and chemical applications. *Angew Chem Int Ed* 53: 4756–4795.

35 Aslan, K., Gryczynski, I., Malicka, J. et al. (2005). Metal-enhanced fluorescence: an emerging tool in biotechnology. *Curr Opin Biotechnol* 16: 55–62.

36 Lim, T.S., Fu, C.C., Lee, H.Y. et al. (2009). Fluorescence enhancement and lifetime modification of single nanodiamonds near a nanocrystalline silver surface. *Phys Chem Chem Phys* 11: 1508–1514.

37 Andersen, S.K.H., Kumar, S., and Bozhevolnyi, S.I. (2016). Coupling of nitrogen-vacancy centers in a nanodiamond to a silver nanocube. *Opt Mater Express* 6: 3394–3406.

38 Lourenco-Martins, H., Kociak, M., Meuret, S. et al. (2018). Probing plasmon-NV^0 coupling at the nanometer scale with fast electrons and photons. *ACS Photonics* 5: 324–328.

39 Gong, J., Steinsultz, N., and Ouyang, M. (2016). Nanodiamond-based nanostructures for coupling nitrogen-vacancy centres to metal nanoparticles and semiconductor quantum dots. *Nat Commun* 7: 11820.

40 Wang, N., Liu, G.Q., Leong, W.H. et al. (2018). Magnetic criticality-enhanced hybrid nanodiamond thermometer under ambient conditions. *Phys Rev X* 8: 011042.

41 Alexander, J.W. (2009). History of the medical use of silver. *Surg Infect* 10: 289–292.

42 Wikipedia (2018). Argyria. https://en.wikipedia.org/wiki/Argyria (accessed 16 April 2018).

43 Rai, M., Yadav, A., and Gade, A. (2009). Silver nanoparticles as a new generation of antimicrobials. *Biotechnol Adv* 27: 76–83.

44 Lara, H.H., Garza-Treviño, E.N., Ixtepan-Turrent, L., and Singh, D.K. (2011). Silver nanoparticles are broad-spectrum bactericidal and virucidal compounds. *J Nanobiotechnol* 9: 30.

45 Lu, A.H., Salabas, E.L., and Schüth, F. (2007). Magnetic nanoparticles: synthesis, protection, functionalization, and application. *Angew Chem Int Ed* 46: 1222–1244.

46 Wang, Y.X.J. (2011). Superparamagnetic iron oxide based MRI contrast agents: current status of clinical application. *Quant Imaging Med Surg* 1: 35–40.

47 Dutz, S. and Hergt, R. (2014). Magnetic particle hyperthermia – a promising tumor therapy? *Nanotechnology* 25: 452001.

48 Plakhotnik, T., Aman, H., Zhang, S., and Li, Z. (2015). Super-paramagnetic particles chemically bound to luminescent diamond: single nanocrystals probed with optically detected magnetic resonance. *J Phys Chem C* 119: 20119–20124.

49 The Nobel Prize in Physiology or Medicine for 2003 – Press Release. *Nobelprize.org*. Nobel Media AB 2014. http://www.nobelprize.org/nobel_prizes/medicine/laureates/2003/press.html (accessed 16 April 2018).

50 Storey, P. (2006). Introduction to magnetic resonance imaging and spectroscopy. In: *Magnetic Resonance Imaging: Methods and Biologic Applications* (ed. P.V. Prasad), 3–57. Humana Press.

13

Nanodiamond-Enabled Medicine

Equipped with specific targeting ability (Chapter 4), inherent biocompatibility (Chapter 5), and versatile imaging capability (Chapters 7–12), surface-functionalized *fluorescent nanodiamonds* (FNDs) have a promising role to play in the magical land of medicine. We mean the medicine of this generation and the future. It seems fitting to have an overview here, specifically, in the areas of precision medicine and nanomedicine where *nanodiamonds* (NDs) have found their place.

Precision medicine, best described as "prevention and treatment strategies that take individual variability into account" [1], may not be a new concept but is boosted mainly by big data becoming available (and accessible), emerging in recent years as a current trend in modern medicine. In his annual State of The Union address in 2015, the U.S. President Obama launched a new *Precision Medicine Initiative* and stated that the goal of such an initiative was "to give all of us access to the personalized information we need to keep ourselves and our families healthier" [2]. From that point onward, although not exactly the equivalent, precision medicine is synonymous with *personalized medicine*. The goal, to put it bluntly, is to provide the right drug at the right dose to the right person. With the participation of nanomedicine, the statement may be expanded to include "in the right place and at the right time".

Nanomedicine is both young and technological, cohesively integrated from multiple disciplines. A nicely drafted definition is provided by the *European Science Foundation*: "Nanomedicine uses nano-sized tools for the diagnosis, prevention, and treatment of disease and to gain increased understanding of the complex underlying patho-physiology of disease. The ultimate goal is to improve quality-of-life" [3]. A comprehensive study of the nanotechnologies in nanomedicine over a 10-year period has concluded three major areas of applications, including (i) diagnostics, sensors, and surgical tools used outside the patients, (ii) imaging agents and monitoring technologies applied at the cell level and up to the whole body, and (iii) nanomaterials and devices for drug delivery and therapeutics [4]. As unveiled in the previous chapters, FNDs and their nanohydrids pose a remarkable prospect as a competent candidate for tasks (i) and (ii) stated above. For the sake of drug delivery and therapeutics (task (iii)), NDs (both fluorescent and nonfluorescent) are useful as a biocompatible and nontoxic platform for nanomedicine, which is aimed at delivering only the amount of active ingredient needed for treating the exact spots that are infected at the molecular or cellular level over a designated period of time necessary for curing disease without overdoses or side effects. We start the discussion in this chapter with NDs as therapeutic carriers.

Fluorescent Nanodiamonds, First Edition. Huan-Cheng Chang, Wesley Wei-Wen Hsiao and Meng-Chih Su.
© 2019 John Wiley & Sons Ltd. Published 2019 by John Wiley & Sons Ltd.

13.1   NDs as Therapeutic Carriers

Therapeutics is a branch of medicine that deals with the treatment of disease and the action of remedial agents [5]. Nanotechnology is becoming an integral part of this medical development because nanoparticle-based therapeutics hold the promises to overcome biological barriers, deliver hydrophobic drugs and biologics, and selectively target cancer cells [6]. However, in general, nanoparticles themselves may be highly reactive due to their small sizes and large specific surface areas (Chapter 4). Interactions of them with biomolecules in cells, tissues, and even extracellular environments can sometimes trigger a sequence of unexpected, undesired, or lethal effects (Chapter 5). These dynamic characteristics determine the biocompatibility of the nanoparticles as well as the efficacy of the intended treatments. Therefore, despite their potential benefits, the approved cases of nanoparticle-based medicines for clinical use are few and far between [7, 8]. The inherently biocompatible NDs are indeed in the right position to offer a possible solution to overcome these hurdles and challenges.

Detonation nanodiamonds (DNDs) and *high-pressure high-temperature nanodiamonds* (HPHT-NDs) are two most popular types of diamond nanoparticles used in the life sciences research [9]. They both are capable of binding with biomolecules and bioactive molecules (such as drugs) as a result of the interactions between the functional groups on their surface and the molecules of interest (Chapter 4). For acid-treated NDs as an example, most functional groups derivatized on the particles can carry charges in aqueous solution and the overall charge density depends on the pK_a values of these groups and the pH of the solution. The forces involved in the carrier–cargo interactions encompass electrostatic, hydrogen bonding, hydrophobic, and van der Waals forces. These acid-treated NDs stand out as an appealing drug carrier for their high loading capacities and the abilities to protect and retain the inherent therapeutic effects of the noncovalently bound molecules. Equally important, if not more, is the versatile surface chemistry that has enabled the development of novel delivery methods for optimal loading, specific targeting, and controlled release of cargo molecules on the ND surface for therapeutic treatments. NDs have been harnessed for the delivery of many classes of molecules, including small molecules, peptides and proteins, and nucleic acids, with a major focus on the use of chemotherapeutic agents for biomedical applications [9]. While some studies have covalently conjugated drug molecules with surface-functionalized NDs, which may impose some challenges on drug release, the majority of the present research has employed physical adsorption procedures.

Aside from the surface properties, the structure of the nanoparticles is another factor in favor of NDs for therapeutic applications. As discussed in Section 2.3.3, DNDs are synthesized by shock wave compression using explosive compounds such as 2,4,6-trinitrotoluene (TNT) and 1,3,5-trinitro-1,3,5-triazinane (RDX). With proper explosive mixture ratios, the shock wave can produce NDs with an average size of 4–5 nm in diameter for the primary particles. However, DNDs always agglomerate to form covalently bound clusters during the synthesis, with many nanometer-sized cavities present in the interiors of the clusters (Figure 2.7), as revealed by *transmission electron microscopy* (TEM) [10, 11]. These cavities are disordered in structure and size, allowing the particles to be filled with assorted molecules. Similar to mesoporous silica nanoparticles [12], they are useful as drug delivery devices.

The most unique feature that makes NDs stand out among other nanoparticle-based drug carriers is, perhaps, their fluorescence property. FNDs made of HPHT-NDs are appealing in this aspect because their fluorescence is bright and their spectroscopic properties are largely unaffected by the attachment of various types of therapeutic agents to the surface. Once properly conjugated with the therapeutic agents, FNDs can be used to measure, monitor, and even alter the biochemical processes within cells. They have the potential to serve as a treatment tool (e.g. with the carried drugs), and more significantly, a preventative monitoring device (e.g. for detecting pre-cancerous changes). Possessing several key properties necessary for clinical applications, i.e. stability and compatibility in biological environments as well as scalability in production, surface-functionalized FNDs are emerging as a powerful and versatile drug delivery platform [13, 14].

13.2 Drug Delivery

Drug delivery at the nanoscale is an interdisciplinary field that spans across chemistry, biology, and medicine. A variety of organic and inorganic nanoparticles has been tested as the drug carriers, including lipid-based vehicles (such as liposomes and micelles), polymeric nanoparticles (such as hydrogels and dendrimers), metal and metal oxide nanoparticles (such as gold or iron oxides), silica nanoparticles (such as amorphous or mesoporous silica), and carbon nanostructures (such as fullerenes, nanotubes, graphenes, and NDs) [15, 16]. To be a suitable drug delivery device, the carrier should possess the following characteristics: (i) sufficient loading capacity in proportion to the weight of the carrier, (ii) versatile binding with bioactive molecules, (iii) functional mechanism for targeted release, and (iv) high-sensitivity tracking by noninvasive methods. With good biocompatibility, large specific surface areas, high bioconjugation ability, and unique magneto-optical properties as discussed in the previous chapters, FNDs comply with all these requirements.

13.2.1 Small Molecules

Doxorubicin (DOX), an apoptosis-inducing drug widely used for cancer chemotherapy, is the first drug tested with NDs as the carriers [17]. Containing an amino group (Figure 13.1a), the molecule is fluorescent and therefore can be used as a *theranostic* (therapeutic and diagnostic) agent [18]. Ho and coworkers [17, 19–24] made considerable efforts to develop ND-based therapeutics with the DOX molecule. Instead of using DND agglomerates as shown in Figure 2.7, the experiments adopted monodisperse DNDs (2–8 nm in size) produced by wet ball milling [11]. Cytotoxicity tests through quantitative *real-time polymerase chain reaction* (RT-PCR) along with the DNA fragmentation assays all confirmed the innate biocompatibility of the monodisperse DND particles. To load DNDs with DOX, the researchers took advantage of the fact that DNDs could easily agglomerate in high-concentration salt solution (e.g. NaCl) to form noncovalently bound clusters. These DND clusters were then bound with DOX through ionic interactions between the negatively charged carboxyl groups on DNDs and the positively charged amino group of DOX (Figure 13.2a and b). It was found that the DND

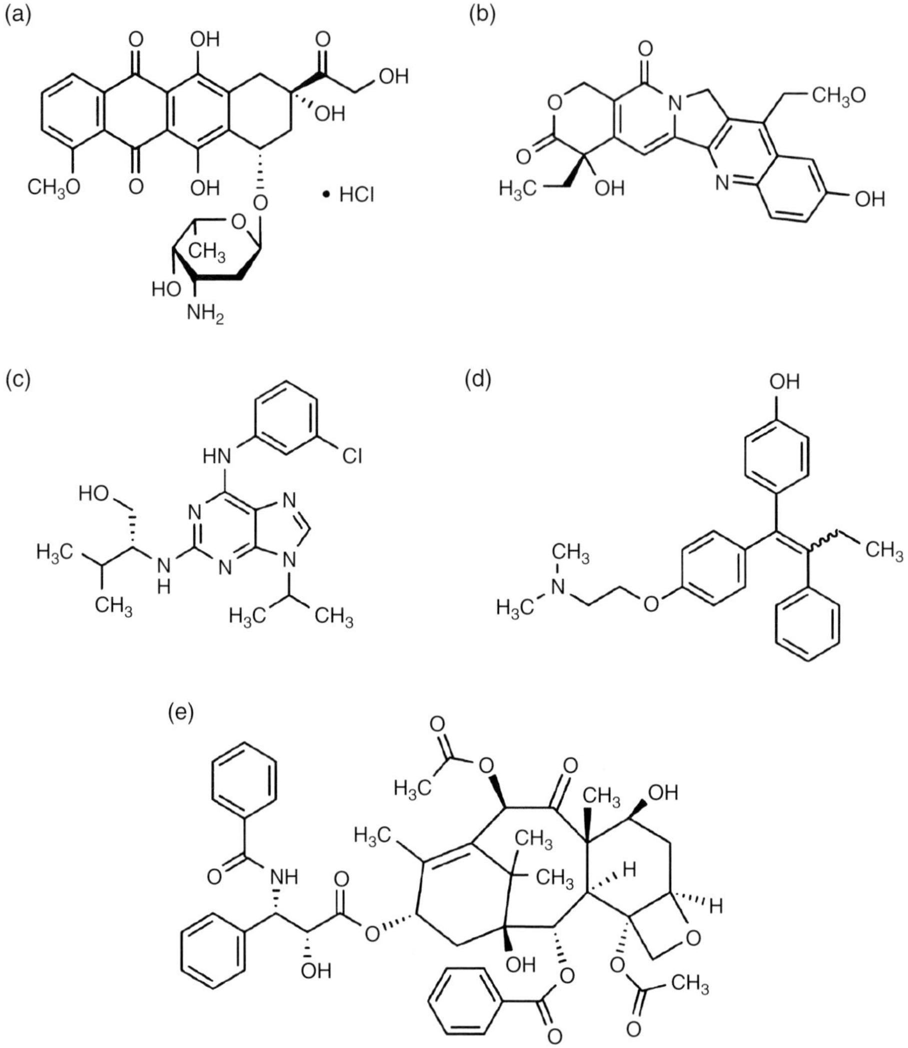

Figure 13.1 Chemical structures of some commonly used drugs loaded on NDs. (a) Doxorubicin, (b) 10-hydroxycamptothecin, (c) purvalanol A, (d) 4-hydroxytamoxifen, and (e) Paclitaxel.

hydrogels loaded with DOX could efficiently deliver the drugs into living cells such as murine macrophages and human colorectal carcinoma cells by endocytosis. The same group further investigated the rate of DOX–DND movement into cells by using DNDs physically bound with fluorescently labeled poly-L-lysine observable by confocal fluorescence microscopy. Cytotoxicity and genotoxicity assays proved that the DOX–DND complexes prompted the cell death, showing clearly that DND could serve as an effective carrier therapeutically significant for both systemic and localized drug delivery [25].

The findings that the salt addition facilitates the loading of DOX into the DND clusters and conversely the salt removal triggers its release suggest a simple switching mechanism that has great potential for use in medical practice. In a supplementary study,

(a)

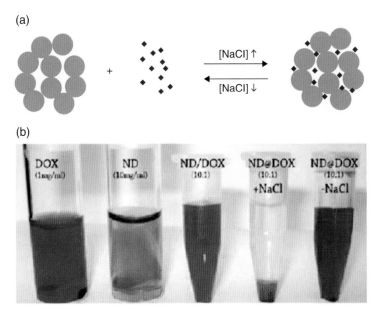

(b)

Figure 13.2 (a) Schematic drawing and (b) photograph of NaCl-mediated loading and release of doxorubicin. Blue spheres denote DNDs and red dots denote DOX. *Source:* Adapted with permission from Ref. [17]. Reproduced with permission of American Chemical Society.

both theoretical and experimental, Adnan et al. [21] looked into the influence of pH on the degree of the DOX loading and found that DOX was only able to bind with DNDs at a high pH (pH > 10). The amount of the bound DOX increased with the ionic strength and basicity of the environment, a trend in line with the strengthening of the electrostatic interactions [26]. Similar studies were carried out by other research groups using DNDs or HPHT-NDs surface-coated with polyethylene glycol to achieve drug delivery and slow release of DOX [27, 28]. Indeed, the delivery strategies thus developed all showed the transport of DOX into cancer cells, as confirmed by confocal fluorescence microscopy and flow cytometric analysis utilizing the intrinsic fluorescence of DOX. However, as the fluorescence properties of DOX depended sensitively on its local environment [29], FNDs were additionally used as references to support their findings [28]. Following the same approach, Wang et al. [30] performed targeting therapy of cancer cells using transferrin-conjugated FNDs as the DOX carrier and proved successful delivery and release of DOX at the targeted sites.

Inspired by the studies of DOX, Guan et al. [31] loaded DNDs with cis-dichlorodiamineplatinum (II) ($Pt(NH_3)_2^{2+}$, CDDP), another widely used anticancer drug. The loading was made by surface functionalization of DNDs with carboxyl groups, followed by ionic interactions of the surface-bound $-COO^-$ with the Pt(II) of CDDP. It was found that CDDP could be released from the composite in phosphate-buffered saline (PBS) of pH 6.0 at a rate twice faster than that of pH 7.4. This pH-responsive release property implied a reduction of toxic side effects. For example, the CDDP–DND composites may release a small amount of CDDP during their circulation period in the blood (pH 7.4) but deliver a larger amount to acidic compartments such as lysosomes (pH < 6) in cells.

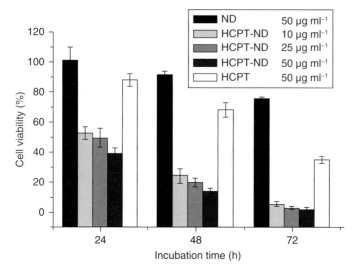

Figure 13.3 Viability of HeLa cells treated with DNDs, HCPT-DNDs, and HCPT alone at different concentrations and time points. *Source:* Reprinted with permission from Ref. [32]. Reproduced with permission of Elsevier.

In another study, Li et al. [32] investigated the dose-dependent behaviors of DNDs as an anticancer drug delivery vehicle. The researchers demonstrated that 10-hydroxyca-mptothecin (HCPT) (Figure 13.1b) was capable of attaching to DNDs through the NaOH-aided solubility enhancement of HCPT at an increased diffusion rate when the drug entered the interior of the covalently agglomerated DNDs. The sustained release of HCPT from the DND agglomerates occurred at pH < 6. The chemotherapeutic effect of the HCPT–DND complexes was found to be higher than that of the stand-alone HCPT by a significant margin (Figure 13.3), suggesting that DNDs were a promising drug delivery platform for cancer therapy.

A main limitation in the systemic administration is the low water solubility of certain drugs. Take two commonly used drugs for liver and breast cancer treatments, Purvalanol A (Figure 13.1c) and 4-hydroxytamoxifen (Figure 13.1d) for example; they both are sol-uble in dimethyl sulfoxide (DMSO) but have a poor solubility in water. Chen et al. [33] chose these two as the model compounds to study the improvement of the drug solubil-ity by complexation with acid-treated, monodisperse DNDs. A visible enhancement in the dispersibility was found for both drugs in the DMSO/water solution (Figure 13.4a–f). The study provided a useful strategy for applying water-insoluble drug molecules to treatment-relevant scenarios through the complexation with DNDs (Figure 13.4g and h).

Someone may ask: Could drug molecules be delivered through covalent conjugation with the carriers? What are the pros and cons of a covalent delivery? Indeed, research-ers in the field have also covalently grafted DOX onto the surface of NDs to avoid premature release and enhance its delivery. This was made by covalent conjugation of the particles with DOX and the cell-penetrating peptide TAT, a HIV trans-activator of the transcription protein, in sequence by carbodiimide chemistry (Section 4.2.2) [34]. The same approach was further carried out by Zhang et al. [35] who presented a mul-timodal DND drug delivery system for the targeting, imaging, and enhanced treatment

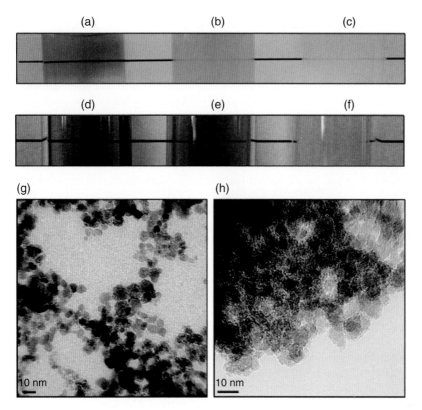

Figure 13.4 Dispersion of Purvalanol A and 4-hydroxytamoxifen (4-OHT) in water before and after complexation with monodisperse DNDs. Vials were prepared against background, and the reduction in turbidity mediated by the DNDs was confirmed under the following conditions: (a) 1 mg ml^{-1} DND in 5% DMSO; (b) 1 mg ml^{-1} DND, 0.1 mg ml^{-1} Purvalanol A in 5% DMSO; (c) 0.1 mg ml^{-1} Purvalanol A in 5% DMSO; (d) 1 mg ml^{-1} DND in 25% DMSO; (e) 1 mg ml^{-1} DND, 0.1 mg ml^{-1} 4-OHT in 25% DMSO; (f) 0.1 mg ml^{-1} 4-OHT in 25% DMSO. (g, h) TEM images of pristine DNDs (g) and 4-OHT–DND complexes (h). *Source:* Reprinted with permission from Ref. [33]. Reproduced with permission of American Chemical Society.

of Paclitaxel (PTX) (Figure 13.1e), a drug commonly used to treat ovarian, breast, and lung cancers. The novelty of this work is that it performed heterofunctionalization of DNDs by grafting fluorescently labeled PTX–DNA conjugates as well as monoclonal antibodies for epidermal growth factor receptor (EGFR) on the same DND surface for specific targeting and imaging purposes. The integration of therapeutics and diagnostics provides a transition from conventional medicine to contemporary personalized and precision medicine, as discussed at the beginning of this chapter.

13.2.2 Proteins

Beyond small molecules, NDs can also play a significant role in protein therapeutics [36]. In a typical application, protein molecules are first loaded on the surface of the carrier, transported, and then released at a specific target site. As pointed out in Section 4.2, care must be taken to conserve their functionalities when loading the biological macromolecules onto NDs. A spectroscopic study of bovine serum albumin (BSA) on DNDs showed that most structural features of the protein were preserved, although the adsorbed BSA might have undergone some minor conformational changes due to the protein-surface interactions [37]. In another study for lysozyme physically anchored on HPHT-NDs (~100 nm in diameter), the hydrolytic activity of the adsorbed protein was found to be retained but reduced to 15–70% of the activity of free lysozyme in solution [38]. Similarly, for α-bungarotoxin, a neurotoxin from *Bungarus multicinctus*, the protein could maintain its bioactivity even after physical adsorption onto carboxylated HPHT-NDs [39]. The activity was confirmed by blocking the membrane protein, α-7-nicotinic acetylcholine receptor, expressed in oocytes from *Xenopus laevis*. The same conclusion was reached for rabbit anti-mouse antibodies covalently immobilized on DNDs [40].

Insulin was the first recombinant human protein therapeutic developed in the 1980s [41]. The therapeutics employs recombinant insulin synthesized by protein engineering to replace its natural counterpart deficient in diabetes mellitus. The human insulin has a molecular mass of 5808 Da, consisting 51 amino acids as shown in Figure 13.5a. Shimkunas et al. [42] explored the feasibility of using monodisperse DNDs for insulin therapy, where the protein molecules served as a potential promoting agent for wound healing and vascularization for severe burns and other possible conditions. The study demonstrated an efficient method for noncovalent conjugation of insulin to the surface of DNDs and, when exposed to an alkaline environment, the attached insulin could be released from the complexes (Figure 13.5b). The research team confirmed the effective binding and release of insulin from the DND carriers through imaging and adsorption/desorption assays. Both cytotoxicity and RT-PCR analysis revealed that the protein's functionality remained active after desorption but was inactive when adsorbed onto the DND surface, a behavior attributed to the rapid formation of stable insulin aggregates at interfaces [43, 44]. The result demonstrated that DND could play an effective role in insulin delivery. In addition to insulin, the same research group also investigated the feasibility of using DNDs as a protein delivery vehicle for the transforming growth factor beta antibody (anti-TGF-β), which is a potential anti-scarring agent [45]. While the anti-TGF-β-DND complexes were stable in water, the antibodies could be triggered to release upon their incubation in serum-containing media. *Enzyme-linked immunosorbent assays* (ELISA) verified the preservation of the protein activity after release.

Pushing the research further along, Moore et al. [46] developed novel nanoparticle suspensions that could be used in oral surgery as injectable alternatives for the delivery of bone morphogenetic proteins (BMP-2) and low doses of basic fibroblast growth factors (b-FGF). The study evaluated the efficacy of DNDs for simultaneous and targeted delivery of both proteins as well as their ability to hasten localized bone growth. The high adsorption ability of DNDs allowed BMP-2 and b-FGF to be promptly loaded into the DND clusters by physical adsorption, as confirmed by Fourier transform infrared spectroscopy and ELISA. The results successfully demonstrated that DNDs were indeed qualified for the simultaneous delivery of these two functional proteins (BMP-2 and b-FGF), both required for bone healing *in vivo*. All the above studies together support the notion that ND is a useful protein delivery vehicle.

A promising advance in the field is the development of ND-based vaccines. *Nanovaccine*, defined as any vaccines containing nanoparticles, is a new kind of immunotherapy. It has

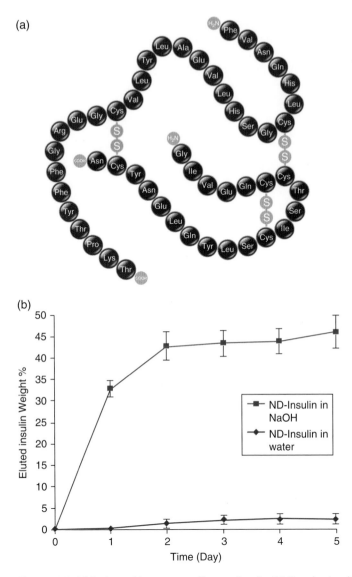

Figure 13.5 (a) Amino acid sequence of human insulin. (b) Five-day insulin desorption test of ND–insulin complexes treated with NaOH (pH 10.5) and water, showing insulin release in an alkaline pH environment. *Source:* Adapted with permission from Ref. [42]. Reproduced with permission of Elsevier.

attracted considerable interest recently because using nanoparticles for vaccine delivery allows the improvement of antigen stability and the enhancement of antigen immunogenicity [47]. Moreover, it enables targeted delivery and slow release of antigens as discussed in the previous sections. Pham et al. [48] demonstrated the use of NDs to facilitate the development of vaccine against H7N9 viruses with the antigen, hemagglutinin subtype 7 (H7). Hemagglutinin is a glycoprotein on the surface of influenza viruses, responsible for both viral attachment and viral/host membrane fusion. It is included in all currently approved human influenza vaccines. The research team first mixed trimeric

H7 with HPHT-NDs (100 nm in diameter) at a weight ratio of 1 : 12 to form noncovalently bound H7–ND complexes and then determined their activities by the hemagglutination assay [49]. A 64-fold increase in the activity was found for H7–ND, compared with that of free trimeric H7. Additionally, the H7–ND complex elicited a significantly higher level of H7-specific antibodies than the free trimeric H7 did after the second and third immunization in mice, as revealed by ELISA and Western blotting. The results suggested an important role of NDs in developing better and more cost-effective nanovaccines than traditional vaccine formulations.

13.3 Gene Therapy

13.3.1 RNA

Gene therapy is a technique that potentially treats a disease at its genetic roots [50]. The primary goal of gene therapy is to introduce exogenous genetic materials into cells to swap out abnormal genes or to enable additional functions. The conventional approach of gene therapy is to transfect cells with a polymer-encapsulated DNA plasmid that can replace a defective gene in the target-cell genome. RNA interference has recently emerged as a new therapeutic pathway that can silence harmful genes by delivering complementary short interfering RNA (siRNA) to the targeted cells [51]. However, the siRNA delivery suffers from many of the same difficulties found in the DNA delivery as discussed in the next section. One of the difficulties is that the actual application of gene therapy to humans must rely on the use of benign and effective carriers for the genetic materials. Researchers have explored an array of materials to address the challenges associated with the delivery. Materials that have been applied include polymers, lipids, peptides, antibodies, aptamers, and small molecules [52]. Most of these compounds ease the delivery of genetic information by encapsulating or condensing nucleic acids into nanosized particles to increase their stability in the bloodstream or facilitate their uptake by cells.

Polyethylenimine (PEI) is one of the most commonly used polymers for nucleic acid delivery. An important feature of PEI is that it contains a high concentration of protonated amines, which makes it suitable for condensing large, negatively charged molecules such as RNA to form polyplexes [53]. The polyplexes first enter cells via endocytosis. Within endosomes, the unprotonated amine groups of PEI absorb protons from the cytosol, leading to an osmotic swelling of the vesicles and an increased influx of Cl^- ions and water. The swelling may eventually cause disruption of the endosomal membrane, which subsequently releases the contents into the cytoplasm. This is known as the *proton sponge effect* [54]. However, the PEI treatment has some toxic side effects, which are directly related to the polymer size and thus limit its practical therapeutic application.

In an effort to establish NDs as a siRNA delivery vehicle, Chen et al. [55] mixed monodisperse DNDs with excess PEI800 to form noncovalently bound DND–PEI complexes, followed by incubation of the complexes with siRNA for the delivery purpose. They chose PEI with a molecular mass of 800 Da (thus, PEI800) for the conjugation to avoid the toxicity effect often accompanied by high-molecular-weight PEI to cells. The strong electrostatic interactions between the oppositely charged DND–PEI and siRNA allowed

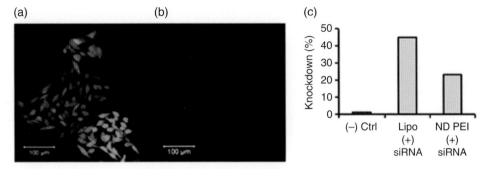

Figure 13.6 (a, b) Confocal fluorescence microscopy and (c) flow cytometric analysis of GFP knockdown in M4A4 cells transfected with GFP-expressing plasmids. Images shown in (a, b) are that of the negative control (a) and the treatment group with ND-PEI + siRNA (b). Lipo stands for lipofectamine, a commonly used transfection reagent. *Source:* Reprinted with permission from Ref. [55]. Reproduced with permission of American Chemical Society.

the particle to act as an effective nucleic acid carrier. Examined with both confocal fluorescence microscopy (Figure 13.6a and b) and flow cytometry (Figure 13.6c), knockdown experiments using human breast cancer cells (M4A4) transfected with green fluorescent protein (GFP) showed an efficiency exceeding 20%.

Research studies along this line continued the coating of NDs with basic amino acids such as lysine [56] and other cationic polymers such as poly(allylamine hydrochloride) (PAH) [57] through covalent immobilization to carry anionic siRNA. Alhaddad et al. [57] demonstrated the feasibility of using PEI- or PAH-coated FNDs to deliver siRNA into Ewing sarcoma cells. In this experiment, the siRNA sequences able to target the oncogene junction EWS-Fli1 in the chimeric mRNA were attached to the PEI- or PAH-coated FNDs with an average size of approximately 50 nm. Cellular uptake of the particles was evidenced by the intrinsic fluorescence of FNDs from the hosted color centers. Confocal fluorescence imaging confirmed the colocalization between the FND vectors and siRNA labeled with fluorescein isothiocyanate (FITC) (Figure 13.7a and b). With the use of FNDs, the researchers were able to evaluate the desorption kinetics of siRNA from the carriers in living cells. For cells cultured at five different time points, the signals of FITC were stronger than that of FNDs at the initial stage, a signature that siRNA was bound to both PEI- and PAH-coated carriers (Figure 13.7c). The FITC intensity, however, drastically decreased in the case of PEI coating at 24 hours, while it was only slightly reduced in the PAH coating, indicating a higher affinity of siRNA for PAH-FNDs. A specific inhibition of the EWS/Fli-1 gene expression was found at the mRNA and protein levels, proving the efficacy of this method.

13.3.2 DNA

Gene delivery is a process to introduce foreign DNA into host cells. It falls into two categories: viral and nonviral [58]. The former has an unrivaled level of gene transfection efficiency. However, it fails to become an all-utility vector due to the problems associated with formulation, storage, gene-carrying capacity, and residual viral elements that can potentially cause insertional mutagenesis, cytotoxicity, immunogenicity,

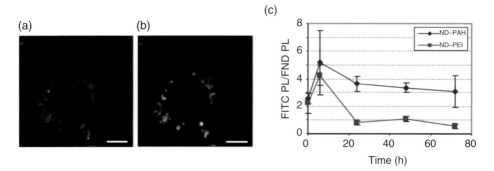

Figure 13.7 Colocalization studies between (a) PEI-FND vectors and (b) siRNA labeled by FITC in NIH/3T3 EF cells. (c) Quantitative estimate of the siRNA release time, using the photoluminescence intensities (PL) of FITC over the whole cell, normalized to that of FND. *Source:* Adapted with permission from Ref. [57]. Reproduced with permission of John Wiley & Sons.

and tumorigenicity. The field of nonviral vectors began as a response to these problems by including biocompatible materials designed and fabricated through innovative synthesis schemes. The preparation of nonviral vectors is relatively easy, less immunogenic and oncogenic, and allows all gene sizes. Nanoparticles have been increasingly popular as a nanoconstruct for the gene delivery [59]. However, because of the complexities involved in the precess, bringing nanoparticle-based gene therapy from the benchtop to the bedside still requires a comprehensive understanding of the advanced drug and gene delivery systems as well as how to apply nanoparticles in the therapy.

Various surface-modified NDs have been proposed as potential gene carriers in the literature [60–63]. Specifically, Zhang et al. [60] demonstrated that DNDs surface-functionalized (either covalently or noncovalently) with PEI800 served well as an effectual plasmid DNA delivery vehicle. The composite material DND-PEI800 showed low cytotoxicity with a transfection efficiency similar to that of DND-PEI25k, which had higher toxicity. A favorable property of the cross-linked ND-PEI800 was that it bound with plasmid DNA through electrostatic interactions and protected the plasmid from degradation in solution. Compared with PEI800, DND-PEI800 mediated a 70-fold upsurge in transfection efficiency while at the same time maintaining good biocompatibility with the cells. Also, the enhancement factor increased to 400 and 800 over that of amine- and carboxyl-terminated NDs, respectively. The transfection efficiency followed the trend of DND-PEI800 > PEI800 > DND-NH$_2$ > DND-COOH > naked DNA. The increased transfection efficiency of ND-PEI800 over ND-NH$_2$ was attributed to the proton sponge effect as discussed previously.

A bottleneck of the nanoparticle-based gene delivery is that the DNA molecules on the carriers must escape from endosomes in order to be available for cells to replicate or express. The proton sponge effect is one of the strategies. Interestingly, with HPHT-NDs as the carriers, another possible mechanism may exist. It was reported by Chu et al. [62] that prickly FNDs made of HPHT-NDs could easily enter cells via endocytosis, followed by a quick endosomal escape to the cytoplasm. Confocal fluorescence microscopy and TEM identified endosomal membrane rupturing to be the major route of the particles escaping from the confinement due to the unusual shape effect [64]. Little cytotoxicity was observed for such cytosolic release. The researchers demonstrated a viable application of the method by transfecting HepG2 cells with the GFP gene.

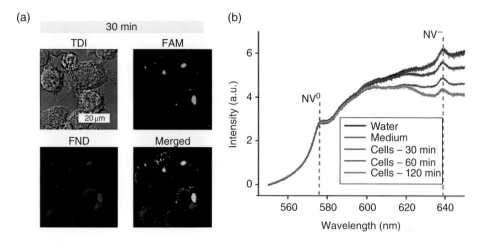

Figure 13.8 Imaging DNA release from the FND–PEI–DNA complexes in IC-21 macrophages.
(a) Confocal fluorescence images of the cells incubated with FND–PEI–DNA for 30 minutes, where DNA was labeled with fluorescein amidite (FAM) before forming complexes with FND–PEI. (b) Fluorescence spectra of NV centers measured by 532 nm laser excitation, measured for FND–PEI–DNA in water, medium, and cells after incubation for 30, 60, and 120 minutes. TDI, transmission/bright field image. *Source:* Adapted with permission from Ref. [63].

A nice piece of work to demonstrate the feasibility of tracking gene delivery in real time with FNDs was conducted by Petrakova et al. [63], who monitored the DNA transfection and payload release in cells over 12 hours using PEI-coated FNDs as the carriers (Figure 13.8a). They observed the events based on the changes in the fluorescence spectra of nitrogen-vacancy centers (including NV^- and NV^0) hosted in the FNDs of 35 nm in diameter. Specifically, the reduction of the fluorescence intensity at 638 nm relative to that at 576 nm served as an indicator for the release of DNA from the PEI-coated FND surface (Figure 13.8b). The changes of the NV^-/NV^0 fluorescence intensity ratios with or without DNA attachment were attributed to the modifications of the surface electric fields coupled with the electronic states of NV centers in the matrix [65]. Their results highlight the potential use of FNDs not only as nontoxic biolabels, but also as non-photobleachable fluorescent nanosensors to reveal complex intracellular events.

It is now well supported by multiple evidences that the advantages of using FND as a gene delivery device are many-fold. First, the nanomaterial is highly fluorescent and thus can be used to monitor the transfection process by optical imaging of live cells. Second, being an electron-dense material, it can be detected by TEM for high-resolution localization (Section 10.4). Third, the spiky and rough shape of the nanomaterial facilitates endosomal membrane rupturing and escape of the carried nucleic acids in cells.

13.4 Animal Experiments

Mice represent the most commonly used model animals for the efficacy testing of drugs. Compared with other model organisms, mice have several advantages: (i) similar (99%) genome to the humans; (ii) good genetic/molecular toolboxes available, and

(iii) small size that allows large scale and high throughput studies [66]. Liu et al. [67] explored the feasibility of using mouse models to validate the ND-based therapy. The researchers first demonstrated that HPHT-NDs covalently linked with PTX could significantly reduce the cell viability of A549 human lung carcinoma cells. The ND–PTX complexes induced both mitotic arrest and apoptosis in A549 cells, in addition to inhibiting tumorigenesis and the formation of lung cancer cells in the xenografts of severe combined immunodeficiency mice. The conjugated chemotherapeutic drug still preserved its anticancer activities, which could induce mitotic blockage, apoptosis, and anti-tumorigenesis in human lung carcinoma cells.

In the treatment of cancer, chemotherapy resistance is one main obstruction. The resistance often leads to the recurrence of tumors and cross-resistance against other chemotherapeutic drugs. Along with the rapid development of nanotechnology, scalable and biocompatible nanotherapies have emerged in an effort to overcome the drug resistances while, at the same time, improving the antitumor drug efficacy. Chow et al. [20] reported that ND-DOX could overcome drug efflux and increase apoptosis in both murine liver and mammary carcinoma cell models. Additionally, the new ND-DOX system exhibited a lower level of toxicity than the standard DOX treatment *in vivo*. Furthermore, compared with DOX alone, ND-DOX significantly prolonged the drug retention in tumor cells with improved safety and efficacy.

Other opportunities for NDs abound, such as in the identification of key extracellular receptors and signaling pathways in cancer research. Researchers in this field are developing targeted drug delivery strategies capable of enhancing efficacy and reducing toxicity in chemotherapy of patients with malignant tumors. For instance, receptor-targeted therapies are now routinely integrated into the breast cancer chemotherapy programs. To address the receptor-targeting efficiency, Moore et al. [68] developed ND-lipid hybrid particles consisting of NDs noncovalently bound with epirubicin (a chemotherapy drug) and then encapsulated in a lipid double layer containing a small amount of biotinylated lipid molecules, as shown in Figure 4.9. Through biotin–streptavidin interactions, the lipid-encapsulated particles were then conjugated with biotinylated antibodies that targeted EGFRs on breast cancer cells and tumors implanted in mice. Their result showed a marked improvement in the treatment efficacy and chemotherapeutic tolerance. The scalability, chemical stability, and biocompatibility were all strongly favorable, suggesting a promising future of the approach for chemotherapeutic delivery applications.

Glioblastoma is the most common and lethal malignant brain tumor today [69]. Because of the difficulty of penetrating the *blood–brain barrier* (BBB) with standard drugs and their poor retention in conventional treatments, researchers have been investigating alternative options. To overcome the obstacles in treating glioblastoma, Xi et al. [24] applied a convection-enhanced delivery method to administer DOX adsorbed on monodisperse DNDs. In this experiment, drugs were delivered through catheter(s) placed stereotactically directly within the tumor mass, around the tumor, or in the resection cavity of glioma-bearing rats to bypass the BBB. The conjugation with DNDs was proven to enhance the DOX uptake and retention in the glioma cell line and in normal rodent parenchyma (cf., captions of Figure 13.9 for details). Furthermore, through the convection-enhanced delivery, the conjugation with NDs successfully localized the toxicity and significantly prolonged the killing efficacy of DOX to the brain tumors.

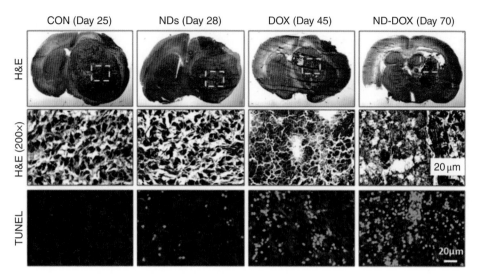

Figure 13.9 Improving glioblastoma therapy efficiency via convection-enhanced delivery of DND-DOX. (Top) Representative images of H&E staining from each treatment group, showing tumor size regression at the indicated day after tumor inoculation in glioma-bearing rats. CON: 0.9% saline; NDs: 5 mg ml^{-1} DNDs, DOX: 1 mg ml^{-1} DOX; DND-Dox: 5 mg ml^{-1} DND and 1 mg ml^{-1} DOX. (Middle) 200× magnification of boxed areas. (Bottom) Apoptotic cell death, measured by TUNEL staining of the boxed areas. Nuclei were counterstained with DAPI (blue). H&E, hematoxylin and eosin; TUNEL, terminal deoxynucleotidyl transferase mediated dUTP nick end labeling; DAPI, 4′,6-diamidino-2-phenylindole. *Source:* Reprinted with permission from Ref. [24]. Reproduced with permission of Elsevier.

Finally, as a footnote, for readers who are interested in delving further into the applications of NDs for personalized and precision medicine, please refer to the review article of Ho et al. [70].

References

1 Collins, F.S. and Varmus, H. (2015). A new initiative on precision medicine. *N Engl J Med* 372: 793–795.
2 The White House, President Barack Obama (2016). The precision medicine initiative. https://obamawhitehouse.archives.gov/node/333101 (accessed 16 April 2018).
3 European Science Foundation's (2005). Forward look nanomedicine: an EMRC consensus opinion. www.esf.org (accessed 16 April 2018).
4 Duncan, R. and Gaspar, R. (2011). Nanomedicine(s) under microscope. *Mol Pharm* 8: 2101–2141.
5 The Oxford Dictionaries (2016). *The Oxford Dictionaries*. Oxford: Oxford University Press.
6 Desai, N. (2012). Challenges in development of nanoparticle-based therapeutics. *AAPS J* 14: 282–295.
7 Bae, Y.H. and Park, K. (2011). Targeted drug delivery to tumors: myths, reality and possibility. *J Control Release* 153: 198–205.
8 De Jong, W.H. and Borm, P.J.A. (2008). Drug delivery and nanoparticles: applications and hazards. *Int J Nanomedincine* 3: 133–149.

9 Mochalin, V.N., Shenderova, O., Ho, D., and Gogotsi, Y. (2012). The properties and applications of nanodiamonds. *Nat Nanotechnol* 7: 11–23.

10 Krüger, A., Kataoka, F., Ozawa, M. et al. (2005). Unusually tight aggregation in detonation nanodiamonds: identification and disintegration. *Carbon* 43: 1722–1730.

11 Osawa, E. (2008). Monodisperse single nanodiamond particles. *Pure Appl Chem* 80: 1365–1379.

12 Slowing, I.I., Vivero-Escoto, J.L., Wu, C.W., and Lin, V.S. (2008). Mesoporous silica nanoparticles as controlled release drug delivery and gene transfection carriers. *Adv Drug Deliv Rev* 60: 1278–1288.

13 Vaijayanthimala, V. and Chang, H.C. (2009). Functionalized fluorescent nanodiamonds for biomedical applications. *Nanomedicine* 4: 47–55.

14 Vaijayanthimala, V., Lee, D.K., Kim, S.V. et al. (2015). Nanodiamond-mediated drug delivery and imaging: challenges and opportunities. *Expert Opin Drug Deliv* 12: 735–749.

15 Koo, O.M., Rubinstein, I., and Onyuksel, H. (2005). Role of nanotechnology in targeted drug delivery and imaging: a concise review. *Nanomedicine* 1: 193–212.

16 Chow, E.K. and Ho, D. (2013). Cancer nanomedicine: from drug delivery to imaging. *Sci Transl Med* 5: 216rv4.

17 Huang, H., Pierstorff, E., Osawa, E., and Ho, D. (2007). Active nanodiamond hydrogels for chemotherapeutic delivery. *Nano Lett* 7: 3305–3314.

18 Kelkar, S.S. and Reineke, T.M. (2011). Theranostics: combining imaging and therapy. *Bioconjugate Chem* 22: 1879–1903.

19 Pierstorff, E. and Ho, D. (2008). Nanomembrane-driven co-elution and integration of active chemotherapeutic and anti-inflammatory agents. *Int J Nanomedicine* 3: 425–433.

20 Lam, R., Chen, M., Pierstorff, E. et al. (2008). Nanodiamond-embedded microfilm devices for localized chemotherapeutic elution. *ACS Nano* 2: 2095–2102.

21 Adnan, A., Lam, R., Chen, H. et al. (2011). Atomistic simulation and measurement of pH dependent cancer therapeutic interactions with nanodiamond carrier. *Mol Pharm* 8: 368–374.

22 Chow, E.K., Zhang, X.Q., Chen, M. et al. (2011). Nanodiamond therapeutic delivery agents mediate enhanced chemoresistant tumor treatment. *Sci Transl Med* 3: 73ra21.

23 Man, H.B., Kim, H., Kim, H.J. et al. (2014). Synthesis of nanodiamond-daunorubicin conjugates to overcome multidrug chemoresistance in leukemia. *Nanomedicine* 10: 359–369.

24 Xi, G., Robinson, E., Mania-Farnell, B. et al. (2014). Convection-enhanced delivery of nanodiamond drug delivery platforms for intracranial tumor treatment. *Nanomedicine* 10: 381–391.

25 Lam, R. and Ho, D. (2009). Nanodiamonds as vehicles for systemic and localized drug delivery. *Expert Opin Drug Deliv* 6: 883–895.

26 Yan, J.J., Guo, Y., Altawashi, A. et al. (2012). Experimental and theoretical evaluation of nanodiamonds as pH triggered drug carriers. *New J Chem* 36: 1479–1484.

27 Zhang, X.Y., Wang, S.Q., Fu, C.K. et al. (2012). PolyPEGylated nanodiamond for intracellular delivery of a chemotherapeutic drug. *Polym Chem* 3: 2716–2719.

28 Wang, D.X., Tong, Y.L., Li, Y.Q. et al. (2013). PEGylated nanodiamond for chemotherapeutic drug delivery. *Diam Relat Mater* 36: 26–34.

29 Karukstis, K., Thompson, E., Whiles, J., and Rosenfeld, R. (1998). Deciphering the fluorescence signature of daunomycin and doxorubicin. *Biophys Chem* 73: 249–263.

30 Wang, D., Li, Y., Tian, Z. et al. (2014). Transferrin-conjugated nanodiamond as an intracellular transporter of chemotherapeutic drug and targeting therapy for cancer cells. *Ther Deliv* 5: 511–524.

31 Guan, B., Zou, F., and Zhi, J.F. (2010). Nanodiamond as the pH-responsive vehicle for an anticancer drug. *Small* 6: 1514–1519.

32 Li, J., Zhu, Y., Li, W. et al. (2010). Nanodiamonds as intracellular transporters of chemotherapeutic drug. *Biomaterials* 31: 8410–8418.

33 Chen, M., Pierstorff, E.D., Lam, R. et al. (2009). Nanodiamond-mediated delivery of water-insoluble therapeutics. *ACS Nano* 3: 2016–2022.

34 Li, X.X., Shao, J.Q., Qin, Y. et al. (2011). TAT-conjugated nanodiamond for the enhanced delivery of doxorubicin. *J Mater Chem* 21: 7966–7973.

35 Zhang, X.Q., Lam, R., Xu, X. et al. (2011). Multimodal nanodiamond drug delivery carriers for selective targeting, imaging, and enhanced chemotherapeutic efficacy. *Adv Mater* 23: 4770–4775.

36 Leader, B., Baca, Q.J., and Golan, D.E. (2008). Protein therapeutics: a summary and pharmacological classification. *Nat Rev Drug Discov* 7: 21–39.

37 Wang, H.D., Niu, C.H., Yang, Q., and Badea, I. (2011). Study on protein conformation and adsorption behaviors in nanodiamond particle-protein complexes. *Nanotechnology* 22: 145703.

38 Nguyen, T.T.B., Chang, H.C., and Wu, V.W.K. (2007). Adsorption and hydrolytic activity of lysozyme on diamond nanocrystallites. *Diam Relat Mater* 16: 872–876.

39 Liu, K.K., Chen, F., Chen, P.Y. et al. (2008). Alpha-bungarotoxin binding to target cell in a developing visual system by carboxylated nanodiamond. *Nanotechnology* 19: 205102.

40 Purtov, K.V., Petunin, A.I., Burov, A.E. et al. (2010). Nanodiamonds as carriers for address delivery of biologically active substances. *Nanoscale Res Lett* 5: 631–636.

41 Carter, P.J. (2011). Introduction to current and future protein therapeutics: a protein engineering perspective. *Exp Cell Res* 317: 1261–1269.

42 Shimkunas, R.A., Robinson, E., Lam, R. et al. (2009). Nanodiamond-insulin complexes as pH-dependent protein delivery vehicles. *Biomaterials* 30: 5720–5728.

43 Li, S. and Leblanc, R.M. (2014). Aggregation of insulin at the interface. *J Phys Chem B* 118: 1181–1188.

44 Lin, C.L., Lin, C.H., Chang, H.C., and Su, M.C. (2015). Protein attachment on nanodiamonds. *J Phys Chem A* 119: 7704–7711.

45 Smith, A.H., Robinson, E.M., Zhang, X.Q. et al. (2011). Triggered release of therapeutic antibodies from nanodiamond complexes. *Nanoscale* 3: 2844–2848.

46 Moore, L., Gatica, M., Kim, H. et al. (2013). Multi-protein delivery by nanodiamonds promotes bone formation. *J Dent Res* 92: 976–981.

47 Zhao, L., Seth, A., Wibowo, N. et al. (2014). Nanoparticle vaccines. *Vaccine* 32: 327–337.

48 Pham, N.B., Ho, T.T., Nguyen, G.T. et al. (2017). Nanodiamond enhances immune responses in mice against recombinant HA/H7N9 protein. *J Nanobiotechnol* 15: 69.

49 Killian, M.L. (2008). Hemagglutination assay for the avian influenza virus. *Methods Mol Biol* 436: 47–52.

50 Naldini, L. (2015). Gene therapy returns to centre stage. *Nature* 526: 351–360.

51 Pai, S.I., Lin, Y.Y., Macaes, B. et al. (2006). Prospects of RNA interference therapy for cancer. *Gene Ther* 13: 464–477.

52 Kanasty, R., Dorkin, J.R., Vegas, A., and Anderson, D. (2013). Delivery materials for siRNA therapeutics. *Nat Mater* 12: 967–977.

53 Dunlap, D.D., Maggi, A., Soria, M.R., and Monaco, L. (1997). Nanoscopic structure of DNA condensed for gene delivery. *Nucleic Acids Res* 25: 3095–3101.

54 Behr, J. (1997). The proton sponge: a trick to enter cells the viruses did not exploit. *Chimia* 51: 34–36.

55 Chen, M., Zhang, X.Q., Man, H.B. et al. (2010). Nanodiamond vectors functionalized with polyethylenimine for siRNA delivery. *J Phys Chem Lett* 1: 3167–3171.

56 Alwani, S., Kaur, R., Michel, D. et al. (2016). Lysine-functionalized nanodiamonds as gene carriers: development of stable colloidal dispersion for *in vitro* cellular uptake studies and siRNA delivery application. *Int J Nanomedicine* 11: 687–702.

57 Alhaddad, A., Adam, M.P., Botsoa, J. et al. (2011). Nanodiamond as a vector for siRNA delivery to Ewing sarcoma cells. *Small* 7: 3087–3095.

58 Mintzer, M.A. and Simanek, E.E. (2009). Nonviral vectors for gene delivery. *Chem Rev* 109: 259–302.

59 Jin, S., Leach, J.C., and Ye, K. (2009). Nanoparticle-mediated gene delivery. *Methods Mol Biol* 544: 547–557.

60 Zhang, X.Q., Chen, M., Lam, R. et al. (2009). Polymer-functionalized nanodiamond platforms as vehicles for gene delivery. *ACS Nano* 3: 2609–2616.

61 Martin, R., Alvaro, M., Herance, J.R., and Garcia, H. (2010). Fenton-treated functionalized diamond nanoparticles as gene delivery system. *ACS Nano* 4: 65–74.

62 Chu, Z., Miu, K., Lung, P. et al. (2015). Rapid endosomal escape of prickly nanodiamonds: implications for gene delivery. *Sci Rep* 5: 11661.

63 Petrakova, V., Benson, V., Buncek, M. et al. (2016). Imaging of transfection and intracellular release of intact, non-labeled DNA using fluorescent nanodiamonds. *Nanoscale* 8: 12002–12012.

64 Chu, Z., Zhang, S., Zhang, B. et al. (2014). Unambiguous observation of shape effects on cellular fate of nanoparticles. *Sci Rep* 4: 4495.

65 Petrakova, V., Taylor, A., Kratochvílova, I. et al. (2012). Luminescence of nanodiamond driven by atomic functionalization: towards novel detection principles. *Adv Funct Mater* 22: 812–819.

66 Vandamme, T.F. (2014). Use of rodents as models of human diseases. *J Pharm Bioallied Sci* 6: 2–9.

67 Liu, K.K., Zheng, W.W., Wang, C.C. et al. (2010). Covalent linkage of nanodiamond-paclitaxel for drug delivery and cancer therapy. *Nanotechnology* 21: 315106.

68 Moore, L., Chow, E.K., Osawa, E. et al. (2013). Diamond-lipid hybrids enhance chemotherapeutic tolerance and mediate tumor regression. *Adv Mater* 25: 3532–3541.

69 Davis, M.E. (2016). Glioblastoma: overview of disease and treatment. *Clin J Oncol Nurs* 20: S2–S8.

70 Ho, D., Wang, C.H., and Chow, E.K. (2015). Nanodiamonds: the intersection of nanotechnology, drug development, and personalized medicine. *Sci Adv* 1: e1500439.

14

Diamonds in the Sky

Ever wondered where to look for *nanodiamonds* (NDs) in nature? Perhaps a more relevant question is: Do they even exist in natural environment? After all, NDs are very tiny and could be exceedingly difficult to find. For example, in order to make up a thickness of a single-strand human hair, one would need to weave 1000 NDs side-by-side, let alone produce a size of a half-carat diamond (as on a humble wedding ring) that would need a mound of 30 trillion (3×10^{13}) NDs. While diamonds appear to be extremely rare on the earth surface (Section 2.3), who would have ever thought that in dark skies over the far deep space, there just might be abundant NDs present on stars and nebulas enough to light up the skies throughout an entire galaxy? How did scientists find it? This is the story of *unidentified infrared* (UIR) emission and *extended red emission* (ERE) of stardust [1].

14.1 Unidentified Infrared Emission

Carbon is the fourth most abundant element in the Galaxy. The possible presence of this element as diamond in space was first proposed by Saslaw and Gaustad [2] in 1969 to account for the observed interstellar extinction curves in the far ultraviolet (UV) region of galactic radiation. Little, if any, attention was paid to this hypothesis until 1987 when Lewis et al. [3] reported the discovery of small diamond grains in primitive meteorites from outside of the solar system (Section 2.4). These diamond grains have an average size of approximately 3 nm in diameter with a concentration as high as 1400 ppm in carbonaceous meteorites (Table 14.1) [4]. It turns out that NDs are the most abundant presolar grains in primitive meteorites and, therefore, are expected to be an abundant component of the stardust in the interstellar medium as well. But how can one find diamonds in cosmic medium? Traditionally, scientists have relied on optical methods to study cosmic objects, for instance, by using telescopes to map their surface landscape and emission spectra. Then, what are the spectroscopic signatures that may cast light on the composition of these interstellar grains?

The search for diamonds in the interstellar medium by spectroscopic means was actively pursued in the 1990s [5–8]. Allamandola et al. [5, 6] made one of the first attempts to identify diamond dusts in dense molecular clouds based upon the broad infrared (IR) absorption band centered around 2880 cm^{-1} (or 3.47 μm). Zooming into the microscale, Hill et al. [7] conducted laboratory studies of the IR absorption bands

Fluorescent Nanodiamonds, First Edition. Huan-Cheng Chang, Wesley Wei-Wen Hsiao and Meng-Chih Su.
© 2019 John Wiley & Sons Ltd. Published 2019 by John Wiley & Sons Ltd.

Table 14.1 Types, abundances, and sizes of stardust in meteorites.

Mineral	Abundance (ppm)	Size (µm)
Diamond	1400	0.002
SiC	30	0.3–50
Graphite	10	1–20
Si_3N_4	0.002	≤1
Oxides	50	0.1–2
Silicates	200	≤1

Source: From Ref. [4].

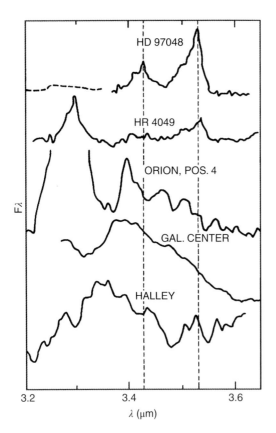

Figure 14.1 Comparison of the 3.43- and 3.53-µm emission bands of *HD 97048* to a number of other astronomical features. *Source:* Reprinted with permission from Ref. [9].

of NDs extracted from Orgueil meteorites. However, no satisfactory matches were found between the astronomical and laboratory spectra. Alternatively, one may try to find NDs in space by examining the IR emission bands from galactic nebulae and circumstellar medium. In doing so, the emission from the *HD 97048* star caught scientists' attention. The emission spectrum of the star exhibited two prominent features at 3.43 and 3.53 µm, distinct from other astronomical bands (Figure 14.1). In a study of the nature of the emission from *HD 97048* and *Elias 1*, Schutte et al. [9]

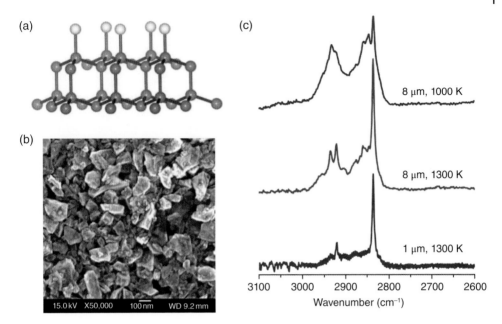

Figure 14.2 (a) Structure of a hydrogenated diamond C(111) surface. Gray spheres denote carbon atoms and white spheres denote hydrogen atoms. (b) Scanning electron microscopy image of gravel-like diamond nanocrystals used in the laboratory studies. (c) Laboratory absorption spectra of hydrogenated ND films prepared in different thickness (1 and 8 μm) and treated at different microwave plasma temperatures (1000 and 1300 K). The spectra were acquired at 300 K. *Source:* Adapted with permission from Ref. [10]. Reproduced with permission of American Chemical Society.

noticed several anomalies in the observations. First, these two emission bands could be detected only under very special conditions (e.g. close to a hot stellar source). Second, the spatial variation of the emission intensities was fairly independent of other emission bands in the same frequency region. Third, both bands always appeared together, suggesting a spectral correlation between the two. An assignment of these emission bands was proposed by the astronomers to be the C–C overtones and some combination bands of highly excited large *polycyclic aromatic hydrocarbon* (PAH) molecules [9]. However, the assignment did not seem to present a satisfactory interpretation for the origin of these UIR emission bands.

In about the same time and without the knowledge of UIR, Chang and coworkers [10] studied the IR spectroscopy and vibrational relaxation dynamics of CHx ($x = 1$–3) on diamond nanocrystal and single-crystal surfaces. Three types of single crystals were used, including C(100), C(110), and C(111) (cf., Figure 14.2a for the hydrogenated C(111) surface). The study was aimed at elucidating the formation mechanism of diamonds by *chemical vapor deposition* (CVD) where atomic hydrogen plays a crucial role (Section 2.3.2) [11]. With a microwave reactor to produce pure hydrogen plasmas, the research team hydrogenated diamond thin films consisting of 100 nm particles of irregular sizes and shapes (Figure 14.2b), and obtained their absorption spectra at 3–5 μm with a Fourier transform infrared spectrometer. Surprisingly sharp absorption features were observed at 2835 and 2920 cm^{-1} (equivalent to 3.53 and 3.42 μm), respectively (Figure 14.2c), which were later confirmed by single crystal experiments to originate from

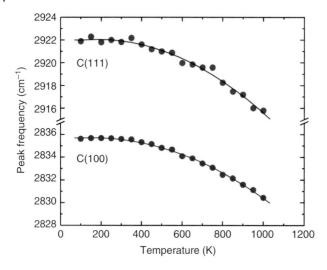

Figure 14.3 Temperature-dependent shifts of the absorption bands of the CH stretching vibrations on {100} and {111} facets of hydrogenated NDs. *Source:* Adapted with permission from Ref. [16]. Reproduced with permission of AIP Publishing LLC.

the CH stretches of H-covered {111} and {100} facets [12, 13]. The appearance of these two prominent absorption features for the hydrogenated NDs was accounted for by the unique H-etching anisotropy of the diamond substrate [13–15]. Further studies of the thermal effect on the CH stretching vibrations showed a band shift of approximately $5\,cm^{-1}$ to the red when the sample temperature was raised from 300 to 1000 K (Figure 14.3) [16].

In 1997, Geballe [17] provided a high-quality IR emission spectrum of *Elias 1* in the Taurus dark cloud from a ground-based observation. In this emission spectrum as shown in Figure 14.4, the spectral feature at $2830\,cm^{-1}$ (3.53 µm) was much better resolved than before, with a full width at half maximum of $11\,cm^{-1}$. The characteristic of this anomalous IR emission band was also confirmed by the spectra of the same object taken by the short-wavelength spectrometer on board of the *Infrared Space Observatory* [17]. Two years later, Guillois et al. [8] pointed out a nearly perfect match between the emission spectra taken from *Elias 1* (and also *HD 97048*) and the absorption bands of CH stretches on hydrogenated ND surfaces (Figure 14.4), claiming "Here we report what we think to be the first unambiguous evidence of the presence of small crystallites of diamond in the dusty envelopes surrounding stars." The claim was supported by further laboratory studies [12, 13, 16]. Based on the spectral redshift of $5\,cm^{-1}$ for the CH stretching vibrations, a temperature of 1000 K was estimated for these circumstellar NDs. The research team also made careful energy considerations to understand the nature of the IR emission process [8, 18].

One may ask: How can the circumstellar NDs display such distinct vibrational features? It is perceivable that diamond grains may be formed and situated in a close proximity to the star, from which they are exposed to intense UV photon fluxes, thus reaching high temperatures. The high temperature and high hydrogen-flux conditions resemble that of the microwave plasma used in the CVD diamond synthesis. Conceivably, the temperatures of these two cosmic environments are sufficiently high to allow hydrogenation and

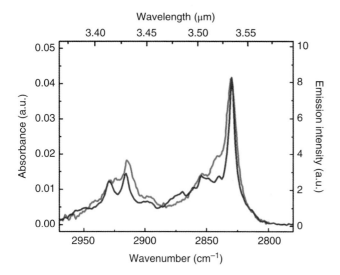

Figure 14.4 Comparison between the infrared emission spectrum of *Elias 1* (green) and the laboratory absorption spectrum (red) of hydrogenated NDs measured at 300 K. Note that the absorption spectrum is redshifted by 5 cm^{-1} in comparison. *Source:* Adapted with permission from Ref. [8].

H-etching of the ND surfaces to occur. The suggestion is independently supported by the observations of H recombination lines in both *Elias 1* and *HD 97048* [19].

Another useful piece of information provided by the observations is the size of the UIR carriers. The information can be deduced from the findings that the 3.53 μm band coincides with the stretching resonance of the CH monolayer on the C(111) single-crystal surface [12], in addition to its width (full widths at half maximum of ~6 cm^{-1}). Therefore, these NDs cannot be too small since a large heterogeneously broadened width would have been resulted for the CH stretches on the nanometer-sized H-covered {111} facets [20]. Analysis of the IR absorption spectra of meteoritic NDs (~3 nm in diameter) [21] and a series of synthetic diamonds (5–700 nm in diameter) [22] has shown that the sharp 3.53-μm band can emerge only when particles have an average size of 25 nm or greater. This derived size is quite different from that of the ND dust in meteorites (Table 14.1), suggesting that these diamond nanoparticles must have been formed in space through different mechanisms [21].

Figure 14.4 represents, arguably, the best agreement ever obtained between laboratory spectra of a solid-state candidate and astronomical features. Taking advantage of the brightness of the IR emissions, Habart et al. [23] and Goto et al. [24] have been able to acquire spatially resolved spectra of the 3 μm bands at *HD 97048* and *Elias 1*, respectively. Compared to the PAH emission, the diamond emission is more centrally concentrated. It should be noted here that the CH stretching vibrations are the only distinct spectral features that can be clearly identified so far based on the IR emission spectra of NDs. The corresponding CH bending vibrations cannot be found in the 6–7 μm range [25] because of the weakness of the absorption and, possibly, the severe band broadening resulting from its coupling with diamond surface phonons resonant in the same frequency region. The absence of this spectral feature stands as a sharp contrast to the prominence of the CH bending of PAHs [26], representing another unique characteristic of the H-terminated NDs.

The remarkable matching in the peak positions, bandwidths, and profiles of the two 3.43- and 3.53-μm features between laboratory and astronomical spectra provides a strong evidence for the existence of NDs in the circumstellar medium. Started as a long-standing mystery [27], the UIR emission bands may very well have originated from hydrogenated diamond nanoparticles formed in astronomical environments rich in energetic hydrogen atoms, a condition similar to that of CVD diamond synthesis in the laboratories. It seems amazing that a fundamental study of hydrogen chemistry on diamond surfaces can lead to the discovery of diamonds in the sky.

14.2 Extended Red Emission

In space's dustiest places, there exists a faint, rose-colored glow called ERE [28]. This kind of light has been found in a wide range of astronomical environments including reflection nebulae, planetary nebulae, emission nebulae, and the diffuse interstellar medium of our galaxy. So, what seems to have caused this odd cosmic light?

A nebula, in all shapes and sizes, is an interstellar cloud consisting of dust, hydrogen, helium, and other ionized gases. ERE was first detected by Cohen et al. [29] in 1975 for a planetary nebula located in the constellation of *Monoceros* roughly 2300 light years from Earth. The nebula has a red color and a characteristic rectangular shape surrounding the star *HD 44179* and called, rightfully, the *Red Rectangle* (Figure 14.5) [30]. A broad emission band in the wavelength range of 500–900 nm appears in the areas of abundant UV light [31], suggesting that the ERE process may be associated with high-energy photons striking on solid materials. The emission could be thought of as in an office-used fluorescent lamp, where the discharged UV light strikes on a thin film of coating materials inside the lamp's tube, the materials in turn emitting visible light. Proposed models for the materials (also known as the ERE carriers) include hydrogenated amorphous carbons [32], PAHs [33], and silicon nanoparticles [34, 35].

Witt and Vijh [28], after compiling decades of research, concluded a total of 13 conditions that must be fulfilled by any candidate to be considered a credible ERE carrier (Table 14.2). No association of ERE with diamonds was made before year 2000

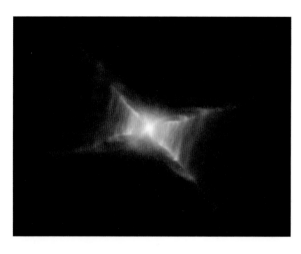

Figure 14.5 The proto-planetary nebula *HD 44179* (also called the "*Red Rectangle*") in the constellation of *Monoceros*. This picture was obtained with the Hubble Space Telescope (courtesy of Hans Van Winckel, Martin Cohen, NASA, and ESA). *Source:* Reprinted with permission from Ref. [30].

Table 14.2 Characteristics of extended red emission and its carrier.

1) Carriers must consist of cosmically abundant refractory elements that are depleted from the gas phase in interstellar space and are capable of forming photoluminescent materials

2) The ERE carrier particles appear to originate in such dust-forming environments as the proto-planetary *Red Rectangle* nebula and in C-rich planetary nebulae

3) Carrier particles must survive under a wide range of interstellar and circumstellar conditions and have a Galaxy-wide distribution

4) The ERE appears to be uncorrelated with the strength of the interstellar 2175 Å extinction feature and the ubiquitous UIR emission bands

5) ERE is a photoluminescence process

6) The photoluminescence occurs in a broad, unstructured band with a peak wavelength varying between 600 and >900 nm, within given sources and from source to source

7) The width and the peak wavelength of the ERE band are positively correlated

8) Efficient excitation of the ERE process requires photons with energies in excess of 7.25 eV

9) No photoluminescence is detectable in the 350–500 nm wavelength range under excitation by UV continuum radiation

10) Under conditions of low radiation density the ERE quantum yield is $\gg$10%

11) The ERE is unpolarized with an isotropic radiation pattern

12) The ERE is observed in the absence of scattering and appears, therefore, not associated with submicron-sized scattering grains

13) The ERE peak wavelength and quantum yield are strongly affected by the density and hardness of the local UV photon field

Source: From Ref. [28].

except that of Duley [36] who proposed that the sharp emission lines observed around 580 nm in the *Red Rectangle* spectra might have originated from the electronic transitions of structural defects similar to those found in the terrestrial diamonds. In 2006, Chang and coworkers [37] proposed that the nitrogen-vacancy (NV) centers in NDs could be a possible carrier of the ERE bands because the physicochemical properties of this carbon-based nanomaterial matched exactly the first four of the 13 conditions set out by Witt and Vijh [28]. Additionally, in later studies [38, 39], the photoluminescence spectra of the NV centers were shown to share multiple similarities with the ERE bands as discussed below.

In their 2006 work, Chang et al. [37] produced carbon atom vacancies in type Ib NDs by repeatedly bombarding the particles with 3-MeV protons in a vacuum to simulate what might have occurred in the interstellar medium. The radiation-damaged NDs were then annealed at 800 °C to form *fluorescent nanodiamonds* (FNDs) (~100 nm in diameter). As discussed in Section 3.3.2, there are two types of NV centers in FNDs produced in this manner, NV^0 and NV^-, and they are characterized by the sharp zero-phonon lines at 575 and 637 nm, respectively (Table 3.2). The intensity ratio is close to $NV^0 : NV^- = 1 : 3$ under continuous laser excitation over the wavelength range of 450–610 nm at the steady state. Despite that the peak wavelength may vary between 610 and 690 nm (Figure 3.5), depending on the wavelength of excitation, the observed photoluminescence spectra are all ERE-like [39].

Figure 14.6a presents a picture of the planetary nebula *NGC 7027* in the constellation of *Cygnus*, obtained by the *Hubble Space Telescope* [40]. Compared with the ERE band observed for this nebula [41], the laboratory spectrum obtained by excitation of FNDs

(a)

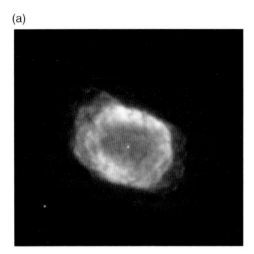

Figure 14.6 (a) The planetary nebula *NGC 7027* in the constellation of *Cygnus*. This picture was obtained with the Hubble Space Telescope (Courtesy of William B. Latter, NASA, and ESA). *Source:* Reprinted with permission from Ref. [40]. (b) Comparison between the observed ERE spectrum of *NGC 7027* and the laboratory fluorescence spectrum of FND excited with 510–560 nm light. *Source:* Reprinted with permission from Ref. [38].

(b)

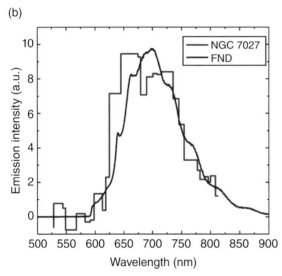

at 510–560 nm shares a close resemblance (Figure 14.6b), leading Chang et al. [37] to hypothesize that FNDs may be responsible for the red emission. The hypothesis is supported by the facts that *NGC 7027* is rich in carbon [42] and, additionally, nitrogen atoms are abundant in the universe and they can be readily incorporated into the diamond lattice as impurities during circumstellar condensation. Therefore, as vacancies in diamonds are created in the laboratory by high-energy proton irradiation under vacuum, the similar defective diamond nanoparticles can potentially form in the astronomical environments where the ERE bands have been observed.

While FNDs meet many of the conditions listed in Table 14.2, a major concern of the above proposal is that the photoluminescence is produced by excitation of the particles with visible light (500–600 nm), rather than UV photons with energies in excess of 7.25 eV (or equivalent to 171 nm), listed as item 8 in the table. To address this issue, Lu et al. [39] have recently conducted experiments with radiation from a

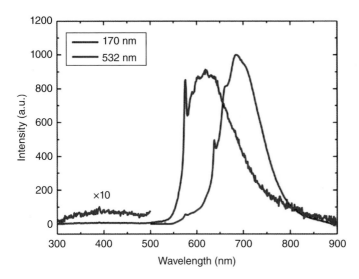

Figure 14.7 Emission spectra of FNDs excited with 170- and 532-nm light at 300 K. The spectra were obtained for the same sample. Note the occurrence of photoionization, which converts NV⁻ to NV⁰ under far-UV excitation. *Source:* Reprinted with permission from Ref. [39]. Reproduced with permission of John Wiley & Sons.

synchrotron source in the wavelength range of 125–350 nm. No emission was observed between 300 and 500 nm under far-UV excitation such as at 170 nm (cf., item 9 in Table 14.2). However, much to their surprise, the photoluminescence spectra were dominated by the NV⁰ emission, in sharp contrast to the excitation with visible photons, where the emission of NV⁻ prevailed (Figure 14.7). The result was attributed to the photoionization of NV⁻ to NV⁰ due to the interband excitation, i.e. the excitation of electrons from the valence band to the conduction band, by the far-UV photons. The energy released from the electron–hole recombination after excitation was sufficient to ionize the negatively charged defects to form electronically excited NV⁰ centers, which subsequently relaxed to the ground state by emitting red photons as in the cases of *cathodoluminescence* (Section 10.3). Assuming that both far-UV and optical photons were involved in the excitation process [43], the researchers were able to reproduce successfully the ERE band observed in *NGC 7023*, a bright reflection nebula in the constellation *Cepheus* (Figure 14.8a) [44], by linearly combining the photoluminescence spectra of both NV⁰ and NV⁻ centers with an intensity ratio of 1 : 2 (Figure 14.8b).

Going one step further, Lu et al. [39] set out to measure the quantum yield (Φ) of the emission under UV excitation. Instead of using FNDs, they recorded the photoluminescence excitation spectra of *fluorescent microdiamonds* (FMDs, diameter ~400 μm) by scanning the excitation energy over 125–675 nm and monitoring the resulting emission signals at 683 nm. The reasons to use micron-sized particles were to avoid the light-scattering problem of FNDs and to ensure that the exciting photons were completely absorbed by the substrate. From an observation of the photoluminescence signals in both intracenter and interband excitation regions, they were able to obtain a quantitative measure for the quantum yield at 170 nm by referring to the well-documented value of $\Phi \approx 99\%$ at 532 nm for the NV⁻ centers (Table 3.2). An overall quantum yield of $\Phi \geq 20\%$

(a)

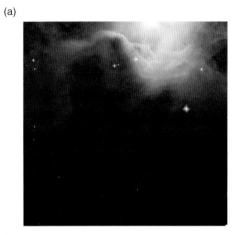

(b)

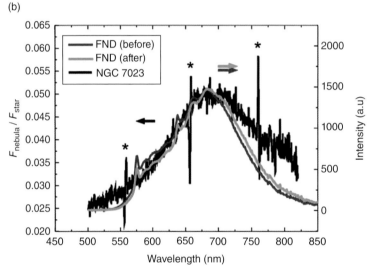

Figure 14.8 (a) The reflection nebula *NGC 7023* in the constellation of *Cepheus*. This picture was obtained with the Hubble Space Telescope (Courtesy of NASA and ESA). *Source:* Reprinted with permission from Ref. [44]. (b) Comparison of a laboratory photoluminescence spectrum and the ERE band from *NGC 7023*. The ERE band was obtained on dividing the nebular spectrum by the spectrum of the illuminating star. The sharp features denoted with asterisks result from incomplete cancellation of night-sky and nebular emission bands. The laboratory spectra were synthesized by combining the photoluminescence signals of NV^0 and NV^- with (green) or without (red) corrections for the interstellar reddening effect. *Source:* Reprinted with permission from Ref. [39]. Reproduced with permission of John Wiley & Sons.

was determined for both FMDs and FNDs, in agreement with the specific characteristic of ERE (cf., item 10 in Table 14.2).

We now turn to the long-standing mystery of the *Red Rectangle*, which is a nebula-emitting ERE with exceptional strength. The high quality of the spectra allows for a stringent test of the theory involving FND as a possible ERE carrier. A notable feature of the *Red Rectangle* is that the spectral profile of the ERE band varies substantially with distance from the central star *HD 44179*. Figure 14.9 shows two spectra taken at the

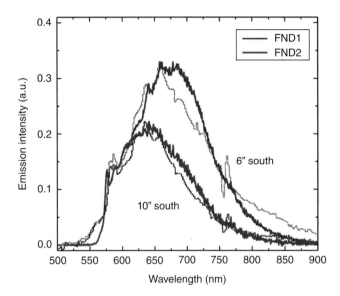

Figure 14.9 Comparison between the observed ERE spectra of the *Red Rectangle* at 10″ south and 6″ south of *HD 44179* and the laboratory fluorescence spectra of FNDs with different contents of NV^0 and NV^-. The fluorescence intensity ratios of $NV^0 : NV^-$ in samples FND1 and FND2 are roughly 3 : 2 and 1 : 0, respectively. *Source:* Reprinted with permission from Ref. [38].

angular distances of 6 and 10 arcsec (6″ and 10″) south, respectively. The peak position of the ERE band shifts from 635 to 660 nm as the distance moves from 10″ south at 6″ south, respectively, within the same object [45, 46]. To understand the shifts, Chang [38] modeled the ERE spectra using the experimental data of NV^0 and NV^- emissions with different intensity ratios and was able to properly reproduce the astronomical observations at these two offset distances, as shown in Figure 14.9. The result seems to indicate an intriguing particle size effect, that is, the ERE-emitting FNDs at 10″ are smaller than those at 6″. The conclusion is supported by the laboratory studies of FNDs over the size range of 10–100 nm that the NV^0 centers prefer to form in smaller diamond nanoparticles than NV^- due to the surface-induced charge state conversion between these two types of structural defects (Section 3.3.2) [47]. It is also in line with the common knowledge that particles of smaller size can be blown away more easily than the larger ones from the illuminating star by radiation pressure [48].

The diamond model can also account for the absence of blue and green photoluminescence in the interstellar medium. As discussed in Section 3.3.3, the observation of the blue and green emissions requires nitrogen impurities to be present as aggregates in the diamond lattice (e.g. H3 and N3 in Table 3.2). However, it is known that the time period required for single substitutional nitrogen atoms to form aggregates in diamond is about 10^9 years at 1200 °C [49]. This condition, clearly, cannot be readily realized in the interstellar or even circumstellar medium. With this simple model, one can additionally attribute the weakness of the ERE bands at the wavelength greater than 900 nm to the small photoluminescence quantum yield (~1.4%) of the V^0 centers even at low temperature (Table 3.2). No evidence has yet be discovered for the association of ERE with molecular ions like C_{60}^+, which shows a prominent absorption band at 957.7 nm [50]. Further discussion of the diamond-based ERE model can be found in Ref. [39].

14.3   Cosmic Events at Home on Earth

Following the mysteries in outer space, as we gaze down on earth, we cannot help wondering if what could happen in skies might just be possible on earth. Granted that we do not have red skies hanging over our heads (good heavens!), but are there stories about NDs on the planet earth that are related to the cosmic events? Where can we find natural NDs on earth, perhaps, after the massive impacts by comets or asteroids? It turns out not only natural NDs are present on earth but they also have been used as a marker for identifying cosmic events, as told in the story of the *Younger Dryas event*.

There were major geological periods throughout history of our planet that have been identified to have widespread effects globally and long-lasting devastations. The most recent one of such major events, also the most well-known by popular culture, is the *Cretaceous-Paleogene* (K-Pg, also called *Cretaceous-Tertiary*) extinction event some 66 million calendar years before the present time (*cal BP*) that had wiped out an estimate of more than 75% living creatures from earth, including the dinosaurs. Originally thought as caused by volcanism and marine regression, it is now broadly accepted that the K-Pg extinction was a result of extraterrestrial impact. In 1980, Luis Alvarez (1968 Nobel laureate in Physics) and his research team discovered, among soot and other glassy spherules typically formed at combustion temperatures, an unusually high content of the rare element iridium (Ir, atomic number 77) in the K-Pg layer of the sediment samples collected from all over the world [51]. Iridium binds nicely with iron; therefore, any iridium present at the time of earth crust formation would have been transported to the earth core with iron, leaving only a trace amount on the earth surface. On the other hand, extraterrestrial comets and asteroids typically show a high level of iridium content that can prevail upon the impact on earth. Therefore, Alvarez's finding of the high iridium content was a clear evidence linked directly to a cosmic impact.

A proposed hypothesis nonconventional as such is bound to invite public scrutiny and met by widespread suspicions and, indeed, it was. After all, it is hard to imagine a rock falling from sky would have wiped out the entire species of dinosaurs, the largest and strongest animals on the surface of earth at the time, reshuffling the entire hierarchy of inhabitant species on the planet that forever changed the course of earth evolution. Then, in 1990, two years after the death of Alvarez, a breakthrough came from the discovery of the giant Chicxulub crater mostly buried under the coast of the Yucatan Peninsula in Mexico. The crater is in an oval shape with an average diameter of 110 mi, about half of it in the ocean, which is believed to be created by an extraterrestrial impactor of a size estimated 6–9 mi in diameter. The finding of the impact site and its geological location, a region of sulfur-rich carbonate rocks, poses strong potential for massive deposit of sulfuric acid in the earth's atmosphere followed possibly by large-scale acid rain, providing undisputable support for the hypothesis of a cosmic impact. In 2010, a panel consisting of 41 scientists worldwide reviewed all reported findings in the past 20 years on the K-Pg event and concluded that it was indeed a cosmic impact of a collisional energy roughly about one million times greater than the energy released by the atomic bombs dropped at Hiroshima and Nagasaki in World War II.

So, what is all this to do with NDs? This brings us to the *Younger Dryas* event. The landmark geological drills of the *Greenland Ice Sheet Project*, GISP (1979–1981) and GISP2 (1988–1993), have produced ice cores ranging from 2000 to 3000 m in depth with diameters of 10–15 cm that contain remarkable geologically rich information of earth history. One of the data derived from these ice cores is the annual average temperature on earth that can be traced millions years back. In general, the earth's temperature fluctuates from year to year, normally in a tiny fraction of a degree Celsius, and oscillates cumulatively in geological scale between warming and cooling trends from one period to another. The present period, *Holocene*, is an *interstadial* and therefore we are in a warming phase. The *stadial*, a cooling phase, right before Holocene is the *Younger Dryas* dated from 12 900 to 11 700 *cal BP*, when the earth's temperature dropped by 2–6 °C mainly in the northern hemisphere. An abrupt climate change was identified in this geological period that, in large parts of Europe, the warmer climate vegetation was replaced by the colder weather vegetation including the *Dryas octopetala*, which was once abundant by the Scandinavian lakes during the glacial period. This is where the *Younger Dryas* (YD) got its name from, along with the other members in the family: the *Older Dryas* and *Oldest Dryas* occurring around 14 000 and 15 070 *cal BP*, respectively. Incidentally, the study of human evolution has documented that *Homo sapiens*, modern human species, have been around since 250 000 *cal BP*.

Devastated mass extinction occurred as the climate changed in this more than 1000 year-long YD period. An estimated of at least 35 mammal genera disappeared from North America alone, including megafauna (e.g. mammoths), camels, horses, mastodons, and many species of small mammals and birds; marking the end of the *Pleistocene period* (2 588 000–11 700 *cal BP*). As a result, the ecosystems in the American and Eurasian continents were significantly changed, which was clearly reflected in, for example, the sharp decline of the Paleo-Americans population and the shift in the Clovis culture to better adapt in the drastically altered environment. In a way, these changes, dramatic as it might have seemed, actually shaped up the landscape of our natural resources and environment today.

An immediate question now is: What may have been the cause(s) that triggered the sudden climate changes of YD?

The geologists have speculated that the causes for the temperature decline was because of the dramatic reduction or shutdown of the circulation in atmosphere that used to carry warm vapor from tropical oceans to North Atlantic. But, why? What would have caused the shutdown? There are hypotheses (more speculations?) of volcano eruptions, wildfires, or meteorite debris among others that were responsible for causing the climate change. But, none of the proposed theories seems to offer satisfactory explanations nor is there geological evidence to support the speculations. In 2007, representing a team of 26 scientists from 16 research groups, Firestone [52] reported to the National Academy of Science in America a body of carefully studied evidence that seemed to support a hypothesis citing extraterrestrial impact as the trigger of YD cooling period. In the paper, the research team presented analysis of the samples collected from 10 Clovis-age sites well documented and dated with clear *YD boundary* (YDB), a thin layer (usually about 5 cm or less) of sediment that had been identified as the onset of YD directly beneath the darkened soil (called *black mat*) during the YD cooling period. The analysis using modern instrumentation covered all unique substances (so-called the *YD event markers*) discovered in the YDB consisting of (i) magnetic microspherules and grains, (ii) iridium and nickel, (iii) charcoal, (iv) soot and PAHs, (v) carbon spherules, (vi) fullerenes and extraterrestrial helium, and (vii) glass-like carbon. Collectively, the analysis results provided a strong support for the hypothesis that the onset of YD was caused by extraterrestrial airburst or impact on earth.

The arguments for a cosmic impact, regardless how credible the evidence seemed to have suggested, were immediately challenged by the opponents of the impact theory. The weakest point of the defense by anyone who believes in the impact hypothesis was the lacking of the impact site. Many have speculated since then that, if a surface impact did happen, it would probably have landed somewhere around the Great Lakes in Michigan, though no craters have been found yet up to date. (Could it be sunken at the bottom of the Lakes?) The cause for YD remained a mystery and so was the cosmic impact theory.

A breakthrough finally came in 2014. In the year's September issue of *The Journal of Geology*, the top story was an article based on a collective effort of 26 researchers from 21 universities in 6 countries reporting the details of evidences supporting the cosmic impact hypothesis [53]. And, the key evidence was, yes, our very own NDs! In addition to the seven burning proxies analyzed in Firestone's paper previously, the 2014 team discovered a rich content of NDs in the YDB layer, not above nor below the layer, but only within YDB. Therefore, NDs were likely formed at the onset of YD and may very well be related to the causes of the long cooling period. Multiple forms of NDs were found: cubic, lonsdaleite-like crystal, n-diamonds, and i-carbon nanoparticles. Furthermore, these NDs were identified to be terrestrial, i.e. carbons originally on earth turned into NDs only after experiencing the changes involving extreme conditions such as high-energy impact. The research team carefully compared the experimental conditions used in today's laboratories for manufacturing NDs, both detonation and HPHT NDs, and ruled out possibilities of wildfires and lightening, the two frequently thought causes. The team further compared the samples obtained from the K-Pg sites and other known impact sites, all showing rich ND contents. A close look into the K-Pg sites led the research team to find that NDs only existed in the boundary layer, not the layer above or below, exactly the same as that of NDs in YDB. The structures of NDs produced at the K-Pg impact appeared to be identical to the NDs in YDB but with a less variety, probably due to the significant difference in the impact energy. The 2014 paper also investigated all other potential causes of the YD cooling brought forward by scientists, including volcanism, sclerotia, and anthropogenic activities, and ruled out all of them. The final conclusion was that it must require a cosmic impact to meet the necessary conditions of exotic temperature and explosive pressure in order to form NDs at the onset of the YD period. NDs proved to be the defining evidence in closing the arguments to finally solving the long-standing mystery of YD. Now the only thing left is to find the impact crater. Someday, perhaps.

As the writing of this book was about to complete, we learned a special documentary, *The Day the Dinosaurs Died*, was broadcasted by BBC Two on 15 May 2017 [54], reporting the current status on the cosmic impact theory of dinosaur's extinction. In addition to the impact theory, described at the beginning of this section, the documentary also announced a two-year Dinosaur crater drill project that had started in 2016 at approximately 19 mi off the shore of Mexico's Yucatan Peninsula. The project was led by Professors Joanna Morgan of the Imperial College London and Sean Gulick of the University of Texas, who have drilled into the inner core of the impact crater 4300 ft beneath the Gulf of Mexico. They expected to piece up frame-by-frame a detailed timeline of the impact event by analyzing the content of the rock cores recovered from the drill, while the world is watching and waiting patiently.

References

1 Kwok, S. (2013). *Stardust: The Cosmic Seeds of Life*. Springer.

2 Saslaw, W.C. and Gaustad, J.E. (1969). Interstellar dust and diamonds. *Nature* 221: 160–162.

3 Lewis, R.S., Ming, Y., Wacker, J.F. et al. (1987). Interstellar diamonds in meteorites. *Nature* 326: 160–162.

4 Davis, A.M. (2011). Stardust in meteorites. *Proc Natl Acad Sci USA* 108: 19142–19146.

5 Allamandola, L.J., Sandford, S.A., Tielens, A.G.G.M., and Herbst, T.M. (1992). Infrared spectroscopy of dense clouds in the C-H stretch region: methanol and "diamonds". *Astrophys J* 399: 134–146.

6 Allamandola, L.J., Sandford, S.A., Tielens, A.G.G.M., and Herbst, T.M. (1993). Diamonds in dense molecular clouds: a challenge to the standard interstellar-medium paradigm. *Science* 260: 64–66.

7 Hill, H.G.M., Jones, A.P., and d'Hendecourt, L.B. (1998). Diamonds in carbon-rich proto-planetary nebulae. *Astron Astrophys* 336: L41–L44.

8 Guillois, O., Ledoux, G., and Reynaud, C. (1999). Diamond infrared emission bands in circumstellar media. *Astrophys J* 521: L133–L136.

9 Schutte, W.A., Tielens, A.G.G.M., Allamandola, L.J. et al. (1990). The anomalous 3.43 and 3.53 μ emission features toward HD 97048 and Elias 1: C-C vibrational modes of polycyclic aromatic hydrocarbons? *Astrophys J* 360: 577–589.

10 Chang, H.C., Lin, J.C., Wu, J.Y., and Chen, K.H. (1995). Infrared spectroscopy and vibrational relaxation of CHx and CDx stretches on synthetic diamond nanocrystal surfaces. *J Phys Chem* 99: 11081–11088.

11 Spear, K.E. and Frenklach, M. (1994). Mechanisms for CVD diamond growth. In: *Synthetic Diamond: Emerging CVD Science and Technology* (ed. K.E. Spear and J.P. Dismukes), 243–304. Wiley.

12 Cheng, C.L., Lin, J.C., Chang, H.C., and Wang, J.K. (1996). Characterization of CH stretches on diamond C(111) single- and nanocrystals by infrared absorption spectroscopy. *J Chem Phys* 105: 8977–8978.

13 Cheng, C.L., Chang, H.C., Lin, J.C. et al. (1997). Direct observation of hydrogen etching anisotropy on diamond single crystal surfaces. *Phys Rev Lett* 78: 3713–3716.

14 Stallcup, R.E. and Perez, J.M. (2001). Scanning tunneling microscopy studies of temperature-dependent etching of diamond (100) by atomic hydrogen. *Phys Rev Lett* 86: 3368–3371.

15 Kuroshima, H., Makino, T., Yamasaki, S. et al. (2017). Mechanism of anisotropic etching on diamond (111) surfaces by a hydrogen plasma treatment. *Appl Surf Sci* 422: 452–455.

16 Lin, J.C., Chen, K.H., Chang, H.C. et al. (1996). The vibrational dephasing and relaxation of CH and CD stretches on diamond surfaces: an anomaly. *J Chem Phys* 105: 3975–3983.

17 Geballe, T.R. (1997). Spectroscopy of the unidentified infrared emission bands. In: *From Stardust to Planetesimals*, ASP Conference Series, vol. 122 (ed. Y.J. Pendleton), 119–128. San Francisco: ASP.

18 Guillois, O., Ledoux, G., Nenner, I. et al. (1998). Excitation processes for the emission of the unidentified IR bands. *Faraday Discuss* 109: 335–347.

19 Van Kerckhoven, C., Tielens, A.G.G.M., and Waelkens, C. (1999). The peculiar 3.43 and 3.53 µm emission features towards HD 97048 and Elias 1. In: *The Universe as Seen by ISO*, vol. 1 (ed. P. Cox and M.F. Kessler), 421–423. Noordwijk: ESA.

20 Chen, C.F., Wu, C.C., Cheng, C.L. et al. (2002). The size of interstellar nanodiamonds revealed by infrared spectra of CH on synthetic diamond nanocrystal surfaces. *J Chem Phys* 116: 1211–1214.

21 Jones, A.P., d'Hendecourt, L.B., Sheu, S.Y. et al. (2004). Surface C-H stretching features on meteoritic nanodiamonds. *Astron Astrophys* 416: 235–241.

22 Sheu, S.Y., Lee, I.P., Lee, Y.T., and Chang, H.C. (2002). Laboratory investigations of hydrogenated diamond surfaces by infrared spectroscopy: implications for the formation and size of interstellar nanodiamonds. *Astrophys J* 581: L55–L58.

23 Habart, E., Testi, L., Natta, A., and Carbillet, M. (2004). Diamonds in HD 97048: a closer look. *Astrophys J* 614: L129–L132.

24 Goto, M., Th, H., Kouchi, A. et al. (2009). Spatially resolved 3 µm spectroscopy of Elias 1: origin of diamonds in protoplanetary disks. *Astrophys J* 693: 610–616.

25 Ando, T., Ishii, M., Kamo, M., and Sato, Y. (1993). Thermal hydrogenation of diamond surfaces studied by diffuse reflectance Fourier-transform infrared, temperature-programmed desorption and laser Raman spectroscopy. *J Chem Soc Faraday Trans* 89: 1783–1789.

26 Van Kerckhoven, C., Tielens, A.G.G.M., and Waelkens, C. (2002). Nanodiamonds around HD 97048 and Elias 1. *Astron Astrophys* 384: 568–584.

27 Joblin, C. (1998). Which carriers for the unidentified IR emission bands? Observations and laboratory simulations. *Faraday Discuss* 109: 349–360.

28 Witt, A.N. and Vijh, U.P. (2004). Extended red emission: photoluminescence by interstellar nanoparticles. *ASP Conf Ser* 309: 115–138.

29 Cohen, M., Anderson, C.M., Cowley, A. et al. (1975). The peculiar object HD 44179 ("The red rectangle"). *Astrophys J* 196: 179–189.

30 Van Winckel, H., Cohen, M., and NASA/ESA. (2004). The remarkable red rectangle: stairway to heaven? https://www.spacetelescope.org/images/heic0408a (accessed 16 April 2018).

31 Schmidt, G.D., Cohen, M., and Margon, B. (1980). Discovery of optical molecular emission from the bipolar nebula surrounding HD 44179. *Astrophys J* 239: L133–L138.

32 Duley, W.W. (1985). Evidence for hydrogenated amorphous carbon in the Red Rectangle. *Mon Not R Astron Soc* 215: 259–263.

33 d'Hendecourt, L.B., Leger, A., Olofsson, G., and Schmidt, W. (1986). The Red Rectangle: a possible case of visible luminescence from polycyclic aromatic hydrocarbons. *Astron Astrophys* 170: 91–96.

34 Ledoux, G., Ehbrecht, M., Guillois, O. et al. (1998). Silicon as a candidate carrier for ERE. *Astron Astrophys* 333: L39–L42.

35 Witt, A.N., Gordon, K.D., and Furton, D.G. (1998). Silicon nanoparticles: source of extended red emission? *Astrophys J* 501: L111–L115.

36 Duley, W.W. (1988). Sharp emission lines from diamond dust in the Red Rectangle? *Astrophys Space Sci* 150: 387–390.

37 Chang, H.C., Chen, K., and Kwok, S. (2006). Nanodiamond as a possible carrier for extended red emission. *Astrophys J* 639: L63–L66.

38 Chang, H.C. (2016). Diamonds in space: a brief history and recent laboratory studies. *J Phys Conf Ser* 728: 062004.

39 Lu, H.C., Peng, Y.C., Chou, S.L. et al. (2017). Far-UV excited luminescence of nitrogen-vacancy centers: evidence for diamonds in space. *Angew Chem Int Ed* 56: 14469–14473.

40 Latter, W.B. and NASA/ESA (1998). Planetary Nebula NGC 7027. https://www.spacetelescope.org/images/opo9811d (accessed 16 April 2018).

41 Furton, D.G. and Witt, A.N. (1990). The spatial distribution of extended red emission in the planetary nebula NGC 7027. *Astrophys J* 364: L45–L48.

42 Kwok, S. and Zhang, Y. (2011). Mixed aromatic-aliphatic organic nanoparticles as carriers of unidentified infrared emission features. *Nature* 479: 80–83.

43 Witt, A.N., Gordon, K.D., Vijh, U.P. et al. (2006). The excitation of extended red emission: new constraints on its carrier from HST observations of NGC 7023. *Astrophys J* 636: 303–315.

44 NASA and ESA (2009). Blushing dusty nebula. https://www.spacetelescope.org/images/heic0915a (accessed 16 April 2018).

45 Witt, A.N. and Boroson, T.D. (1990). Spectroscopy of extended red emission in reflection nebulae. *Astrophys J* 355: 182–189.

46 Ledoux, G., Guillois, O., Huisken, F. et al. (2001). Crystalline silicon nanoparticles as carriers for the Extended Red Emission. *Astron Astrophys* 377: 707–720.

47 Rondin, L., Dantelle, G., Slablab, A. et al. (2010). Surface-induced charge state conversion of nitrogen-vacancy defects in nanodiamonds. *Phys Rev B* 82: 115449.

48 Halliday, D., Walker, J., and Resnick, R. (2010). *Fundamentals of Physics*, 5e. Wiley.

49 Evans, T. and Qi, Z. (1982). The kinetics of the aggregation of nitrogen atoms in diamond. *Proc R Soc A* 381: 159–178.

50 Maier, J.P. and Campbell, E.K. (2017). Fullerenes in space. *Angew Chem Int Ed* 56: 4920–4929.

51 Alvarez, L.W., Alvarez, W., Asaro, F., and Michel, H.V. (1980). Extraterrestrial cause for the cretaceous-tertiary extinction. *Science* 208: 1095–1108.

52 Firestone, R.B., West, A., Kennett, J.P. et al. (2007). Evidence for an extraterrestrial impact 12 900 years ago that contributed to the megafaunal extinctions and the Younger Dryas cooling. *Proc Natl Acad Sci USA* 104: 16016–16021.

53 Kinzie, C.R., Que Hee, S.S., Stich, A. et al. (2014). Nanodiamond-rich layer across three continents consistent with major cosmic impact at 12 800 Cal BP. *J Geology* 122: 475–506.

54 BBC Two (2017). The day the dinosaurs died. http://www.bbc.co.uk/programmes/b08r3xhf (accessed 16 April 2018).

Further Reading

Review Articles

Aharonovich, I., Greentree, A.D., and Prawer, S. (2011). Diamond photonics. *Nat Photon* 5: 397–405.

Barnard, A.S. (2009). Diamond standard in diagnostics: nanodiamond biolabels make their mark. *Analyst* 134: 1751–1764.

Castelletto, S., Li, X., and Gu, M. (2012). Frontiers in diffraction unlimited optical methods for spin manipulation, magnetic field sensing and imaging using diamond nitrogen vacancy defects. *Nanophotonics* 1: 139–153.

Chaudhary, H.M., Duttagupta, A.S., Jadhav, K.R. et al. (2015). Nanodiamonds as a new horizon for pharmaceutical and biomedical applications. *Curr Drug Deliv* 12: 271–281.

Chen, X. and Zhang, W. (2017). Diamond nanostructures for drug delivery, bioimaging, and biosensing. *Chem Soc Rev* 46: 734–760.

Chipaux, M., van der Laan, K.J., Hemelaar, S.R. et al. (2018). Nanodiamonds and their applications in cells. *Small* 14: 1704263.

Drezet, A., Sonnefraud, Y., Cuche, A. et al. (2015). Near-field microscopy with a scanning nitrogen-vacancy color center in a diamond nanocrystal: a brief review. *Micron* 70: 55–63.

Holt, K.B. (2007). Diamond at the nanoscale: applications of diamond nanoparticles from cellular biomarkers to quantum computing. *Phil Trans R Soc A* 365: 2845–2861.

Krueger, A. (2008). Diamond nanoparticles: jewels for chemistry and physics. *Adv Mater* 20: 2445–2449.

van der Laan, K., Hasani, M., Zheng, T., and Schirhagl, R, (2018). Nanodiamonds for *in vivo* applications. *Small* 14: 1703838.

Lai, L. and Barnard, A.S. (2015). Functionalized nanodiamonds for biological and medical applications. *J Nanosci Nanotechnol* 15: 989–999.

Lim, D.G., Prim, R.E., Kim, K.H. et al. (2016). Combinatorial nanodiamond in pharmaceutical and biomedical applications. *Int J Pharm* 514: 41–51.

Liu, J.H., Yang, S.T., Chen, X.X., and Wang, H. (2012). Fluorescent carbon dots and nanodiamonds for biological imaging: preparation, application, pharmacokinetics and toxicity. *Curr Drug Metab* 13: 1046–1056.

Man, H.B. and Ho, D. (2013). Nanodiamonds as platforms for biology and medicine. *J Lab Autom* 18: 12–18.

Merchant, K. and Sarkar, S.K. (2016). Fluorescent nanodiamonds for molecular and cellular bioimaging. *IEEE J Select Topics Quantum Electron* 22: 6802311.

Montalti, M., Cantelli, A., and Battistelli, G. (2015). Nanodiamonds and silicon quantum dots: ultrastable and biocompatible luminescent nanoprobes for long-term bioimaging. *Chem Soc Rev* 44: 4853–4921.

Nagl, A., Hemelaar, S.R., and Schirhagl, R. (2015). Improving surface and defect center chemistry of fluorescent nanodiamonds for imaging purposes – a review. *Anal Bioanal Chem* 407: 7521–7536.

Najeeb, S., Khurshid, Z., Zohaib, S. et al. (2016). Dental applications of nanodiamonds. *Sci Adv Mater* 8: 2064–2070.

Narayan, R.J., Boehm, R.D., and Sumant, A.V. (2011). Medical applications of diamond particles and surfaces. *Mater Today* 14: 154–163.

Nebel, C.E., Shin, D., Rezek, B. et al. (2007). Diamond and biology. *J R Soc Interface* 4: 439–461.

Neburkova, J., Vavra, J., and Cigler, P. (2017). Coating nanodiamonds with biocompatible shells for applications in biology and medicine. *Curr Opin Solid State Mater Sci* 21: 43–53.

Nunn, N., Torelli, M., McGuire, G., and Shenderova, O. (2017). Nanodiamond: a high impact nanomaterial. *Curr Opin Solid State Mater Sci* 21: 1–9.

Passeri, D., Rinaldi, F., Ingallina, C. et al. (2015). Biomedical applications of nanodiamonds: an overview. *J Nanosci Nanotechnol* 15: 972–988.

Plakhotnik, T. (2017). Diamonds for quantum nano sensing. *Curr Opin Solid State Mater Sci* 21: 25–34.

Rosenholm, J.M., Vlasov, I.I., Burikov, S.A. et al. (2015). Nanodiamond-based composite structures for biomedical imaging and drug delivery. *J Nanosci Nanotechnol* 15: 959–971.

Say, J.M., van Vreden, C., Reilly, D.J. et al. (2011). Luminescent nanodiamonds for biomedical applications. *Biophys Rev* 3: 171–184.

Schrand, A.M., Hens, S.A.C., and Shenderova, O.A. (2009). Nanodiamond particles: properties and perspectives for bioapplications. *Crit Rev Solid State Mater Sci* 34: 18–74.

Shenderova, O.A. and McGuire, G.E. (2015). Science and engineering of nanodiamond particle surfaces for biological applications. *Biointerphases* 10: 030802.

Sotoma, S., Epperla, C.P., and Chang, H.C. (2018). Diamond nanothermometry. *ChemNanoMat* 4: 15–27.

Szunerits, S., Barras, A., and Boukherroub, R. (2016). Antibacterial applications of nanodiamonds. *Int J Environ Res Public Health* 13: 413.

Turcheniuk, K. and Mochalin, V.N. (2017). Biomedical applications of nanodiamond. *Nanotechnology* 28: 252001.

Whitlow, J., Pacelli, S., and Paul, A. (2017). Multifunctional nanodiamonds in regenerative medicine: recent advances and future directions. *J Control Release* 261: 62–86.

General References

Banwell, C.N. and McCash, E.M. (1994). *Fundamentals of Molecular Spectroscopy*. College: McGraw Hill.

Born, M. and Wolf, E. (1997). *Principles of Optics*. Cambridge University Press.

Cantor, C.R. and Schimmel, P.R. (1980a). *Biophysical Chemistry. Part I: The Conformation of Biological Macromolecules*. W.H. Freeman.

Cantor, C.R. and Schimmel, P.R. (1980b). *Biophysical Chemistry. Part 2: Techniques for the Study of Biological Structure and Function*. W.H. Freeman.

Cantor, C.R. and Schimmel, P.R. (1980c). *Biophysical Chemistry. Part 3: The Behavior of Biological Macromolecules*. W.H. Freeman.

van Holde, K.E., Johnson, C., and Ho, P.S. (2005). *Principles of Physical Biochemistry*, 2e. Prentice Hall.

Hollas, J.M. (2004). *Modern Spectroscopy*, 4e. Wiley.

Sheehan, D. (2013). *Physical Biochemistry: Principles and Applications*, 2e. Wiley.

Index

Fluorescent Nanodiamonds, First Edition. Huan-Cheng Chang, Wesley Wei-Wen Hsiao and Meng-Chih Su.
© 2019 John Wiley & Sons Ltd. Published 2019 by John Wiley & Sons Ltd.